J. Gasteiger (Ed.)

Software Development in Chemistry 4

Proceedings of the 4th Workshop
"Computers in Chemistry"
Hochfilzen, Tyrol, November 22-24, 1989

Organized by the working group

der GDCh-Fachgruppe Chemie-Information

Springer-Verlag
Berlin Heidelberg New York
London Paris Tokyo Hong Kong

Prof. Dr. Johann Gasteiger
Institute of Organic Chemistry
Technical University of Munich
Lichtenbergstraße 4, D-8046 Garching
FRG

ISBN 3-540-52173-9 Springer-Verlag Berlin Heidelberg New York
ISBN 0-387-52173-9 Springer-Verlag New York Berlin Heidelberg

This work is subject to copyright. All rights are reserved, whether the whole or part of the material is concerned, specifically the rights of translation, reprinting, re-use of illustrations, recitation, broadcasting, reproduction on microfilms or in other ways, and storage in data banks. Duplication of this publication or parts thereof is only permitted under the provisions of the German Copyright Law of September 9, 1965, in its current version, and a copyright fee must always be paid. Violations fall under the prosecution act of the German Copyright Law.

© Springer-Verlag, Berlin Heidelberg 1990
Printed in Germany

The use of registered names, trademarks, etc. in this publication does not imply, even in the absence of a specific statement, that such names are exempt from the relevant protective laws and regulations and therefore free for general use.

Offsetprinting: Color-Druck Dorfi GmbH, Berlin; Bookbinding: B. Helm, Berlin.
2161/3020 5 4 3 2 1 0 – Printed on acid-free paper

Geleitwort

Neben Arbeit und Kapital wird die Produktion, Verteilung und Nutzung von Information in unserer modernen Industriegesellschaft immer mehr zum Produktionsfaktor. Dies gilt insbesondere für innovationsorientierte Forschungs- und Entwicklungsbereiche, wie etwa die Mikroelektronik, die Biotechnologie und die Chemie.

Im weltweiten Informationsangebot behaupten wir mit den Datenbanken BEILSTEIN ONLINE und GMELIN ONLINE, deren Aufbau Schwerpunkt im Fachinformationsprogramm der Bundesregierung ist, unsere traditionelle Kompetenz für den Bereich der stoffbezogenen Fakteninformation. Ergänzend dazu spielen spektroskopische Methoden eine erhebliche Rolle für viele Arbeitsbereiche der Chemie. Mit den kürzlich begonnenen Aktivitäten zum Aufbau eines Informationssystems Spektroskopie auf der Basis von SPECINFO werden entscheidende Verbesserungen zur Automatisierung in der Strukturaufklärung und Analytik erwartet.

Angesichts der Informationsfülle in der Chemie ist, neben der Aufarbeitung und Speicherung, die Entwicklung moderner Werkzeuge zur Erschließung des gespeicherten Wissens von großer Bedeutung. Andererseits ist für eine Vielzahl der bekannten Stoffe das Eigenschaftsprofil nicht annähernd vollständig verfügbar. Durch die systematische Auswertung strukturorientierter Datenbanken können Zusammenhänge zwischen den Stoffdaten ermittelt werden, die sich in Regeln für eine Datenprognose niederschlagen. Hier besteht eine wichtige Zukunftsaufgabe für die Fachinformation, und hier setzt auch das Thema Ihrer Veranstaltung "Software-Entwicklung in der Chemie" an.

Für die Vorträge, die Diskussionen und Präsentationen wünsche ich den Veranstaltern und Teilnehmern viel Erfolg und vielfältige Anregungen für ihre Tätigkeit in ihren Einrichtungen.

Dr. Heinz Riesenhuber

Bundesminister für Forschung und Technologie
der Bundesrepublik Deutschland

Foreword

Besides work and finance, the production, distribution, and the use of information are increasingly becoming production factors in our modern industrial society. This is particularly valid for innovatory fields of research and development such as microelectronics, biotechnology, and chemistry.

In the worldwide supply of information, we are maintaining our traditional competence in the field of factual information on compounds with the databases BEILSTEIN ONLINE and GMELIN ONLINE. The development of these databases is a high priority of the Specialized Information Program of the Federal Republic. In addition, spectroscopic methods play a major role in many areas of chemistry. Recently, action was taken to initiate an Information System of Spectroscopy based on SPECINFO. It is expected that this system will lead to decisive improvements in the automation of structure elucidation and analytical chemistry.

In view of the large amount of information in chemistry, not only the processing and storage of knowledge is of great importance but also the development of modern tools for retrieving it. On the other hand, the set of characteristics for a large number of compounds is far from being completely accessible.

The relationships between the data on compounds can be derived by systematically evaluating structure oriented databases. These relationships can be put down in rules for data prediction. This is an important task for technical information in the future and this is also a focal point of your workshop "Software-Development in Chemistry".

For the lectures, discussions, and presentations, I wish the organizers and participants every success and a great deal of inspiration for their work in their organizations.

Dr. Heinz Riesenhuber

The Federal Minister for Research and Technology
of the Federal Republic of Germany

(translated by Prof. Dr. J. Gasteiger)

Preface

This monograph contains papers presented at the 4th Workshop "Software-Entwicklung in der Chemie" (Software Development in Chemistry) held in November 1989 in Hochfilzen/Austria. It was the 4th workshop organized by the Working Party "CIC - Computer in der Chemie". The proceedings of the first three workshops have all been published in German: Software-Entwicklung in der Chemie 1, 2, and 3; Springer-Verlag, Berlin, 1987, 1988, 1989. As this is the first one to have the proceedings in English some retrospective views on the history of this Working Party and its Workshop are appropriate.

The Division of "Chemie Information" of the German Chemical Society (GDCh) set up a Working Party "CIC - Computer in der Chemie" in December 1985. As a first scientific event this Working Party organized a Workshop "Software-Development in Chemistry" in November 1986 in the tiny alpine village of Hochfilzen/Tirol in Austria.

The main objective of this workshop was to bring together scientists from the various disciplines of chemistry that are actively engaged in developing software or using the computer for the solution of chemical problems. The various activities in organic, inorganic, analytical, and theoretical chemistry are usually presented at highly specialized meetings to experts from their own field. It was thought that a synergistic effect should develop by bringing these specialists together at a meeting that has as a common denominator the use of the computer in chemistry and that this should provide new scientific impulses.

Another intention was to further the development of software for chemistry in the German speaking countries. Even the humblest beginnings of venturing into programming the computer to perform tasks for the chemist should be attracted to assess the state of the art in Germany and then lay a foundation to aim for international standards of excellence. Therefore, the language of the workshop was German.

Only a few dozens of "computer freaks" were expected. Thus we were in for a big surprise when already 135 participants showed up at this first workshop in 1986. The scientific atmosphere was so good, the discussions so intense that it was unanimously agreed to make this workshop an annual event. The second CIC-Workshop was in 1987 again in Hochfilzen, the third one, organized by Prof. G. Gauglitz, in 1988 at the University of Tübingen drawing 265 participants. For the fourth workshop the number of participants had to be limited to 180 with 190 eventually coming again to Hochfilzen.

Concurrent with the success of the workshop the standing of the Working Party "CIC - Computer in der Chemie" within the Division of "Chemie Information" was increasing. To recognize the achievements of the Working Party and the impact of computers on chemistry the Division of the German Chemical Society (GDCh) changed its name in November 1989 to "Chemie - Information - Computer (CIC)". This should lay the foundation for an increase in the promotion of the computer in chemical information and chemistry in general.

The 4th workshop in 1989 encompassed 28 oral presentations, 27 posters, and 26 demonstrations of a variety of software products from academic, industrial, and commercial sources.

The proceedings contain those contributions (oral and poster presentations and some software demonstrations) for which manuscripts were submitted. The papers are grouped according to their topics. As in the previous workshops, main features were the processing of structural information, the building of databases, the processing of spectroscopic data, and molecular modelling. A new major topic was the processing of information on chemical reactions. We also succeeded to open our perspectives to new developments in computer science, and attracted presentations on the simulation of structures and processes. On the other hand, we have to realize that some areas are still not well represented at this workshop. Chemometrics is one such topic that should deserve more attention in the future.

A highlight of the workshop was the evening address by Prof. J. Brickmann on "Art and Computer" with its slide and video show. We all went away impressed and moved by realizing that we have met a person that has managed to combine work and hobby and has brought that to perfection. Clearly, a printed medium is too poor to give only the slightest hint at what is going on at the forefront of computer graphics and computer art. Thus we refrain from presenting any contribution of this apogee.

The success of any meeting depends on the collaboration of many persons. The scientific committee, Prof. J. Brickmann, Prof. G. Gauglitz, Dr. C. Jochum, Dr. R. Neudert, Dr. V. Schubert, Prof. K. Varmuza, Dr. J. H. Winter, and Prof. D. Ziessow were instrumental in attracting contributions and shaping the program. The Division of "Chemie-Information" of GDCh and the Working Party "EDV in der Chemie" of the Austrian Chemical Society (GÖCh) helped in spreading the news of the meeting and in the organisation. Computer hardware was provided by SUN Microsystems, München, Digital Equipment, Frankfurt, Silicon Graphics, Geneva, and W & S-Computersysteme, Innsbruck. The Workshop was held under the auspices of the Bundesminster für Forschung und Technologie (BMFT) of the Federal Republic of Germany. Sponsorship came from the Federation of European Chemical Societies (FECS) and its Working Party "Computational Chemistry".

Last not least, most of the work (in organizing and running the meeting) rested on Ms. Tanja Pietraß who obtained assistance by Ms. Claudia Hofmann. In the final and hectic days my entire research group - V. Dimpfl, W. Hanebeck, W. D. Ihlenfeldt, K. Rafeiner, P. Röse, C. Rudolph, K.-P. Schulz, M. Wagener, W. Witzenbichler - provided every necessary help.

As in previous years again valuable on-site assistance came in the village of Hochfilzen from the mayor, J. Bergmann, and from Ms. H. Gfäller, Ms. Perterer, and J. Arnold. The music band of Hochfilzen welcomed us and made us feel at home from the very beginning.

I can only thank these persons and institutions for making the 4th CIC-Workshop again a big success.

Johann Gasteiger

Table of Contents

1. Representation of Chemical Structures; Factual Databases

H. Maier, D. Walkowiak
Chemical Substructure Search on CD-ROM   1

H. Nöth, E. Striedl
Documentation of Special Bond Types and Evaluation of Factual
Data from "B-Base"   11

J.L. Wisniewski, L. Goebels, A. Lawson
AUTONOM: Automatic Generation of IUPAC-Names from Structural Input   19

L. Domokos
Keys to the Beilstein Database (A Ring Searching Algorithm)   31

E. Lang, T. Förster, C.-W. von der Lieth
The Integration of the Cambridge Crystallographic Data Files into the
Relational Information Network of the German Cancer Research Center   43

A. Nebel, G. Olbrich, R. Deplanque
The Gmelin Information System - The Connection between
Handbook and Database   51

J. Gasteiger, W.D. Ihlenfeldt
The WODCA System (An Integrating Environment for the Chemist)   57

2. Structure and Properties; Bridging the Gap

K. Bley, I. Ugi
Computer-Assisted Analysis of Qualitative Structure/Activity
Relations of Organic Molecules   67

W. Degen
Prediction of the Threshold Soot Index for Hydrocarbon Fuels with
Randic's Topological Indices   75

R. Brüggemann, B. Münzer
A Graph Theoretical Method to Estimate Substance Date   85

B. Heinrich, E. Schreiner, D. Spielbauer, M. Wolperdinger, H.-U. Wagner
PIMO, a Program Visualizing HMO Results by Producing Transferable
Graphics Output   97

J. Barthel, H. Popp, G. Schmeer
The Calculation of Electrolyte Solution Properties with the Help of the
ELDAR Data and Method Bank Exemplified by Electrolyte Conductance   101

3. Molecular Modelling

W. Heiden, M. Schlenkrich, C.-D. Zachmann, J. Brickmann
Triangulation of Molecular Surfaces   115

U. Höweler
MOBY: Molecular Modelling on the PC   123

B. Krieg, P. Keller
ROCOCO: Reference Online Library for the Computer Aided Construction of
Molecular Geometries. Computer Aided Construction of Realistic Molecular
Models Using a Knowledge Base (I)   127

N. Reuter
MMGEO - A Versatile Tool for MMX Users   141

4. Spectral Data; Acquisition and Interpretation

A.N. Davies, H. Hillig, M. Linscheid
JCAMP - DX, A Standard   147

P. Haas, W. Robien
Automatic Interpretation of 2D-NMR-Spectra [1]   157

H. Lohninger
CDDS - A Personal Computer Based System for Automated Interpretation
of GC/MS-Analyses   165

W. Werther, K. Varmuza
EDAS-MS - Exploratory Data Analysis of Mass Spectra   175

W. Hanebeck, K. Rafeiner, K.-P. Schulz, P. Röse, J. Gasteiger
Towards the Automatic Generation of a Mass Spectrum from the
Structure of a Compound   187

G. Gauglitz, S. Weiß
Factor Analysis of Spectral Data from Chemical Reactions   197

S. Ebel, J.S. Kang, W. Windmann
Normalization of in Situ-Spectra in Thin Layer Chromatography   207

S. Ebel, C. Urban, S. Windmann
Algorithms for Use in Purity Control of Drugs Using a UV-Spectroscopy   221

S. Goudetsidis, G. Hägele
DSYM-PC A Novel Program to Simulate HR-NMR Spectra for Spins I=1/2 on
IBM-Compatible Computers of PC/XT/AT Type   233

5. Chemical Reactions and Synthesis Planning

J. Bauer
IGOR2: A Program System for Generating Chemical Reactions and
Structures   237

E. Zass
Reaction Databases in a University Chemistry Department –
Online or In-house?   243

E. Blurock, T. Strelow
Towards Synthesis Planning Aids Through Databank Analysis   255

M. Wagener, J. Gasteiger
Implementation of Synthesis Strategies in PROLOG   265

P. Röse, J. Gasteiger
EROS 6.0, A Knowledge Based System for Reaction Prediction –
Application to the Regioselectivity of the Diels-Alder Reaction   275

R. Moll
CARSA (Computer-Assisted Research in Synthesis and Application)   289

6. Simulations of Structures, Reactions and Properties

J. Kinkel, H.J. Ederer, K.H. Ebert
Modelling of Polymer Gel Formation and Gel Reactions with Monte
Carlo Methods for 3-Dimensional Networks   297

K. Nicklas, J. Böcker, M. Schlenkrich, P. Bopp, J. Brickmann
Molecular Dynamic Simulation of the Interface Aqueous Ionic
Solution / Lipid Membrane   311

B. Speiser
Electroanalytical Simulations. 10. The Simulation of Fast Second
Order Reactions in Electrochemical Systems   321

M. Klein, G. Baier, O.E. Rössler
The Rashevsky-Turing System: Two Coupled Oscillators as a Generic
Reaction-Diffusion Model   331

D. Fröhlich, G. Gauglitz
Photoreactions in Solids – Experiment and Simulation   345

7. Computer Science; New Methods and Their Applications

K. Zercher, B. Radig
The Role of Machine Learning in Knowledge Acquisition   353

M. Tusar, J. Zupan
Neural Networks   363

M. Otto, U. Hörchner
Application of Fuzzy Neural Network to Spectrum Identification   377

H. Armitage, A. Khuen, D. Ziessow
Tools for Automatic Program Generation   385

A. Mechsner
Automatic Translation from FORTRAN to "C"   397

8. Miscellaneous Subjects

A. Bielecki
LABORG - Laboratory Optimisation by the Elimination of Weaknesses
and the Implementation of LIMS   405

J. Kammerer
C - LIT   413

A. Müller, E. Striedl
Possibilities and Limitations of Combined Database Enquiries Modelled
on "Environmental Significance of Aluminum"   417

CHEMICAL SUBSTRUCTURE SEARCH
ON CD-ROM

Dr. Helmut Maier, Dirk Walkowiak

Softron
Gesellschaft für techn.-wiss. Software
Rudolf-Diesel Str. 1   8032 Gräfelfing

Abstract: Does the CD-ROM with its high storage capacitiy but rather slow random access times offer the opportunity to search large scale chemical structure data bases with a PC in acceptable search times? This article briefly describes the problems and introduces the S^4 substructure search system which is a positive answer to that question.

1. Introduction

To be brief, yet not very precise, the problem of chemical substructure searching consists in solving the following task: Retrieve all molecules from an (often very large) set of fullstructures which, in a certain sense, shall contain a predefined query (sub)structure.
Regarding the atoms of the molecules as knodes and the bonds as edges the problem becomes equivalent to the known "Partial Graph Problem" in graph theory. Unfortunately, this theory has not yet

produced a satisfactorily efficient algorithm for this kind of problem.

In the past, various algorithms for chemical substructure searching have been developed which provide acceptable search times on large scale structure files (up to 10 million structures) if powerful mainframe computers are used. CAS-Online, DARC, MACCS and HTSS are examples for this.

Since PCs have continually become more efficient within the last years, it is an obvious idea to implement substructure search systems also on desk-top or work station computers. In certain cases, it is desirable to be independent of host systems.

In principle, the following characteristics of a PC limit the implementation of a structure search system:

a) Still today, a PC cannot compete with a mainframe computer regarding MIPS and FLOPS;
b) a PC has less hard disk capacity;
c) random access to hard disks is slower on a PC than on a mainframe.

Yet for modern PCs, those assumptions are only partially true:

ad a) Concerning pure integer arithmetics, a PC with a 20MHz phase 80386 processor performs only ten times slower than an IBM-3090/150. Taking the typical overhead of a mainframe multiuser operating environment into account, the elapsed times are even comparable.

ad b) Hard disk drives with a capacity of 650 Mbyte comparing well to a 3380 disk of an IBM mainframe are already available for PCs.

ad c)   Fast PC hard disks today have random access times of about 18msec, i.e. they are also comparable to mainframe disks.

However, fast hard disk drives with a large capacity are still rather expensive compared with the costs of a PC itself.

Example: MAXTOR, 570 MB formatted, 16ms, 30.000 h MTBF, final price about DM 27.000,--.

Since large structure files (CAS, Beilstein) are only needed for read access by the user, the CD-ROM as a medium of more reasonable costs suggests itself.

2. CD-ROM

The CD-ROM is a reasonably prized read-only storage medium for large scale data of any kind. Below, the most important characteristics are listed in an abbreviated form:

- Storage capacity about 550 Mbyte,
- random access time about 500 to 800 msec,
- data transfer on the PC bus about 60Kbyte/sec (measured with C and Pascal programs),
- high production costs for the master disk, about DM 7.000,--,
- low production costs for the copies, about DM 10 to DM 20,
- low drive costs, about DM 1.200,--, controller included.

The main technical difference between a magnetical hard disk drive and a CD-ROM drive is the random access time: it is about 50 (!) times slower on a CD-ROM.

This explains why applications using frequent disk I/O suddenly require inacceptable elapsed times when the hard disk is replaced by a CD-ROM drive the software architecture remaining unmodified. Algorithms and software systems performing excellently on magnetical disks may fail totally when used unmodified on CD-ROM.

3. Structure Search Systems and CD-ROM

It is not the intention of this paper to discuss the architectures of the various structure search systems. Yet the following, if vague, statement can be made. Most structure search systems (CAS, DARC, MACCS) carry out a structure search within two steps on a large scale structure file:

Step 1 ("screening") Selected properties of the query structure are used as necessary conditions for the fullstructures to be an answer or not. Thus, the structure file is divided into two sets, a (more or less precise) super set of all possible "hits" and its complementary set of definite "non-hits".
Commonly used properties are, for instance:

- Atom- or bond-centered fragment codes of the structures,
- characteristic ring fragments,
- characteristic chain fragments.

All those properties are suitable which may be characterized as follows: If a property applies to a given substructure it must also apply to a fullstructure containing this substructure. Technically, the screening process is, in most cases, a search on sorted index files leading to random access operations. Here, as a rule, the amount of random access increases with the logarithm to base 2 of the index size.

For example, searching one out of 1 million records requires about 20 random access operations (provided only one record can be kept in memory at the same time).

The screening, therefore, often is an I/O- intensive process.

Step 2 ("ABAS") The super set qualified in step 1 contains at least all answers to the query, however, depending on the accuracy of the screening, possibly many more structures. In order to eliminate the false candidates, a so called ABAS (atom by atom search) step is carried out. During this procedure, all candidates from the super set are matched in an "atom by atom, bond by bond" manner against the query.

In general, the backtracking algorithms used are rather CPU-intensive. Unfortunately, however, it turns out that this step is still "I/O-bound" because the time spent to access the fullstructures randomly from a "slow" storage medium by far dominates the pure computational time.

This is due to the fact that the candidates are distributed all over the whole file, so there is little probability to reach several candidates with one random access.

For example, take a structure file consisting of 1 million structures, each structure stored in 300 bytes, i.e. roughly 160 structures on a 47 Kbyte track (IBM 3380 disk drive), 80 of which can be read using one I/O operation. That means, that even for a relatively large super set of 12.500 candidates, in the worst case, the ABAS will require one disk I/O per structure!

Drawing a conclusion, it is obvious that the architecture of a structure search system as described above is not suitable for a "slow storage medium" such as CD-ROM.

In the following section we will describe the architecture of S^4-CD-ROM (S^4 = SSSS for Softron Substructure Search System). It will be shown that S^4 is able to substitute fast sequential access for time consuming random access. Thus, search times achieved by other systems only on mainframes can now be achieved on a PC.

4. Architecture of S^4-CD-ROM

The S^4 system consists of a screening and an ABAS, but there are two main differences to the systems described above. First, the screening uses variable fragment codes instead of a fixed fragment set. It concentrates on one single fragment in the query trying to make it as selective as possible.

Second, the ABAS requires a special <u>atom file</u> instead of reading from the primary structure file.

In the fileloading phase, the atom file is prepared in such a way, that the number of random accesses during a substructure search is minimized.

Let us first describe the resources needed by our CD-prototype implemetation of S^4.

a) The CD-prototype contains about 350.000 heterocycles from the Beilstein-Handbook H to EIV. For each of the 350.000 structures the following data are stored:

- Connection table,
- coordinates
- chemical name
- BRN (Beilstein Registry Number),
- molecular formula.

This <u>structure file</u> occupies 130Mbyte, that is an average of 370 bytes per structure. The structures are stored in the BRCT-format (i.e. Beilstein Registry Connection Table). S^4 needs this file for structure display only, not for searching.

b) Beyond the above primary data, S^4 needs about 1.4 times the size of the primary data as secondary data (atom file, bundle files). As mentioned above, S^4 uses the so called <u>atom file</u> for the ABAS. It consists of all atoms (not H) of all structures sorted in a certain order. Since each structure, on the average, contains 22.6 atoms (not H), the atom file consists of 7.9 million atom entries. Each entry comprises the complete constitutional information of the structure containing this atom. In principle, this means that the connection table of each structure is stored, on the average, 22.6 times. This high redundancy is important for S^4. Of course, the atom file is stored in a highly compressed format, thus requiring only about 90 Mbytes for the 7.9 million connection tables. This means that a complete connection table, on the average, has been condensed to 12 Byte <u>including</u> a 3 byte molecule number. Thus, a 23 Kbyte block holds nearly 2000 structures!

c) From the atom file, a set of (up to) 11 <u>bundle files</u> must be generated which are used by the screening of S^4. A set of 11 bundle files requires about the same disk space as the whole atom file (about 90 Mbytes for the 350.000 file).

All in all, S^4 needs 310 Mbyte for 350.000 structures, all secondary data included. The coordinates for an accurate display of the hits, the chemical names and molecular formulas are already contained in this number. Since not more than 1 Kbyte storage capacity per (organic) structure is required by S^4, an S^4-searchable CD-ROM can hold up to 550.000 structures.

In order to understand why S^4-CD-ROM provides search times within the range of seconds even on large scale files, its functionality shall be explained briefly.

<u>S^4 Screening</u> Within this step, S^4 qualifies the candidates for the ABAS step. S^4 searches the bundle files in ascending order and generates most selective fragment codes of the query. With each used bundle file the fragment codes become more precise. Depending on the query, it is even possible to achieve an <u>exact</u> fragment code making a subsequent ABAS step obsolete!
All bundle files are stored on the CD-ROM, but the screening can be sped up by storing some or all of them on a magnetic disk. The S^4 system includes an installation program which enables the user to store selected bundle files on one or more magnetic disks (e.g. network drives). In a configuration file, S^4 can be instructed how many screening levels it is actually going to use and where the bundle files are stored. The screening can be interrupted at any time which results in a possibly larger (but complete!) super set.

The screening passes all generated fragment codes of the query to the ABAS. Each fragment code determines a unique <u>consecutive</u> section in the atom file containing the candidates in sequential order.

<u>S^4 ABAS</u> The ABAS scans all sections of the atom file which have been qualified by the screening. The sections are sorted by the screening in ascending order. "Reasonable" queries lead to few, big sections, some queries may lead to many, small sections. The positioning of a section may require a random access, reading a section requires only sequential access to the atom file. Recalling from section 4.b) that one single read operation can load a large amount of structures into memory, the advantages of the S^4 file system become obvious.

If the screening was exact for a section, the ABAS need not process the candidates but only reads the molecule numbers. Otherwise, all "false" candidates are eliminated by the ABAS.

It may also happen that different sections contain atoms of the same hit. These "multiple" hits are excluded by the ABAS in a final sorting step.

The S^4 approach to chemical substructure searching has shown that it is worth putting strong emphasis on the fileloading process, where the data is prepared for a minimized random access during the search phase.

This is the actual reason for S^4 being so efficient not only on the classic magnetic disks, but also on a more sequential medium like the CD-ROM.

DOCUMENTATION OF SPECIAL BOND TYPES AND EVALUATION OF FACTUAL DATA FROM "B-BASE"

H. Nöth, E. Striedl

Institut für Anorganische Chemie,
Ludwig-Maximilians-Universität München,
Meiserstraße 1, D-8000 München 2, FRG

Computer assisted handling of chemical structures - the international language of chemistry - has become very comfortably within the last 30 years. Problems arise by leaving the field of classic valence bond type structures and incorporating metal complexes or cluster compounds for instance. This report shows the concept for documentation of Lewis adduct compounds, transition metal complexes with π bonds including d- or f-orbitals and cluster compounds within the factual database "B-Base". Furthermore it describes the evaluation of factual data for visualisation via graphical comparison. A new feature of ChemBase software, now applicated to "B-Base", enables the linkage of substructure and graphs of shift ranges.

INTRODUCTION

Computer assisted handling of chemical information has become more and more important within the last 30 years, to a large extent caused by the much improved performance of the available hardware. Meanwhile chemical structures - the international language of chemistry - are graphically handled very comfortably even by Personal Computers [1]. Problems will only arise by leaving the field of classic valence bond type structures and incorporating metal complexes or cluster compounds for instance [2].

The constructive work on **B-Base** within the last two years has resulted in a database, which at present contains approximately 9,200 document units, covering primary literature up to the middle of 1988 [3]. Corresponding to the multifarious chemistry of the element Boron, **B-Base** includes compounds which differ considerably from the domain of known connection table based structure handling programs - the covalent bond type based organic chemistry.

STRUCTURAL REPRESENTATION OF SPECIAL BOND TYPES

Lewis Adduct Compounds

The following abridged description of the documentation of Lewis adduct compounds within the factual database **B-Base** shows a concept, being developed with regards to one of the key questions of structure-NMR-shift-correlation in the field of boron chemistry: What is the coordination number of the central boron atom considered?

The formation of adduct compounds of boron containing compounds with the solvent leads to a change of the coordination number at the boron atom. But structural descriptions of these compounds in reports vary from incorrect 3-coordinated boron compounds (only mentioning solvents) to mesomeric structures without any distinction between different bond types.

As illustrated in Figure 1A the bond between two components, sometimes marked by an asterisk, must be represented as a single bond. Assigning a **wavy "Either bond"** emphasizes the adduct character, but signifies no further difference. Formal **charges** must be applied to the related atoms according to the automatic supplying of hydrogen atoms, because structures like 1B, which would otherwise be wrong, will result. Mesomeric structures like 1C are better documented without emphasizing using the "Either bond".

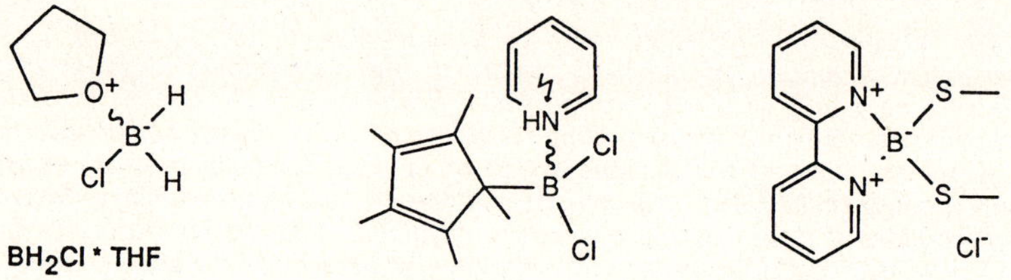

Figure 1: Documentation of Lewis adduct compounds in **B-Base**.

A similar problem for structure-shift correlations arises from different notation of compounds with pp-π back donation like the example in Figure 2. In these cases we try to achieve uniform representation within **B-Base** according to the established interpretations.

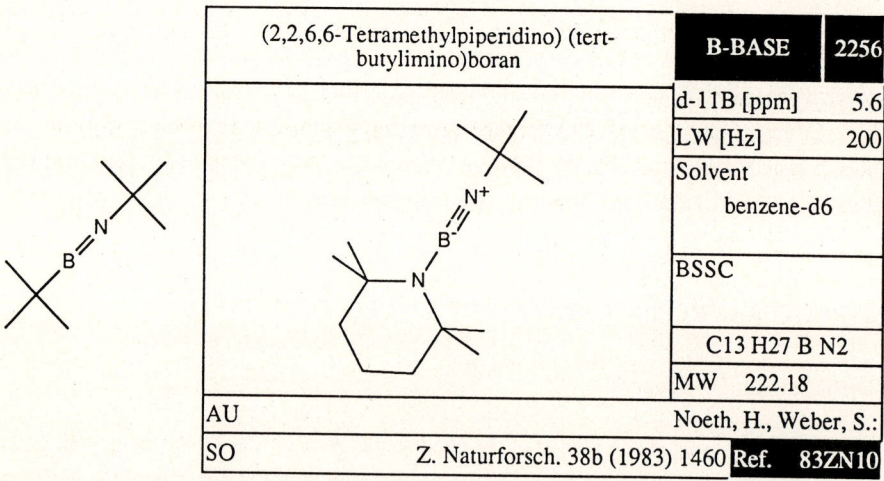

Figure 2: Documentation of pp-π back bonding within various publications.

D- or F-π Bonds within Transition Metal Complexes

Today Chemical Abstracts Service (CAS) has no structure conventions applied to transition metal complexes [4] which separate σ- from π-bonding interaction between metal atoms and ligands. Even the projected GMELIN-ONLINE will treat all metal to ligand connectivities as ordinary covalent bonds [5].

Our approach to this problem is pointed out by Figure 3: within **B-Base** the user gets *graphical information* about where a π-bonding interaction is located. These dashed lines result from a "**stereo down bond**" displayed at the stereo display mode "Alternate". Because they are simple optical attributes (coded as a single bond in the connection table), these "π-bond types" cannot be searched for in the database. Therefore a user of **B-Base** executes a rough preselection by the program's search options and then selects the desired compounds via visual examination.

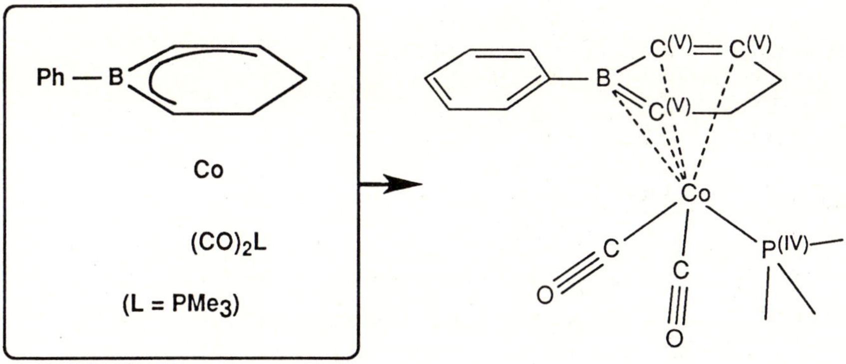

Figure 3: Redefining valences and assigning "stereo down" bond types to metal-ligand π-bonds results a **graphical distinction between σ- and π-interaction**.

Cluster Compounds

In addition, this concept has been expanded for cluster compounds as depicted in Figure 4.

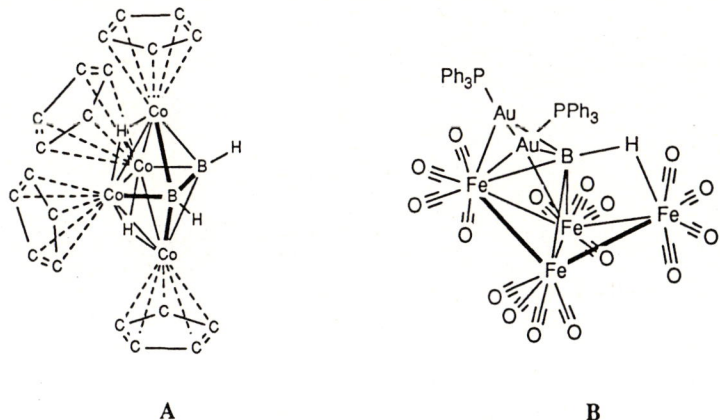

A   B

Figure 4: Representative sample structures for cluster compounds within **B-Base**.

EVALUATION OF FACTUAL DATA FROM "B-BASE"

Above all, using **B-Base** simplifies a fast comparison between substructures and related shift values from NMR-experiments. One further aspect is the possibility of a comparative overall access to published boron NMR data. Visualization of shift ranges for various classes of boron compounds via bar graphs appears to be an obvious application in order to obtain a general view of the content of **B-Base**.

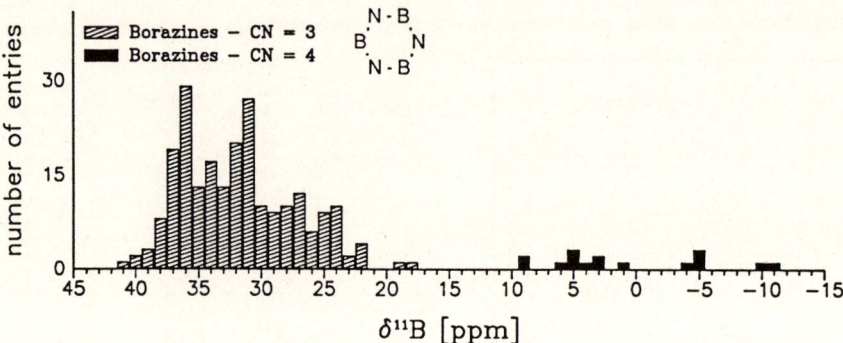

Figure 5: An impressive verification of a well-known fact: borazines show little tendency for increasing the coordination number of their boron atoms.

Figures 5 and 6 show plots which indicate the numbers of compounds with comparable nuclear shielding within a defined class of compounds. Graphs like these will be used for detecting remarkable deviations from the profile. The analysis of extremes leads in most cases to a modification of the incorrect coordination number at the boron atom of interest, brought about by incomp-

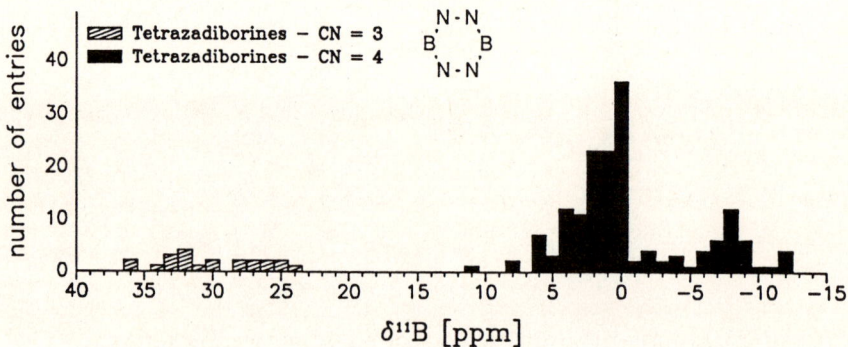

Figure 6: The evidence of a reverse situation within the class of tetrazadiborines supports the hypothesis of a poor cyclic mesomeric pp-π back donation.

lete or ambiguous publication. On the other hand a lot of documentation units require a specification of the second coordination sphere as the example of Figure 7 shows.

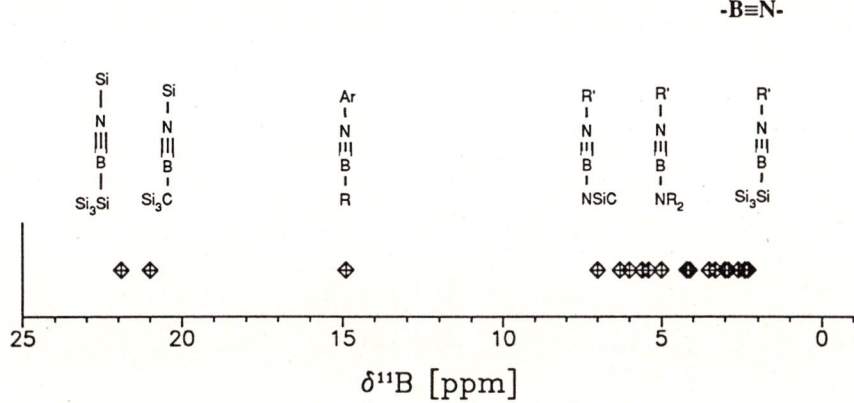

Figure 7: The impact of the second coordination sphere on the ^{11}B-nuclear shielding: R, R'= alkyl-, Ar = aryl-substituents.

Every new synthesized compound needs to be compared to similar or analogous classes of compounds. Input of structure and related ^{11}B-NMR shifts of such compounds into **B-Base** enables

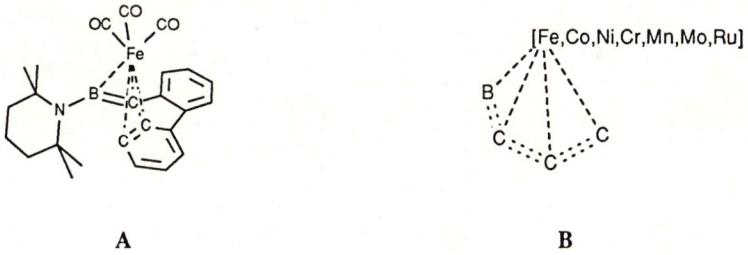

Figure 8: The object of analysis (A), and a feasible query (B) which should achieve a widespread variety of hits.

direct comparison to the whole database content in many ways. Figures 8 and 9 illustrate an example, that supports by documentary evidence the novel character of compound 8A.

LINKAGE OF SUBSTRUCTURE AND SHIFT RANGES WITHIN B-BASE

According to a new feature of ChemBase software [7] the shift ranges of discrete substructures can be obtained directly via display of their graphs. This new enhancement of **B-Base** is demonstrated by Figure 10. In practice it is accessible by using a specially designed form, named

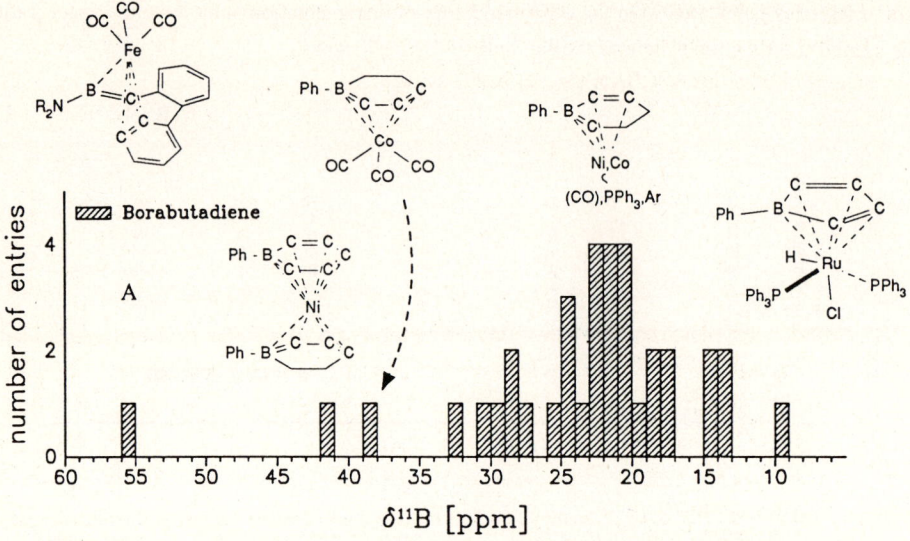

Figure 9: The graph illustrates the novel character of the new compound 8A.

B-Base-Graph. The form incorporates one database field, defined for displaying META-Files and one box for the name of the related META-File. To prevent casual access, this box is hidden by the label box **B-Base-Graph**.

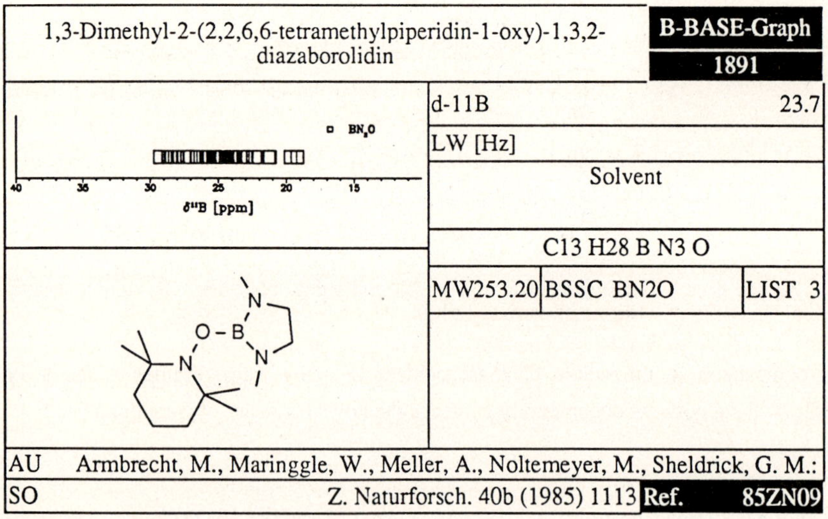

Figure 10: **B-Base-Graph** links each documentation unit with a graph, illustrating the shift range of the compound's boron environment

META-Files for this purpose may be made by the following procedure:
- Running a substructure search within **B-Base** [3].
- Saving a Data File (ASCII) of the ^{11}B-NMR-shifts.
- Generating a HPGL-Plot-File [8] from the Data File by a suitable program.
- Transforming the Plot-File with FROMHPGL.EXE [7] into a *.MET-File.

CONCLUSION

The concept, presented here, for the documentation of Lewis adduct compounds fills an essential gap within **B-Base**. The expansion to metal complexes and other compounds with partial bond orders may be considered as an acceptable compromise - but additionally encodable bond types will be useful to the whole field of inorganic chemistry.

Graphical descriptions of shift ranges - well-known from ^{1}H- or ^{13}C-NMR-Spectroscopy - represent a further service for boron chemists, the more so because within **B-Base** they are integrated into each documentation unit.

---

1. a) *W.T. Wipke*, "Evolution of Molecular Graphics". In **Graphics for Chemical Structures**; Warr, W.A. (Editor); **ACS Symposium Series** 341: Washington, DC, 1987; p. 1.
   b) *D.E. Meyer*, "Use of Microcomputer Software to Access and Handle Chemical Data". In **Chemical Structures**; W.A. Warr (Editor); Springer-Verlag: Berlin Heidelberg New York, 1988; p. 251.
2. *G.G.V. Stouw*, "Potential Enhancements to the CAS Chemical Registry System". In **Chemical Structures**; W.A. Warr (Editor); Springer-Verlag: Berlin Heidelberg New York, 1988; p. 211.
3. a) *H. Nöth* and *E. Striedl*, Mitteilungsblatt, Ges. Dtsch. Chem., Fachgruppe Chemie Information, ISSN 0178-4927, **15** (1989) p. 3-10.
   b) *H. Nöth* and *E. Striedl*, "**B-Base** - eine ^{11}B-NMR-Faktendatenbank für den Personal Computer". In *Software-Entwicklung in der Chemie 3*; G. Gauglitz (Editor); Springer Verlag, Berlin, Heidelberg, New York (1989) p. 101-108.
   c) *E. Striedl*, Thesis, University of Munich, in preparation.
4. *A.W. Ryan* and *R.E. Stobaugh*, "The Chemical Abstracts Service Chemical Registry System IX. Input Structure Conventions". *J. Chem. Inf. Comput. Sci.* 1982, **22**, p. 22-28.
5. EXCERP Programm EXC3D; Input 3-dimensional Structure Conventions for GMELIN-ONLINE database; Manual for excerptors, August 1989.
6. *G.E. Herberich* and *E. Raabe*: J. Organomet. Chem. **309** (1986) 143.
7. Version 1.3 of ChemBase, August 1989; Molecular Design Ltd., San Leandro, CA 94577.
8. HPGL: *H*ewlett *P*ackard *G*raphics *L*anguage.

AUTONOM: AUTOMATIC GENERATION OF IUPAC-NAMES FROM STRUCTURAL INPUT

J.L. Wisniewski, L. Goebels, A. Lawson

Beilstein Institute, Varrentrappstrasse 40-42,
D-6000 Frankfurt/Main, FRG

Abstract: An algorithm has been developed for the computer generation of organic compound names from structure diagram input. The structure diagrams of compounds are input via a graphic interface and the names are generated purely on the basis of the resulting molecular connection tables. The system, when fully implemented and installed, will be used in the production of both the Beilstein Database and the Beilstein Handbook. It is also expected that a PC-version of AUTONOM will provide the chemical community with a new valuable tool. This paper describes the general design of AUTONOM and presents a detailed analysis of software solutions adopted during the work on the system.

INTRODUCTION

The need to precisely describe information on the structure of chemical compounds spans written and oral communication in a multi-disciplinary, international environment. The communication relies upon the use of chemical names, structural diagrams and formulas. The structural diagram conventions are established as an international standard [1] and transcend language barriers among chemists; however, to nonchemists they are only interesting shapes and strange configurations which convey little understandable information. Names which can accurately describe the composition and format of the structure are still vital for a wider audience. They are also vital for institutions producing chemical information, such as Beilstein or CAS, as indexing tools and are important search key fields for their databases.

Nevertheless, a complete, comprehensive grammar for systematic chemical nomenclature does not exist to date. The system of recommendations [2] which have been developed by the Commission On Nomenclature of Organic Chemistry of the IUPAC has not become a universal standard, mainly because of the complexity of the recommended rules and frequent ambiguity in name assignment. Additionally, the rules are by no means in a form that may be directly translated into computer-orienatated algorithms.

Neither Beilstein nor CAS follow, in full extension, the recommendations while preparing their indexes. Both institutions revised the IUPAC system and created their own 'systematic' nomenclature and one has not to be a nomenclature specialist in order to notice the acknowledged fact [3] that the two nomenclature systems are far apart in numerous cases.

The main driving force for this divergence was the desire to create a really unambiguous system which would translate unique graphic structure representations into equally unique names and vice versa. It became obvious that such a bi-directional non-synonym translation would be necessary for any successful implementation of computerized chemical retrieval applications. It also became obvious that such a task can only be conducted by a computer program [4].

Several organizations have reported computer programs that converse names into structures. The earliest efforts in this area were undertaken as early as 1974 by CAS [4]. Later Beilstein reported the use of the VICA name-to-structure conversion system to assist in translating the Beilstein Handbook names into structure records for the Beilstein database [5]. The implementation of a similar program on a personal computer has been recently reported by a group from the University of Hull [6,7,8].

To date there are no reports of a general program in the reverse direction, i.e., from structures to names. The first naming program for a limited class of compounds restricted only to alloys, co-polymers, mixtures, and additional compounds was reported by Vander Stouw et al. in 1976 [9]. Then in 1981, CAS presented a paper [10] on plans for complete automatic generation of organic compound names. The paper discussed the design of the algorithm; however, there has been no subsequent report of the implementation of such an algorithm. Recently, Meyer and Gould [11] have confirmed a feasibility of creating a microcomputer-based program that would accept structure input and generate an IUPAC-like sytematic names.

The present paper describes the AUTONOM (AUTomatic NOMenclature) computer system, which is the latest product of the Beilstein Institute's Research Division, aimed at providing the chemical community with a fully automated 'structure-to-name' translator. AUTONOM analyzes structure diagrams of organic compounds entered via a graphic interface and generates IUPAC compatible names purely on the basis of the molecular connection tables of the input structures. Figure 1 illustrates the use of the program in a PC-based implementation. The given structure was drawn using a mouse, and the AUTONOM name was returned (as shown) within 8 seconds.

AUTONOM 6-(8-Bromo-10,11-dihydro-5#H!-dibenzo[#a!,#d!]cyclohepten-10-yl)-5-(3-methyl-but-2-enyloxy)-5,6-dihydro-1#H!-pyrazin-2-one

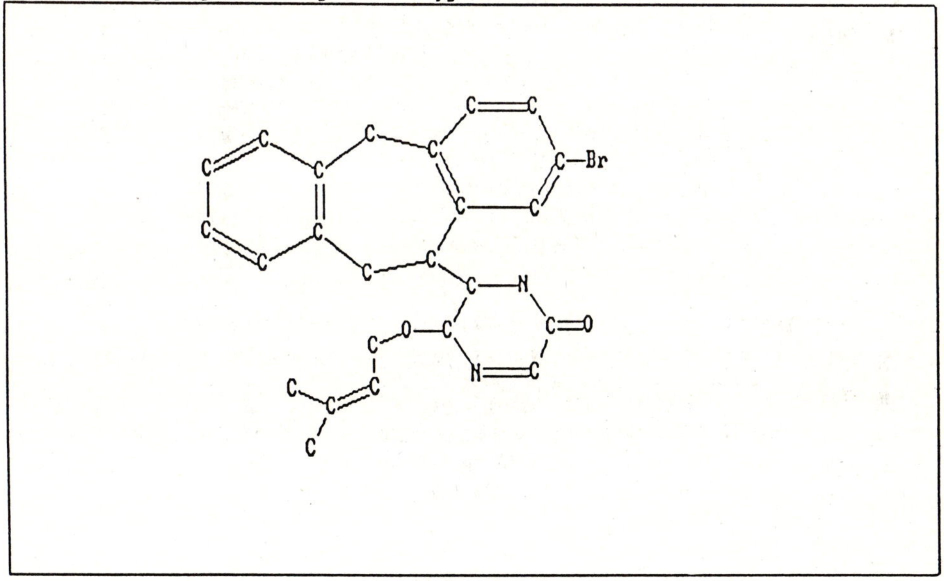

Fig. 1. Name and structure output screen from AUTONOM.

AUTONOM is a general purpose system, which means that is not limited to specific classes of compounds.

THE ALGORITHM

The algorithm applied in AUTONOM names a structure in the same manner as would a nomenclature specialist, by identifying candidate parent structural fragments and then successively applying the appropriate IUPAC nomenclature principles to eliminate less-preferred candidates, which then become substituents on the selected parent fragment. The algorithm operates on a systematic step-by-step basis; while a chemist can often apply nomenclature rules without conscious effort, the algorithm of AUTONOM has no such ability and must systematically collect numerous items of information about the structure during the processing cycle.

The complete naming cycle is logically divided into five phases:

* functional group recognition
* ring system recognition
* parent structure selection
* name tree creation
* name assembly

During the first four phases, the algorithm operates on objects that are independent of textual name fragments. These objects fall into three distinct classes, namely functional groups, chains and ring systems. The complete description of all objects present is derived purely from the connection table of the structure being processed. The object selection and identification is itself a very complex routine and is invoked on a regular basis during each of the four phases. The text name fragments are generated and appropriately ordered and assembled only during the very last phase of the naming cycle.

1. Functional Group Recognition.

Acyclic portions of the structure are scanned to identify substructures bearing functional group characteristics, such as hetero atom compositions with unsaturated bonds. A table-driven approach with an atom-by-atom search mechanism is used. Each candidate functional group substructure unit is checked against the pre-defined sorted list of functional groups. On completion, a full list of all the functional groups present is returned and then sorted in nomenclature priority order. The highest ranking groups are then used in the parent structure selection phase to generate a set of candidate parent fragments.

The term 'highest ranking' derives from the IUPAC recommendations [2] and is supposed to set clear conditions on selection of the so called principal functional group, which later (in the completed name), is cited as the suffix. Unfortunately, the formulation of the recommendation (p.86 in [2]) is ambiguous and both CAS and Beilstein re-defined their own recommendations, particulary in the area of acids and their derivatives. AUTONOM implements so called 'rank+multiplicity' order during selection. It can be briefly exemplified by the following scenario: two ester derivatives of a carbothioic acid group, one ester of a carboxylic acid group and a free sulfonic acid group have been localized during the search cycle. The IUPAC recommendations rank free acids highest in the list. This would then demand choosing the sulfonic acid as the principal group and the term 'sulfonic acid' as suffix in the name. It would also mean transforming the both ester derivatives of carbothioic acid and the ester of carboxylic acid into prefixes (ester part name+'thiocarboxy' and ester part name+'carboxy'). AUTONOM follows the IUPAC recommendations as far as rank order of acids is concerned, but additionally, mainly in order to simplify and shorten the name, intersects the rank principle with 'maximum multiplicity' principle. This means, for the above example, choosing the carbothioic acid group as the principal group, multiplying it and transforming the ester of carboxylic acid as well as sulfonic acid into prefixes (in the latter case 'sulfo').

The functional group recognition is not and cannot be restricted to a single run process. The list of localized functional groups in the initially constructed form usually does not stay in this form throughout the whole naming cycle. The initial form is only important for choosing the principal group and the following parent selection phase. Later, after the parent structural fragment had been selected, the list must usually be modified. This also might be the case for the structure from the discussed scenario. The structural unit of the esterified carboxylic acid:

From parent fragment ---> $R_0-C(=O)-O-R_1$ <---

needs no modification if further bonding to the parent fragment is directed as above. However, for the reversed direction, i.e.,

$$\text{---> } R_0\text{---}C(=O)\text{---}O\text{---}R_1 \text{ ---> to parent fragment}$$

the group must be split into two other groups, namely (-O) 'oxy' and (= O) 'oxo'. The splitting takes place sometime later during the name tree creation phase after the parent fragment has been localized and the position of the group, in relation to the parent, has been established.

2. Ring System Recognition.

A dual approach has been adopted for the task:

- look-up dictionary access to trivial name ring systems via an atom-by-atom search mechanism

and

- a strictly algorithmic generation of ring system names, which is based only on the analysis of their internal composition.

The searching is conducted in a previously prepared dictionary containing trivial name ring connection tables together with prescribed numbering and names. In order to speed up the search process only ring connections and atom characteristics, (with no information on bonding), are stored. This way no bond restoration is necessary, which saves time spent on bond denormalization and unscrambling. Additionally, by using an appropriate hashing method the complete set of trivial name ring sytems has been grouped in small pots with hash codes as the addresses. At the time of writing, the dictionary of ring systems contained 2496 rings in 388 different pots, which gives on average around 7 rings per pot. The distribution of rings in pots is not uniform, however the maximum number of rings in a single pot lies below 25 rings in the worst case. The current storage space of ca. 1,5 Mb is planned to be significantly reduced by applying a compression algorithm [12].

AUTONOM finds all the ring systems in the input structure then applies exactly the same hashing algorithm for every localized ring system and conducts searching only in the corresponding small pot instead of the complete dictionary.

The searching differs only slightly from well known substructure searching methods [13]. Instead of looking for a single substructure in the set of structures, it looks for a set of substructures in a single structure. The whole process might descriptively be coined an 'infrastructure search'.

The connection tables of the dictionary ring systems are stored together with the the arrays of fixed locants.

For non-symmetrical systems each atom can be numbered only with one fixed locant. For a symmetrical system this is unfortunately not the case. Depending on the number of symmetry axes, a single atom can be numbered with many locants. In the following example of anthracene, all marked atoms can theoretically be numbered with one:

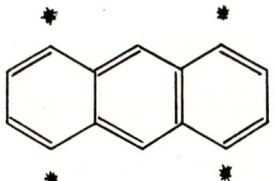

Which of the marked atoms finally gets the one is decided in accordance with the IUPAC rules and normally only this atom is enumerated with one, which guarantees that the rest of the atoms will get the lowest possible locants. When deciding on enumeration, the following structural factors are considered:
- principal groups,
- 'indicated' hydrogens,
- multiple bonds in compounds whose names indicate partial hydrogenation (cycloalkenes, pyrazolines, and the like)
- number of prefixes,
- prefixes in alphabetical order of citation.

There is a wide spectrum of ring system classes for which dictionary access is replaced with purely algorithmic ring identification. The classes include monocyclic, bicyclic, spirocyclic and dispirocyclic alkanes, heteromonocyclics named by the Hantzsch-Widman method , replacement ('a' terms) nomenclature heteromonocyclics, heterobicyclics, spiroheterocyclics and dispiroheterocyclics.

The ring sytems, no matter which identification model has been used to identify them, do not include linear assemblies. The assembly identification for both models is done algorithmically. The current version is capable of identifying any assembly composed of up to 8 identical ring systems.

3. Parent Structure Selection.

Once functional groups and ring sytems are identified, the algorithm proceeds with finding parent chains (chains bearing functional groups or chains bearing no functional groups, which are acceptable as parents only when a function expressible as suffix is not present in the structure). Then potential candidate structural fragments (rings, chains or functional parents if localized) are selected and ranked according to the relevant nomenclature principles. Seniority of ring systems is decided by applying altogether 12 various criteria. In general they are in agreement with the corresponding IUPAC recommendations (p.101 in [2]). The principal chain(s), i.e, the chain(s) upon which the nomenclature and numbering are based, is selected by applying successively 10 various criteria formulated by the IUPAC recommendations (p.97 in [2]).

The highest ranking rings or/and chains are then compared and parent fragment is selected by application of the folowing sequence of principles:
- a) greatest number of the principal functional groups cited as a suffix,
- b) preferred hetero atom content,
- c) a ring is preferred to a chain
- d) the greatest number of substituents,
- e) lowest locants for the substituents,
- f) lowest locants for the substituent cited first as prefi- in alphabetical order.

The other fragments automatically become substituents on the parent structural unit.

4. Name Tree Creation.

The parent fragment selected in the preceding phase is established as the root of the name tree, while the other structural units become branches of the name tree. While starting at the root of the tree and travelling in upward direction, mutual relationships among the branches (e.g. type of bonding, locants of connections, indicated hydrogen locants for ring sytems, etc.) are established and data concerning these relationships are added to the branch descriptors. Concurrently, new branches are introduced; they include newly identified substituent chains and new functional groups generated during the group splitting.

The full description of all the possible objects and their mutual relationship is complete once the tree has been completely traversed in an upward direction. The structure is now divided into substructural units and, the units have been related to the data records containing the full information on the units and their mutual relationships.

At this moment the ring name descriptors are not yet fully complete, e.g. 'cyclopropa[3,4]pentaleno[1,2-d]dioxole' is stored instead of the complete '2a,2b,2c,5a,5b,5c-hexahydro-cyclopropa[3,4]pentaleno[1,2-d]dioxole'.

The missing 'hydrogenation' prefix will be algorithmically generated in the next phase (design and programming of the 'hydrogenation' routine belonged to the most complex task of the AUTONOM project). The correct generation of the prefix can only be successful after the complete set of sub-branches (substituents) of the ' cyclopropa[3, 4]pentaleno[1,2-d]dioxole' branch is identified and described.

Functional groups are described with suffix and prefix forms, e.g. 'carboxylic acid/carboxy'. In the next phase, the algorithm will decide, by examining the principal or non-principal status tag of the functional group, whether it should be 'carboxy' as a prefix, or 'carboxylic acid' or 'oic acid' as a suffix.

For optimization reasons the upward traversal is done only once, with no recursive calls, and the sub-branches are fully known and identified only at the very end of the name tree creation phase.

5. Name Assembly

Starting at the highest branch and traversing the name tree downward, the name fragments for the branches which have been visited are generated and stored in the correct order in a name fragment stack. Then these fragments are combined, by applying the proper chemical nomenclature semantics and syntax into longer fragments and whole process repeats itself until the final name is obtained. Several name assembly routines handle such operations as alphabetization, multiplication, punctuation, vowel deletion, superscript and italic string placing, etc.

Finally, an intelligent text string replacement algorithm takes care of (IUPAC-allowed) non-systematic, traditional nomeclature, e.g the transient name fragment '(1-oxo-ethoxy)' must be replaced with "acetoxy", which is preferred by both CAS and Beilstein . It is worth noting at the moment, that this simple string replacement has important consequences as far as the syntax is concerned. The name fragment '(1-oxo-ethoxy)' would be multiplied with bis-, tris-, tetrakis, etc., multiplication affixes, while the resulting 'acetoxy' is multiplied with di-,tri-, tetra-, etc., affixes. It might sound trivial, but the implementation at this level is by no means trivial.

This process should not be confused with a similar string exchange which can (if required) be implemented in a user-defined shell, at a much later stage of name assembly, where language and usage considerations can be taken into account (propanoic -> propionic, indole -> indol (German), etc.).

SUMMARY

Programming AUTONOM to its current state was a substantial task resulting in 17,500 PASCAL program lines and 87 various routines and functions. It runs on an IBM-AT (or compatible, also 80386-based, naturally) with a minimum of 512 kB of RAM and a hard disk with at least 1,5 MB free for the storage of the dictionary of trivial name ring sytems. On average, depending on the complexity of the structures, AUTONOM names up to 17 structures per minute. Tests conducted on random samples from the Beilstein database indicate that the program achieves expert status in ca. 61 % of the cases tested at the time of writing, where expert status is defined as identical output compared to the Nomenclature Dept. of Beilstein. At the moment AUTONOM cannot handle sterochemistry, which accounts for the great majority of the "missing" 39 % of truly expert status. Input structures are limited to 255 non-hydrogen atoms.

The first complete PC version of AUTONOM as well as the peer mainframe version for production of the Beilstein Database and the Beilstein Handbook are expected to be available in the course of the next year.

ACKNOWLEDGEMENT

The development of the AUTONOM project is generously supported by the Bundesministerium für Forschung und Technologie (BMFT). The authors and the Beilstein Institute gratefully acknowledge this support.

REFERENCES

1 Chemical Structures: The International Language of Chemistry, (1987) Leeuwenhorst Congress Center, Noordwijkerhout, The Netherlands
2 IUPAC, nomenclature of organic chemistry, sections A, B, C, D, E, F and H. (1979) Pergamon, London
3 Laboratory of the Government Chemist Conference on Chemical Nomenclature. (Nov. 1987) London
4 Vander Stouw GG, Elliot PM and Isenberg AC (1974) J Chem Doc 14:185
5 Jochum C (1986) American Chemical Society 192nd Meeting, CINF Abstract 128
6 Cooke-Fox DI, Kirby GH, Rayner JD (1989) J Chem Inf Comput Sci 29:101
7 Cooke-Fox DI, Kirby GH, Rayner JD (1989) J Chem Inf Comput Sci 29:106
8 Cooke-Fox DI, Kirby GH, Rayner JD (1989) J Chem Inf Comput Sci 29:112
9 Vander Stouw GG et al. (1976) J Chem Inf Comput Sci 16:213
10 Mockus J, Isenberg AC, Vander Stouw GG (1981) J Chem Inf Comput Sci 21:183
11 Meyer DE, Gould SR (1988) Am Lab 20(11):92
12 Wisniewski JL (1987) J Inf Sci 13:159
13 Willet P (1987) J Chemometrics 1:139

KEYS TO THE BEILSTEIN DATABASE
(A ring searching algorithm)

László Domokos

Beilstein Institut, 6000Frankfurt/M, Varrentrappstr 40.

Abstract: The experiences with a retrieval algorithm for ring systems are described. The program was tested with 3 million structures of the Beilstein database. The ring searching is a fast and accurate replacement for name fragment searching and gives additional possibilities for structure retrieval from large structure files.

A significant advantage of online databases in comparision with the traditional handbooks is the large variety of different ways of accessing the data. While a handbook can normally be accessed via 1-2 keys, the number of keys used with a database is practically unlimited.

The Beilstein Handbook of Organic Chemistry, for example, has two major keys. The first one is the Beilstein system itself, according to which the compounds are ordered in the Handbook, the second key is the alphabetically ordered register of the chemical names. A third, but not a unique key is the molecular formula used as a primary sort criteria in the formula register of each volume. The recently developed SANDRA(1) program provides an easy access to to the Handbook through the chemical structure. To find a complete set of compounds according to other criteria is practically impossible, unless one reads the book as a whole (ca. 230.000 pages).

The Beilstein online database, which is certainly one of the world's most complex databases, provides several dozens of keys of various types:

- structural : full- and substructure , molecular formula, atom counts, Lawson number (correlated with the Beilstein system), chemical names, fragments of chemical names, etc.
- numerical values or ranges of values of chemical and physical properties : density, viscosity, boiling point, etc.

- text values (description) of chemical and physical properties : derivatives, raman spectrum, etc.
- keywords for chemical and physical properties : electrochemical behaviour, magnetic data, electronic spectrum, NMR data, etc.
- reaction data : starting material, reagents, by-products, etc.
- bibliographical references : author, journal, publication year, patent number, etc.

In addition to this wide range of access possiblities, the Beilstein Institute and the online hosts of the database are steadily considering new useful keys to enhance the usage of the immense amount of information stored in the database.

The most flexible and powerful way of finding a required compound, or compounds containing a given substructure or a class of substructures,is certainly the full- and substructure searching. The complex algorithm and the considerable amount of computer resources required by substructure searching, makes it the most expensive and time consuming search in all chemical databases. Therefore chemists try, if possible, to replace the structure search with a simple search for chemical name fragments to retrieve the required compounds. Unfortunately, because of the lack of a unique well defined nomenclature an exact result cannot be expected. Another problem with name searching arises when not all compounds have names in the database. A strictly algorithmic name generation, offered for example by AUTONOM (2), that directly derives names quickly and reproducably from the chemical structures, might solve both of these problems.

A large and important class of substructures is represented by the rings. To solve the problem of the above mentioned ambiguities of the name fragment searching for the well defined class of ring fragments, a prototype program was developed for searching complete ring systems within the database. The aim was to investigate the use of the technique and to find the optimal way for final implementation.

The software consists of two major parts: the coding part, during which all structures of the database are appropriately coded and the retrieval part. The major steps of the coding are:

- reading the structure
- extracting the ring systems of the structure
- coding uniquely each extracted ring system
- storing the generated codes and several additional descriptors of the ring systems in the database
- building up the indices for efficient retrieval

The steps of retrieval part include:

- definition of a query for the ring system to be searched
- unique coding of the query ring system with the same algorithm as above
- searching for the code(s) using the indices
- displaying the retrieved compounds.

Some data about the system:

- 3.123.407 structures of the Beilstein file were used
- 2.790.870 structures contained at least one ring system
- all bonds of the ring structure are considered as single bonds
- number of attached hydrogen atoms, number of other substituents and the multiplicity of the ring system within the structure can be defined in the query
- non-carbon atoms can be defined either with exact positions or without the the positions
- the test program uses a simple character input for the query definition. Final implementation could use a PC based graphic input program, for example Molkick
- program language : PL1
- the structures and the ring codes are loaded into the Beilstein Institute's ADABAS database
- hardware : IBM-3090/150

EXAMPLES

1. example

In this example a naphthalene ring skeleton with exactly 7 attached hydrogen atoms and with one substituent at any position was searched for. The required ring system must be present in the structure exactly 4 times. The 7 hydrogens and one rest imply 5 double bonds, i.e. an aromatic ring system.

```
Number of atoms   =  10
bonds             = rs66   ( rs stands for ring system )
hetero atoms      = none
Number of hydrogenes = 7
Number of rests      = 1
Multiplicity         = 4

————————> Number of Hits = 43
```

Some hits : (BRN stands for "Beilstein Registry Number)

1. BRN =   77635   2,4,2',4'-Tetra-<1>naphthyl-3,3'-diphenyl-
                     2H,2'H-<2,2'>bichromenyl
2. BRN = 373696   **Tetra-<1>naphthyl-oxiran**
5. BRN = 383127   2,4,2',4'-**Tetra-<1>naphthyl**-3,3'-diphenyl-
                     4H,4'H-<4,4'>bichromenyl
6. BRN = 383318   **Bis-<2,4-di-<1>naphthyl**-3-phenyl-2H-
                     chromen-2-yl>-peroxid
7. BRN = 601457   **Bis-<1,3-di-α-naphthyl-imidazolidinyliden-(2)>**
8. BRN = 604160   * * * NO NAME * * *
9. BRN = 635409   4,4',5,5'-**Tetra-(naphth-2-yl)**-2,2'-azoimidazol
10. BRN = 728643   1,1,2,2-**Tetrakis-<naphthyl-(2)>-hydrazin**
11. BRN = 728672   1,2-**Di-<naphthyl-(1)>-**
                     1,2-**di-<naphthyl-(2)>-hydrazin**
13. BRN = 770880   **Tetra-naphthyl-(1)-hydrazin**
15. BRN = 876259   **Tetrakis-(1-naphthamido)-pyrazin**
20. BRN = 2032716   **Tetra-α-naphthoxyethen**
27. BRN = 2550547   **Tetrakis**(β-**naphthyl**mercapto)ethen
39. BRN = 3023967   <**Di-α-naphthyl-borinsaeure**>-**anhydrid**
41. BRN = 3077831   **Tetra**thioorthokohlensaeure-β-**naphthyl**ester
43. BRN = 3114772   α-**Naphthyl**-β-**naphthyl-ketazin**

Looking at the hits one can see the very good correlation with "tetra naphthyl". However, it can also be seen that the name fragment search query of "tetra* & naphth*", where "*" is a symbol for truncation and "&" for a proper adjacency operator, would not be satisfactory. Hit numbers 6,7,8,11,39 and 41, for example, do not satisfy the query. To formulate a substructure search query to retrieve this set of compounds would be rather complicated. On the other hand, when using substructure searching, it is possible to formulate a question more precisely by, for example, defining the exact positions of substituents.

It must be mentioned here that in these examples only one chemical name is listed even if synonyms exist, and many of the chemical names are from a datapool, from the "short file" (3), which has not been checked yet by the Beilstein editors.

2. example

Same as before but 6 hydrogens, 2 substituents and 3 occurrence.

Number of atoms =   10
bonds   = **rs66**
hetero = none

Number of hydrogenes   = **6**
Number of rests   = **2**

```
Multiplicity           = 3

----------> Number of Hits = 75

Some hits :

   1. BRN =   77500   6-<4-Hydroxy-<1>naphthylazo>-1-<3-(4-
      hydroxy-<1>naphthylazo)-benzyl>-2-<3-(4-hydroxy-<1>naphthy..
   2. BRN =  375533   <2-Hydroxy-<1>naphthyl>-bis-<4-hydroxy-
                     <1>naphthyl>-<1,3,5>triazin
   3. BRN =  375573   4,4',4''-<1,3,5>Triazin-2,4,6-triyl-tri-
                     <1>naphthol
   4. BRN =  375606   Bis-<2-hydroxy-<1>naphthyl>-
                     <4-hydroxy-<1>naphthyl>-<1,3,5>triazin
   5. BRN =  381873   2,4,6-Tris-<2-aethoxy-<1>naphthyl>-
                     <1,3,5>trithian
   7. BRN =  382920   Phosphorsaeure-tris-<2-piperidinomethyl-
                     <1>naphthylester>
  11. BRN =  733441                          * * * NO NAME * * *
  12. BRN =  736697                          * * * NO NAME * * *
  18. BRN =1837219   p-Amino-trisazonaphthalin
  20. BRN =1838465   2,7,13-Tris(β-hydroxy-α-naphthylazo)triptycen
  23. BRN =1845959   Bis-(1-naphthol-2-azo)-2,3-naphthalin
  55. BRN =2912077   Tri-2-<1-methyl-naphthyl>-boron
  63. BRN =2933599   tris-(2-Hydroxy-1-naphthylidenanilin)-silicium-
```

3. example

```
Number of atoms  = 20
bonds   = rs65656
hetero  = *

Number of hydrogenes = any
Number of rests      = any
Multiplicity         = any

----------> Number of Hits =    20

Some hits :

   1. BRN =2000198                    * * * NO NAME * * *
   2. BRN =2109228    10,12-Dihydro-indeno<2.1-b>fluoren
  10. BRN =2531793    5-Methyl-10,12-dihydro-indeno<2,1-b>fluoren
  15. BRN =3061595    11-Phenyl-10.12-dihydro-indeno<2.1-b>fluoren
  16. BRN =3100397    11-Methyl-10.12-dioxo-10.12-dihydro-
                     indeno<2.1-b>fluoren; Isophthalacon
  20. BRN =3110180    10,12-Dioxo-11-(2-carboxy-phenyl)-10,12-
                     dihydro-indeno<2,1-b>fluoren
```

4. example   searching for cuban skeleton

```
Number of atoms =   8
bonds   = rs444,2-7,3-6
hetero  = none


   ――――――――> Number of Hits = 57
```

Some hits :
```
   1. BRN =1878949    4-Cyan-kuban-1-carbonsaeure
   2. BRN =1879431    Methyl 4-cyano-1-cubancarboxylat
   3. BRN =1884091    4-(Carbomethoxy)-1-cubancarbonsaeure
```
   5. BRN =1901366   **Pentacyclo-<4,2,0,0$^{2.5}$,0$^{3.8}$,0$^{4.7}$>-octan**
```
   6. BRN =1923320    Cuban-d
   7. BRN =1925071    Bromcuban
   8. BRN =1990376    Cuban-1,4-dicarbonsaeure-dimethylester
  11. BRN =2082107    1,2,3,4,6,7-Hexadeuterocuban
  12. BRN =2089315    Octadeuterocuban
  13. BRN =2110348    4-Aminomethylcuban-1-carbonsaeure
  14. BRN =2133962    t-Butyl-4-bromcubanperoxicarboxylat
  15. BRN =2167028           * * * NO NAME * * *
  16. BRN =2176416    Di-t-butyl-cuban-1,4-diperoxycarboxylat
  20. BRN =2261963    4-Methylcubancarbonsaeure
  23. BRN =2329478    Cubancarbonsaeure
  24. BRN =2330763    4-Brom-1-carboxy-kuban
  25. BRN =2333624    Cuban-1,4-dicarbonsaeure
  30. BRN =2407360    Octaphenylcuban
  34. BRN =2455107    4-Bromcubancarbonsaeuremethylester
  36. BRN =2497023    Cuban
  49. BRN =2695327    4-Hydroxymethylencubancarbonsaeuremethylester
  57. BRN =2977837           * * * NO NAME * * *
```

This example shows how difficult it is to find the proper structures with a name fragment search. Hit nr.15 does not possess any name in the database, in hit nr.1 cuban is spelled with "k" instead of "c", hit nr.2 could be located with right truncation, hit nr.11 with left truncation, hit nr.14 with both left and right truncations (unless the words are correctly fragmented and inverted), hit nr.5 has a systematic name, etc.

5.example

All separate **six membered carbon rings** with any number of hydrogens and substituents have been searched for. This query is probably the "worst case" because of the extremely large number of hits. Consequently, the elapsed search time was extremely long, 217 seconds, and 10.892 disk accesses were necessary.

Queries with very large number of hits are reasonable when the hit lists are to be further processed using additional selection criteria. For example, find all compounds containing a naphthalene skeleton but not a six membered carbon ring; or find all compounds containing a six membered carbon ring and at least 4 halogen atoms and with a melting point of at least 120 Celsius., etc.

```
Number of atoms = 6
bonds    = rs6
hetero = none
```

                            Time at start = 16:54:02 : 941
                            Time at end   = 16:57:39 : 306

————————> Number of Hits = **1566510**   !

Some hits :

1. BRN = 1001 (S)-5-**Benzoyl**oxy-2-**phenyl**-7-<o^3,o^4,o^6
   -tri**benzoyl**-o^2-(tri-O-**benzoyl**-α-L-rhamnopyranosyl)-β
2. BRN = 1002 O]-**Benzyl**-N]-**benzyl**oxycarbonyl-tyrosyl=>isole
   ucyl=>glutaminyl=>asparaginyl=>S-**benzyl**-cysteinyl=>prol....
etc.

6. example

The same query as above but this time the number of hydrogens is restricted to 5, and the number of substituents is equal to one, i.e. a phenyl ring was searched for.

```
Number of atoms = 6
bonds    = rs6
hetero = none

Number of hydrogenes = 5
Number of rests      = 1
Multiplicity         = any
```

                            Time at start = 16:58:03 : 570
                            Time at end   = 16:59:23 : 933

————————> Number of Hits = **722709**

Some hits :

1. BRN = 1001 (S)-5-**Benzoyl**oxy-2-**phenyl**-7-<o^3,o^4,o^6
       -tri**benzoyl**-o^2-(tri-O-**benzoyl**-α-L-rhamnopyranosyl)-β
2. BRN = 1002 O]-**Benzyl**-N]-**benzyl**oxycarbonyl-tyrosyl=>isole
   ucyl=>glutaminyl=>asparaginyl=>S-**benzyl**-cysteinyl=>prol....

7. example

The same as the previous example but with a multiplicity of 12.

```
Number of atoms =   6
bonds      = rs6
hetero = *

Number of hydrogenes  = 5
Number of rests       = 1
Multiplicity          = 12   <—— !
                              Time at start = 17:00:15 : 934
                              Time at end   = 17:00:16 : 189

————————> Number of Hits = 73
```

Some hits :
1. BRN =   16009 O^1,O^3,O^4,O^5,O^6-**Pentabenzoyl**
   -O]^2-<O^3,O^4,O^6-**tribenzoyl**-O^2-(**tetra**-O-**benzo**
3. BRN = 635642   * * * NO NAME * * *
4. BRN = 773596   2-**Diphenyl**amino-4,6-**di**-<N-<4,6-**bis**-(**dibenzyl**
                    amino)-s-triazin-2-yl>-anilino>-s-triazin
6. BRN = 955510   **Triphenyl**phosphine-palladium
9. BRN = 1280123   2,3,4,6-**Tetra**kis-(**tri**-O-**benzyl**-galloyl)-
                    β-D-methyl-glucosid
13. BRN = 1633388   1,3-**Bis**(2,3,4,6-**tetra**-O-**benzyl**-α-D-gluco
                    pyranosyloxy)-1,1,3,3-**tetraphenyl**-disiloxan
16. BRN = 1661682   α-Cyclodextrin-2,2',2',2'',2'',2''',3,3',3',3'
                    ',3'',3''''-**dodecabenzoat**-6,6',6',6'',6'',6'''-hexa-p-toluols...
27. BRN = 2551431   1,4-**Bis**-(2,3,4,5,6,7-**hexaphenyl**cycloheptatri-
                    1,3,6-enyl)-benzol
35. BRN = 2827190   2,2,4,4,10,10,12,12,19,19,21,21-**Dodekaphenyl**-
                    15,15,24,24-tetramethyltrispiro<5,1,5,3,5,3>undekasiloxan

These kind of searches are very fast, the elapse times measured at normal working load of the machine were, within one second. Hit nr. 6. has an icorrect name.

8. example

The same as above, except that the number of hydrogens is equal to 6, no substituents are allowed, no restriction for multiplicity. i.e. the query corresponds to the full structure of benzene not considering charges and isotopes.

```
Number of atoms = 6
bonds    = rs6
hetero   = none
```

```
Number of hydrogenes  = 6
Number of rests       = 0

───────> Number of Hits = 34

Some hits :
   1. BRN = 969212    Benzene
   2. BRN =1848242    Phenyl-Radikal
   3. BRN =1848658    Phenylkation
   4. BRN =1848659    Deuteriobenzol
   7. BRN =1920388    Benz-in (Arin)
   8. BRN =1920389    <T>-Benzol
   9. BRN =1920819    1-13C-Benzol
  10. BRN =1920828    $^{14}$C-Benzol
  12. BRN =1923021    Benzol<1,6-$^{14}$C>
  13. BRN =1923419              * * * NO NAME * * *
  14. BRN =1924403    1,2,3,5-Tetradeuteriobenzol
  15. BRN =1924607    1,2,4,5-Tetradeuteriobenzol
  16. BRN =1926513    Pentadeuteriobenzol
  17. BRN =1928753    Hexatritiobenzol
  18. BRN =1939404    <1,2-13C>-3,4,5,6-d4-Benzol
  19. BRN =2037211    1,2-Dideuteriobenzol
  23. BRN =2043846    Benzol-14C
  24. BRN =2070583    p-Benzyn
  26. BRN =2079482              * * * NO NAME * * *
  27. BRN =2237356    <1-13C,1-2H>Benzol
  28. BRN =2345576    ($^{12}$C)-Benzol
  29. BRN =2348070    1,2,3,4-Tetradeuterobenzol
  30. BRN =2358531    13C-Benzol
  31. BRN =2359015              * * * NO NAME * * *
  32. BRN =2425357    Benzol
  33. BRN =2428744    1,3,5-$^{13}$C(3)-Benzol
  34. BRN =2498055    p-Deuterophenylradikal
```

9. example

```
Number of atoms =  12
bonds   = rs66,2-11-12-5
hetero  = none

Number of hydrogenes  = 8
Number of rests       = 2
                         Time at start = 18:13:35 : 916
                         Time at end   = 18:13:36 : 554

───────> Number of Hits = 6

Some hits :
```

1. BRN =2284499 2-Benzyl-3-
            phenyl**benzo<5,6>bicyclo<2.2.2>octatrien**
 2. BRN =2292871   * * * NO NAME * * *
 3. BRN =2420589 1,4-Dihydro-1,4-**etheno-naphthalin**-
            2,3-dicarbonsaeure
 4. BRN =2517037   * * * NO NAME * * *
 5. BRN =2566795 2,3-**Benzo**-7,8-bis-<trifluormethyl>-
            bicyclo<2.2.2>-octatrien(2,5,4)
 6. BRN =2646941 2,3-Dicyan-5,6-**benzo-bicyclo<2.2.2>octa**trien

Here we have 6 hits. 2 of them without names, and all the 4 retrieved names have different forms.

10. example

Number of atoms = 7
bonds   = **RS63**
hetero = O 2 N 5
                        Time at start = 09:02:48 : 542
                        Time at end   = 09:02:48 : 559

————————> Number of Hits = **19**

Some hits :
  1. BRN = 574074   * * * NO NAME * * *
  2. BRN = 781750   4,6,6-Trimethyl-3-acetyl-3,4-**epoxy**-tetrahydro-
     piperidon-(2)
  3. BRN = 911446   1-Methyl-4-phenyl-3,4-**epoxypiperidin**
  4. BRN = 980007   * * * NO NAME * * *
  5. BRN =1012980   3-Methoxycarbonylmethyl-1-methyl-4-phenyl-
     5,6-dihydro-2-**pyridon-epoxid**
  6. BRN =1075057   1-Methyl-4-phenyl-3,4-**epoxypiperidin**
 15. BRN =1119315   3,4-**epoxy**-1-benzyloxycarbonyl-**piperidin**
 16. BRN =1140735   * * * NO NAME * * *
 17. BRN =1159702   N-Tosyl-2-trichlormethyl-4,5-**epoxy-piperidin**

CONCLUSIONS:

The above and numerous other examples illustrate the following advantages of the method:

  - very fast retrieval
  - it can be used as a very efficient replacement and enhance-
    ment for name fragment searching of rings
  - it does not replace a substructure search, but opens addi-
    tional perspectives by quick formulation of new type of
    queries for complex ring structures
  - it could be a key to a new door of Beilstein online

REFERENCES:

1. A. Lawson, SANDRA, Springer Verlag, (1987)
2. S. Wisniewski, L.Goebels, A.Lawson, Beilstein AUTONOM: Automatic Generation of IUPAC-Names from Structural Input. 4.Workshop Software Entwicklung in der Chemie, 22-24. Nov. (1989). Hochfilzen.
3. Beilstein Brief, Vol.1.,Nr.1. Springer Verlag, (1986)

THE INTEGRATION OF THE CAMBRIDGE CRYSTALLOGRAPHIC DATA FILES INTO THE RELATIONAL INFORMATION NETWORK OF THE GERMAN CANCER RESEARCH CENTER

Elke Lang[1], Thomas Förster[2], Claus-Wilhelm von der Lieth[1]

Depts. of Spectroscopy[1] and Biostatistics[2], German Cancer Research Center;

Abstract: The Cambridge Crystallographic Data Files (CCDF) represent the largest collection of published crystallographic data. The CCDF, originally implemented in a hierarchical data base design, have been converted to a relational architecture in order to make data access compatible with the concepts developed for the spectroscopic information system, SPEKTREN II, of the German Cancer Research Center (DKFZ), i.e.: the procedures for handling chemical structures and related user-interface software. In addition, the 3-D data of the CCDF can be stored, modified and used to generate starting conformations for Molecular Modeling calculations.

1. INTRODUCTION

The detailed knowledge of the 3-D structure and stereochemistry of molecules is usually an essential key to a better understanding of their biological activities. The Cambridge Crystallographic Data Files (CCDF) (1,2) are worldwide the largest collection of 3-D structures of organic molecules as determined by X-ray crystallography. At the German Cancer Research Center (DKFZ) effective software to retrieve identical and similar molecules on the basis of their topological representation has been developed for the relational spectroscopic data bases of SPEKTREN II (4,5,6) system. The SPEKTREN II retrieval is supported by a graphics oriented user interface and allows one to formulate queries in a much more flexible way than is possible with the original CCDF system. The aim of this work was to convert the CCDF, which were originally implemented in a hierarchical data base design, into a relational architecture (3) offering compatibility with SPEKTREN II. So, we are able to utilize the advantages of the SPEKTREN II retrieval and have easy access to the CCDF. This combination will prove to be very useful both for molecular modeling and for augmenting the spectroscopic data bases with 3-D structural information.

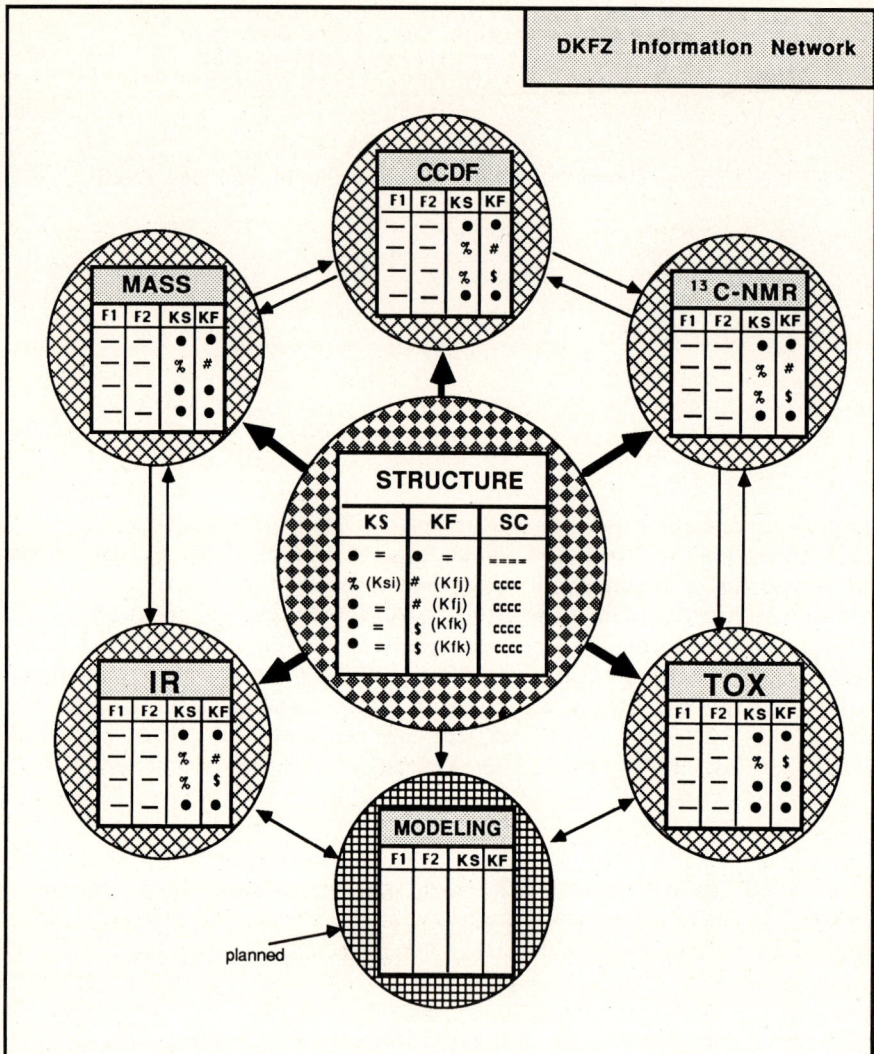

Figure 1:

The Information Network of the German Cancer Research Center (DKFZ) containing spectroscopic, crystallographic and toxicologic information. The different databases are joined together by a canonical description of the chemical structure.

KS: numerical key for the constitution as generated by the MORGAN algorithm.
KF: numerical key for the configuration as generated by the SEMA algorithm.
SC: similarity codes.
F1, F2... : stored information.

2. MATERIALS

2.1 The Original CCDF

The CCDF are a hierarchically organized collection of 3-D coordinates and certain chemical as well as physical properties gained by crystallographic measurement of substances. The data are prepared by abstracting a large, representative portion of the literature. So the CCDF provide a nearly complete overview on the substances for which structural information has been obtained by X-ray diffraction. Currently the CCDF contain about 75,000 entries.
Each entry in the CCDF contains the following information:
- a topological representation of the molecule representing the chemical connectivities for the corresponding structure. The codes that describe structure are derived from the topological description and allow substances to be retrieved by (sub-)structure search based on predefined substructure representing flags;
- chemical and physical information such as sum formula, density;
- bibliographic information (authors' names, literature source, date of publication);
- experimental parameters indicating the reliability of the coordinates;
- 3-D coordinates and related information which can be used to generate 3-D plots in various display modes and to initiate energy calculations or molecular dynamics for modeling purposes.

2.2 The Information Network Installed At The German Cancer Research Center And Some Of The Retrieval Facilities

An information network has been built up in the central spectroscopy department of the German Cancer Research Center based on a relational data base design. It includes data bases with mass, ^{13}C-NMR and IR spectra and a data base containing toxicity-related parameters. The different data bases can be coupled with the aid of the chemical structure (see Figure 1)
A topological description (bond list, atom types and bond types) of molecules, as it is used in the SPEKTREN II system, has proven to be an efficient way to store chemical structures in a computer-readable form. Therefore, this description is used in many chemically-oriented information systems.
 The MORGAN algorithm (9) is implemented to archive a canonical description of chemical structures, which is needed to identify molecules with identical constitution in the various data bases. An extension of that algorithm, SEMA (10), is used to perceive different configurations of molecules.
 For substructure and similarity searches several kinds of substructure codes, e.g. describing ring systems, types of ring fusion, heteroatoms and acyclic chains, have been generated. The coding has been designed in such a way that the codes can be searched in different modes of 'sharpness', i.e. considering the properties with more or less generality (5).

General information

Bibliographic information

REFC	Author	Year	Journal	CASno.	..

Substance information

REFC	MORGan number	Comp. name	Abs. formula	Connection table	...

experimental data

REFC	R-Factor	Tolerance	Temperature	...

sum formula

REFC	Sum formula code

3-D information

crystallographic parameters

REFC	Space Group	Bravais Lattice	...

coordinates

REFC	x-Coordinates
REFC	y-Coordinates
REFC	z-Coordinates

topologic information

REFC	Crystal connectivity

REFC	Atom labels

Figure 2: overview on the CCDF tables and their logical ordering into categories

Since the topological description of molecules is that part of the information which is common to both the CCDF and the spectroscopic data base pool, one is able to use the retrieval possibilities originally developed for SPEKTREN II, if the specific CCDF format of the topological description is transformed to that of SPEKTREN II.

2.3 Hardware And Software

The data base pool of the DKFZ has been implemented on a mainframe IBM-3090 with the VM/SP HPO operating system. The SQL/DS (IBM) data base system offers an interactive language, ISQL, for the creation and maintenance of tables as well as for querying existing tables. Additionally, SQL may be called from several host languages. This concept allows one to combine the data base manipulating features of SQL with the language elements of the host language to operate on data objects. Due to the relevance of 'number-crunching applications' which use the contents of the data base pool, FORTRAN has been chosen as host language, providing high performance in calculation programs.

3 THE INTEGRATED CCDF IMPLEMENTATION

3.1 The Relational Data Base Design

The contents of the relational CCDF tables have been organized in a logical way to speed access to closely related data. The data have been grouped into the following categories (see Figure 2):

General Information
- bibliographic information
- substance information
- experimental data
- sum formula

Three-dimensional Information
- coordinates
- crystallographic parameters
- topological information

Each category defines a table, where the columns represent individual items or properties and the rows represent data sets for individual molecules.
The data items of different tables can be joined by search arguments which are common to all tables. The key attribute REFC (reference code of the CCDF entries) identifies unambiguously a specific entry (whereas several entries may have the same structure) and allows one to couple information stored in distinct tables for a given entry. The key attribute MORG of the substance table is assigned to the MORGAN code and is a unique constitution identification within the whole information network. It serves to join the CCDF tables to the structure tables of the information network. The user need

not specify the data path or any internal parameters that depend on the storage organization, but can retrieve records or sets of records simply by describing the desired properties. This allows one to retrieve groups of entries possessing some common properties regardless of which tables are concerned. The desired query arguments can be combined from column contents according to practical needs.

Creating indices based on the key attributes of the tables guarantees high performance of the queries even if they are nested, relate to several tables, or link together information from various regions. Compared to the hierarchical data model the relational data base design provides high flexibility in querying and updating data bases as well as in altering their structures. Adding new rows to the tables does not require rewriting and sorting of the whole data base file but only some simple insert commands. Alteration of the structure of tables, as might be desired when adding further information, affects only the tables involved. Therefore, the benefits of the relational data base vs. the hierarchical design provide many possibilities for using the contents of the data base pool for further investigations.

3.2 The User Interface

General information can be interactively retrieved using the query language ISQL. Searching for identical and similar structures requires input of the topological information of a molecule. The construction of a molecule's constitution is interactive (each step is supervised by check routines and can be modified/undone by the user), a powerful command language allows one to construct the templates in the same manner as normally done on paper (4). The retrieved structures are displayed graphically on the screen, can be rotated and - if desired - stored in a data format which can be read directly by various other programs, e.g. for the calculation of physico-chemical parameters.

3.3 Retrieval Examples

Figure 3a and 3b show two typical examples for the retrieval of similar structures as they are represented on the screen. At the top left the constitution of the search template is displayed. It has been generated interactively with the aid of the user interface. The other 3 molecules are retrieved entries from the CCDF. Figure 3a shows 3 dibenzo crown ethers. Figure 3b shows 3 steroid skeletons. In both cases arbitrary substitutents were allowed to be matched. Double bond identifications are not stored in the CCDF and therefore not drawn. The capital letter string is the CCDF reference code.

4. SUMMARY AND DISCUSSION

We have shown that it is possible - with a reasonable amount of work - to link together several data bases containing chemical information using a

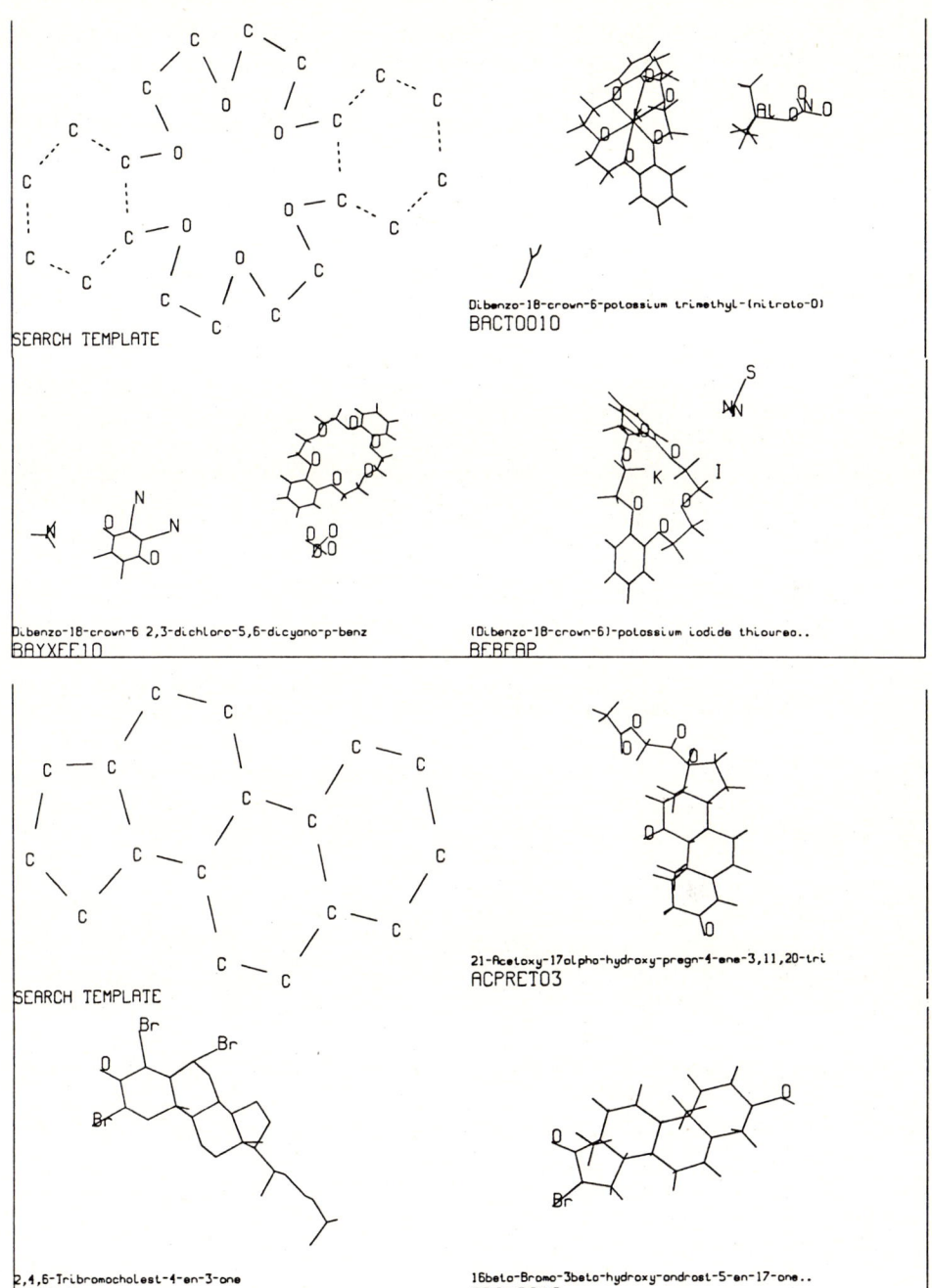

Figure 3: Hardcopies of two retrieval examples. (3a: dibenzo crown ether, 3b: steroid skeleton). At the top left of each screen page the search template is displayed, in addition, three entries possessing the desired structure found in the database. Name and CCDF reference code of each entry are indicated.

relational data structure and the topological description of a molecule. The same could be done for any other data collection containing a complete topological description of the chemical structure.

Our implementation has the advantage that the existing software of SPEKTREN II for structure coding, structure-based retrieval, graphics display and molecular modeling can be applied to the CCDF data. The structure information of a certain chemical structure is stored in a dedicated set of 'structure tables', irrespective of what other kinds of information about the structure are available. These tables are being searched during a substructure retrieval whereby several levels or degrees of similarity can be sought through the use of specific SQL string operators.

Since the spectroscopic and the crystallographic data bases are now linked together in this efficient way, an important tool has been developed which augments the spectroscopic data base also with 3-D structural information. For studies of conformations the conversion of the CCDF into a relational data structure allows us to benefit from the flexible structure-based retrieval. One aim of this approach is the study of the influence of substituents on conformation.

The found structures retrieved from the CCDF can immediately be stored on disk, manipulated with modeling software and used as input for various calculation programs such as MM2 (11), AMPAC (12) and AMBER (13). In order to handle and compare the various calculated conformations in an efficient way, it will be necessary to establish a special data base containing their structural und physicochemical properties.

References:
1   Allen FH et al (1979) Acta Cryst B35:2331
2   Taylor R (1986) J Mol Graph 4:123
3   Lang E (1989) Diplomarbeit, Universität Heidelberg
4   Förster T, Atzberger B, v d Lieth CW, (1988) Comput Chem 12:267
5   Förster T, v d Lieth CW, Opferkuch HJ (1989) Git Fachz Lab 33:318
6   Förster T, v d Lieth CW In: Gasteiger J (ed) (1987) Softwareentwicklung in der Chemie 1, Springer Berlin
7   Date CJ (1981) An Introduction to Data base Systems, Addislon-Wesley
8   SQL/DS Manuals, IBM Inc.
9   Morgan H (1965) J Chem Doc 5:107
10 Wipke WT, Dyott TM (1974) J Amer Chem Soc 96:4834
11 Stewart J J P, (1987) QCPE 527
12 Burkert U, Allinger NL, (1982) Molecular Mechanics
    ACS Monograph 117, Washington DC, USA
13 Weiner SJ, Kollman PA, Case DA, Singh UC, Ghio C
    Alagona G, Profeta S jr and Weiner P
    (1984) J Am Chem Soc106: 784

THE GMELIN INFORMATION SYSTEM
THE CONNECTION BETWEEN HANDBOOK AND DATABASE

A. Nebel, G. Olbrich, and R. Deplanque

Gmelin-Institut für Anorganische Chemie

Varrentrappstr. 40-42

6000 Frankfurt/Main 90

Abstract: The Gmelin Information System consists of the Gmelin Handbook of Inorganic Chemistry and the Gmelin Database. Currently, the Gmelin Handbook is actually created without computer data processing; the procedure of entering data into the database started at the beginning of 1989. Both components will be integrated into the Gmelin Information System. Therefore, the primary information for the handbook production (information about compounds, facts, and bibliographic data) will be delivered from the Gmelin Database. This information and careful study of the original literature forms the basis of the handbook. The distinctive characteristic of the handbook is the evaluation from the view of a scientist of published data concerning a special compound. Within this process, the data from the Gmelin Database will be edited in some cases and after completing a handbook project these data will be uploaded from the handbook department. From this point of view, the connection between the handbook and the database is a process of changing information from one component of the Gmelin Information System to the other. The main qualities of both the handbook, which are described above, and of the database (availability of on-line factual information) should be able to be retained. As an interface between the handbook and the database we have developed the Gmelin Electronic Card Index. It is the card file for the handbook production and employs methods of electronic data processing. The development was based on the hypertext software product CAMS4. This system is a means of presenting information of all kinds as a sequence of pages. Pages are organized into a logical structure called a knowledge base with the program. The pages are connected in the form of a tree structure. The system provides a number of functions in order to access a dBase III Plus compatible data file. Further developments from the CAMS4 hypertext system for the Gmelin Electronic Card Index include a flexible generator of templates and functions for direct access to an ADABAS database under VAX/VMS. The flexible generator for templates is necessary to present the contents of the Gmelin Database (815 fields) in individual form for special handbook projects. This template acts as a filter on the database for downloading, working, and uploading.

1. INTRODUCTION

The Gmelin Information System will be the integration of the Gmelin Handbook of Inorganic and Organometallic Chemistry and the Gmelin Database. Upon implementation of the whole information system, the Gmelin Database will be coupled with the Handbook. Therefore, the database will have different functions. First of all, the on-line retrieval of factual data about a specific compound will be available. The user can obtain bibliographic information as well as references to the Gmelin Handbook, but creation of search criteria with bibliographic data is not possible. The second function of the database is the delivery of primary data for a project in the handbook department. This is the principal part of the integration of the handbook and the database; at the same time, their distinguishing qualities will be retained. The connection of the handbook and the database can be described in detail as follows. The primary information of a project in the handbook department is a catalog of compounds, published facts, and citations. During the initial stages, the chief editor must obtain a complete overview by means of reviewing statistical data available from the database. The next step is downloading of the complete data package into the computer system of the Gmelin Handbook department. The chief editor then distributes the citations to the respective authors. By reading the information from the database, the author can rapidly get an overview of his special handbook writing task. The handbook production then starts with the study of the original literature and the excerption of information relevant to Gmelin. All the excerpted data will be stored in a special text system that will be able to build the logical connection to the respective citation. In some cases, the data which are downloaded from the Gmelin database will be edited, thus, an author must be able to add a new datum to a fact field; this information will in turn be described in detail in the handbook. In addition, an author may find further pertinent literature that must be added to the database. After studying the entire catalog of information in a certain area, citations and their text blocks with excerpted information will be collected by means of different search and sort criteria. Afterwards, such packages can be exported to the text editor system for writing of the scientific manuscript. Uploading of database contents from the handbook department will be accomplished in two steps. The first one is back writing of the data package from the disk of an author into the database of the handbook department. In this procedure, a control will be established on the basis of the regularities for entering data into the database. The second step is uploading into the database. After this entire procedure, which again illustrates the integration of the handbook with the database, an on-line user of the Gmelin Database will be able to obtain evaluated information on a specified compound. The interface between database and handbook is the Gmelin Electronic Card Index. The development of this software is described in the following sections.

2. THE COMPUTER SYSTEM IN THE HANDBOOK DEPARTMENT

As the central database computer and for managing the user files, we have had a VAX 6310 installed. In addition to this central unit, there are seven to eight LAVC (Local Area VAX Cluster) units. All the computers are connected via Ethernet. One LAVC consists of one VAXstation 3100 with 16 Mb of memory and a 300 Mb hard disk, as a boot node for the handbook authors. One cluster comprises five to ten VAXstation 3100 units, each with 8 Mb memory. These workstations are diskless. The application software to be implemented is ADABAS (Software AG) as the database management system and DECwrite (Digital Equipment Corporation) as the scientific text editor. The development of the Gmelin Electronic Card Index was based on the hypertext system CAMS4 (Juniper Systems). Initially, this program shell was available on personal computers under MS-DOS. In the next section, the main features will be described in detail.

3. THE HYPERTEXT SYSTEM CAMS4

The user can build a knowledge base with the CAMS4 system. A knowledge base is a collection of pages. Each page is linked to the other pages in a tree structure and a given page can be directly linked with up to ten successive pages. It may be helpful to picture these page connections as a structure in the form of a tree. If one chooses to move along a particular branch, it eventually divides into other branches, and so on, until the farthest branches at the periphery of the tree are reached. When knowledge is laid out in a tree structure, it is the outer branches that hold the most specific information. The constructor may choose among three different types of pages in order to build a knowledge base. Every page has a number, a subject line, and an area which can display lines of text and pictures. The page number may be used to directly obtain a particular page, bypassing the route specified by the designer of the knowledge base. The subject line is normally used for identification of the page content, especially when using the search function. Within the page area, normal text as well as bitmaps can be displayed. The **text page** includes these three components and the page link editor, where links to other pages can be established. On the **calculator page**, the constructor has the possibility of using an additional program with 36 steps. This feature enables operations with mathematical functions, for which the user provides the initial values. A range of standard functions is available and results can be moved to other pages of the knowledge base. A special group of functions allows direct access to a dBase III Plus compatible data file. Database fields can also be displayed and edited on the page area. The contents of 200 global variables can be stored in an external ASCII file and read from this file into the knowledge base. Furthermore, process control can be turned over to an external program. The third type of page is the **graph page**, which allows the user to display x and y values in different formats. The data for creating a graph may be set by the designer of the knowledge base or entered by the user.

The main modules of the CAMS4 program shell are the **constructor program** for building a knowledge base and the **runtime program** to read and work with the knowledge base.

4. THE DEVELOPMENT OF THE GMELIN ELECTRONIC CARD INDEX

4.1 Concept

Based on the program shell under MS-DOS, described in detail in the previous section, the first step of the development was to put up the system under VAX/VMS. Results from a detailed system analysis showed the necessity of presenting the contents of the Gmelin Database to the handbook department in an individual format. An author needs only a small portion of the 815 data fields for writing a particular article or section. During conventional handbook production, there is a complete inversion of the data structure from the primary information (citation-compounds-facts) to the handbook (compound-fact-citation). The data structure in the Gmelin Database is exactly the same as in the handbook. By using the Gmelin Database for delivery of the primary information, the author must not spend time for inverting the necessary data. Because of these points, we have decided to develop a flexible template generator. With the aid of this template, a subset of records can be downloaded into a data file on the disk of the user. The main feature of this template is its action as a filter for an ADABAS database. After the downloading process, the user can work with the template on his own data file. Evaluated and edited data can then be uploaded through the template into the database of the handbook department.

4.2 The Program Modules of the Gmelin Electronic Card Index

The prototype software package includes three modules. The communication to the database management system ADABAS was obtained by means of a special data dictionary. The program **"fileprep"** allows creation of this data dictionary, which can be accessed by the other modules. It is based on the "field definition table" of ADABAS. The main function of this module is to add long names to the structure of the underlying database structure. The module **"paint"** is the template generator. On an area comparable to that of the page area of CAMS4, the designer is able to place descriptive text and fields from the database. By pressing one button, a pop-up menu shows the list of field names. After a selection process involving moving through the list with up and down arrows, the field is placed at the cursor position. The display format of a field may be edited. All the different ADABAS field types can be displayed, including multiple fields and periodic groups. A template can be stored under a specific name after creation. The third module, "**screen**", enables the user to directly access the database through the template. In figure 1 an example for a template with bibliographic data and some flagffields is shown. Within this module, there are different functions for

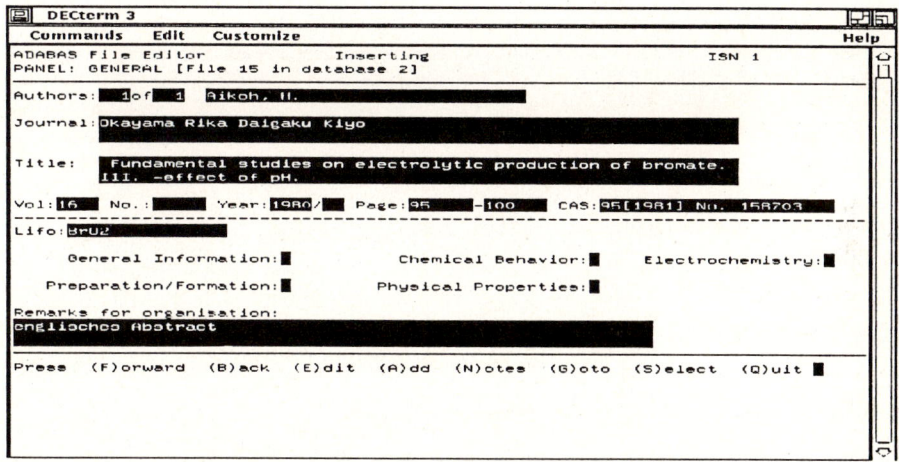

Fig. 1 Template during direct access to an ADABAS database

working on the database and on the data file after downloading. In addition to the functions for looking through the database sequentially, the user is able to build complex search criteria for working on a subset of records. These functions are the result of a detailed system analysis [1, 2, 3] in the handbook department. During the subsequent downloading process, 10 different index files were created. The system analysis [1, 2, 3] showed the necessity of sorting the data in different ways. With this arrangement, the production of the handbook can begin with study of the database contents and the original literature. There is a text window within the screen module, which can be opened with a special "hot key". Text that is entered here will be stored in a separate text file. The connection between the actual record and the text block is accomplished by means of the "Interne Satz Nummer" of ADABAS. An export function, which will be implemented in a future version of the software, allows the transfer of data (database and text window) into the DECwrite text editor system.

Further developments of the Gmelin Electronic Card Index are in progress in order to take advantage of the full range of hypertext capabilities, for example, creation of a tree structure with templates.

References:

1. EDV-unterstützte Erstellung des Gmelin-Handbuchs, Pflichtenheft, Batelle-Institut e.V., Frankfurt a.M., 1986

2. Gmelin-Handbucherstellung (EHE Gmelin), Planungsunterlagen, Version 2.0, Softron GmbH, München, 1987

3. EHE-Gmelin, Elektronischer Karteikasten, Systemanalyse, Gmelin-Institut für anorganische Chemie, Frankfurt/Main, 1988

The WODCA System

An Integrating Environment for the Chemist

J. Gasteiger and W.D. Ihlenfeldt

Organisch-chemisches Institut, Technische Universität München,
Lichtenbergstr. 4, D-8046 Garching, West Germany

Abstract: The WODCA system was developed to serve as the central module of an integrating environment for the interactive handling of programs for synthesis planning and reaction prediction. The program combines both flexibility and user-friendliness by providing fully mouse- and menu-driven input as well as a powerful command language. The program contains a data base management subsystem for access to starting materials catalogs, reads and writes various structure formats and interfaces smoothly to a set of display programs, input facilities and other utilities.

Purpose

The WODCA system (Workbench for the Organisation of Data for Chemical Applications) was developed as the frontend to SESAM (System for the Excogitation of Syntheses for Aliphatic Molecules), a synthesis planning program currently under development. The use of the system goes beyond this single application. It has emerged to be a generally useful tool center for the handling of chemical information originating from various external sources or programs of our research group such as the EROS reaction prediction series [1].

Design Goals

The system should primarily provide a user-friendly man/machine interface to SESAM and EROS. The handling should be as simple as possible, using mouse and menu driven input wherever feasible. On the other side the experienced user must not be prevented from entering highly advanced requests which are not separable into simple choices. Portability was a major concern, too. And last: our research group has created in the last decade a large stack of FORTRAN routines for the handling of molecule information which was to be utilized wherever possible. This dictated the use of the internal molecule data structure of EROS.

Design Concept

The first decision was to limit the full system to a UNIX environment. This restriction makes it easy to split the system into a set of relatively small programs which communicate with each other using pipes, which appear to FORTRAN programs as normal files. The splitting into small programs makes the single programs smaller and easier to modify, avoids mixing of

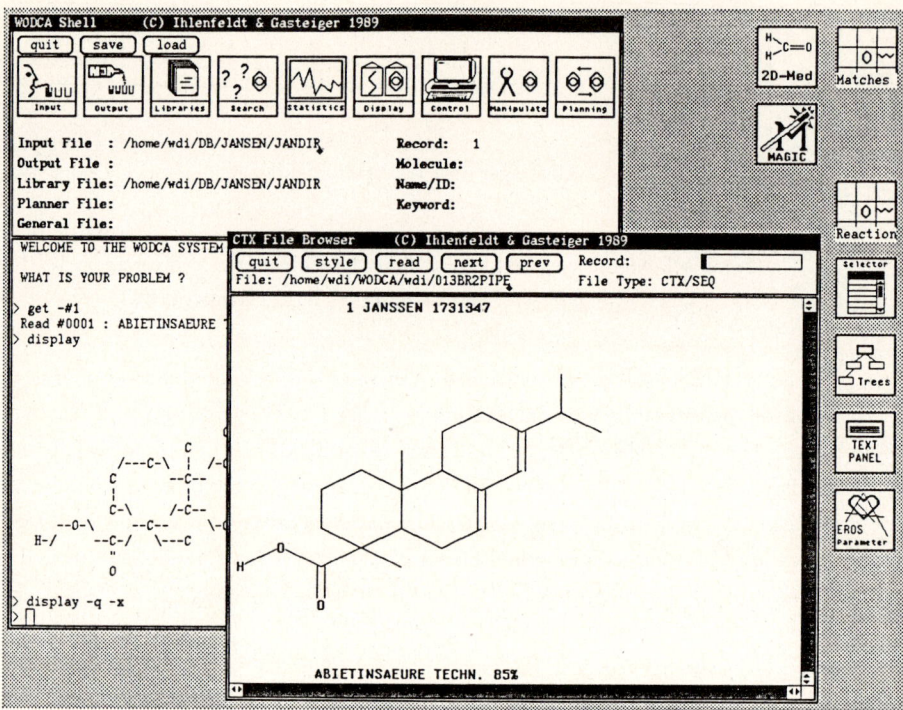

Figure 1: Typical WODCA environment after startup. The WODCA main program is running in the TTY subwindow on the lower left side. Two molecule editors, three display programs for memory, match lists and reaction have been started as well as the reaction tree display and the file tool. The text panel is a macro utility and part of the WODCA shell program. The EROS parameter utility is a control center for the EROS program. A compound from the JANSSEN Chimica catalog has been loaded and is displayed on the peripheral memory content display program.

languages (the WODCA main program is written in FORTRAN, the peripheral programs in C) and separates computation from graphical input and output. The WODCA main program is surrounded by a graphics shell program which reacts to mouse events. Only alphanumeric input is processed directly by the main program. The main program, which performs the major part of computational activity and has by far the largest source code size, does not have any requirements for input or output other than those via standard FORTRAN I/O units and is thus highly portable. It may be used as a standalone system in connection with a simple alphanumeric terminal, if no comfort is expected. Only peripheral programs use graphical input or output. These programs make heavy use of the SunView graphics library. Because they are much smaller than the central program, the effort to reprogram them using other graphics systems is minimized.

Input and Output Operations

WODCA reads and writes a wide variety of file formats. The CTX (ClearText, ASCII keyword coded information representation) format used internally for all programs in our research group

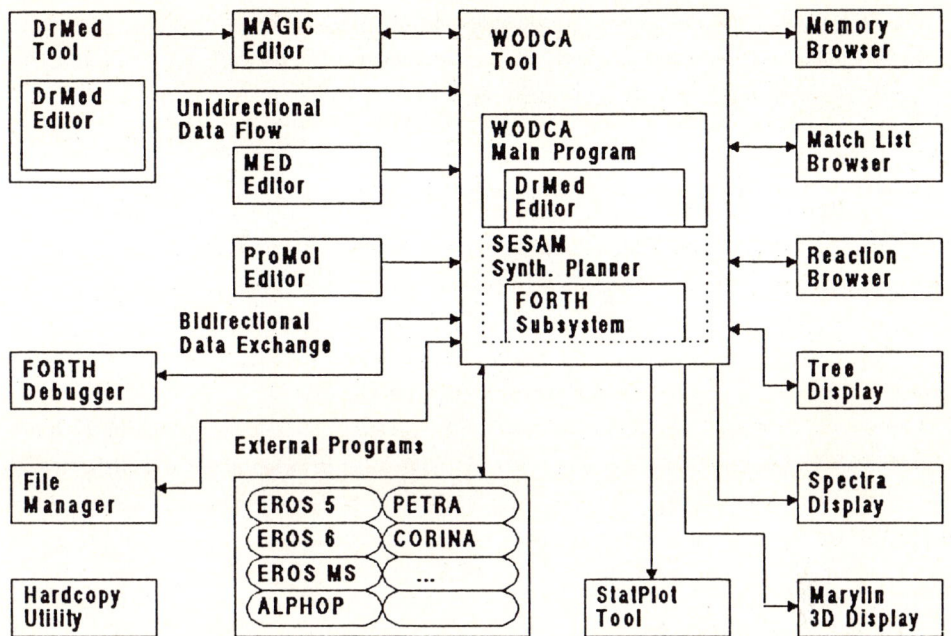

Figure 2: Data flow in the WODCA environment

is the default format and is used to transfer molecule data to and from the peripheral and external programs in the WODCA environment. Other file formats readable are:

— JCAMP

— MDL Molfile

— SMILES

— ALPHOP

— binary

ALPHOP is a compact ASCII molecule coding format which is used for the input of molecules by users equipped with alphanumeric terminals only. File formats which can be written:

— JCAMP

— SMD

— MDL Molfile

— SMILES

— ALPHOP

— binary

Molecules and ensembles of molecules from files of any type can be combined to ensembles, splitted to single molecules or sent to an editor. Files may be scanned for records with arbitrary indices or for molecules with specific properties.

Graphics Display

The WODCA main program has for reasons of portability no graphic capabilities beyond semi-graphic molecule displays using alphanumeric display characters from the ASCII set. Graphical input and output is performed by peripheral programs running in parallel. This is of no severe consequence to the display speed. A few Kbytes of data are transfered fast enough to another program not to have any noticeable impact on the response time. By default three graphics display programs using SunView as graphics system are running: a memory content browser which normally displays the current internal WODCA memory content, a reaction display which shows selected reactions from a reaction tree and a match list browser which gives an overview to the hits of a catalog search. These programs (actually, it is the same program running thrice in parallel with different startup options) are powerful itself: they can be uncoupled from the main program at any time and can be used for example as file scanners.

Analysis of Reaction Trees

Reaction trees produced by the programs of the EROS series (EROS 5, EROS 6 and EROS/MS) can be constructed from the output files of these programs. They are displayed graphically by a specialized display tool running in parallel. The user may freely zoom in and out in these trees, sort them and select reactions, ensembles or single molecules from any tree node for a close-up view.

Data Base Management

WODCA provides a large set of operations for the access to data files containing information on commercially available starting materials or other information. The data base files are basically normal CTX molecule information files formatted for direct access for which parallel files for speeded access have been created. No commercial data base management system is needed. The searches possible on these data bases include:

— direct search for names, name fragments, also using wildcards and full UNIX regular expressions

— phonetic search for names, name fragments

— name fragment proximity searches

— CAS number searches

— full structure search

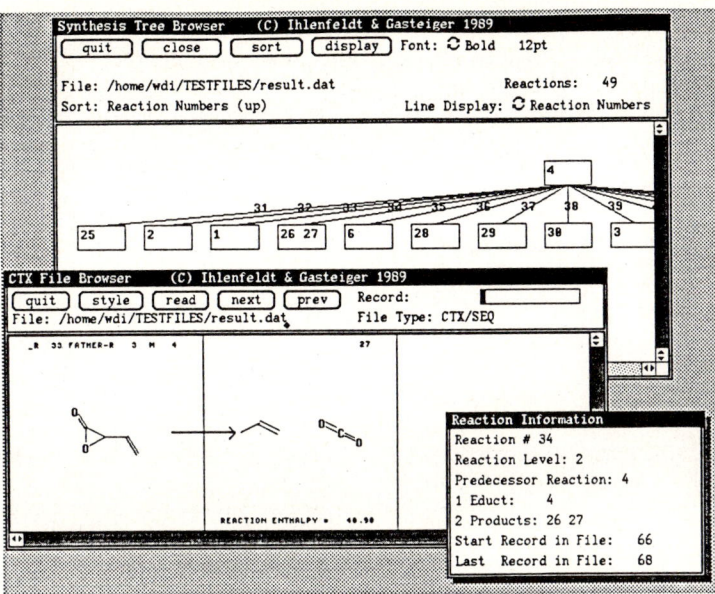

Figure 3: Reaction retrieval from the result file of a EROS 5.1 run using the tree display tool and the reaction display program.

— substructure search, including wildcard atoms, atom lists, special bond types, substructures to be excluded, separated substructures with and without overlay in the target

— atom count searches, including ring counts, aromatic ring counts, atom type counts, atom class counts, molecular weight, price per mol, isotope labelling etc.

— hash code transformation searches

— strategic bond severing combination searches

Match lists may be edited, combined using logical operations, sorted, stored, retrieved etc. The match lists can be displayed for visual control with the aid of a match list browser program running in parallel.

Some of the search options mentioned above deserve further explanation. These operations are mechanisms used also by the SESAM program made available to the human user. A hash code transformation search accepts a molecule or ensemble as input, subjects the input to one of several transformations and calculates, with an extremely small chance of collision, an unique number from the transformation product regardless of internal numbering, Kekulé structure etc. This number has been computed with the same transformation before for all molecules of the data base. The search for data base entries with the same transformation product is extremely fast. The typical search time in a 10,000 entries database including the transformation for the input is less than a second on a diskless ethernet engine with a 680xx microprocessor. The transformations are predefined and cover certain kinds of chemical similarity, which may

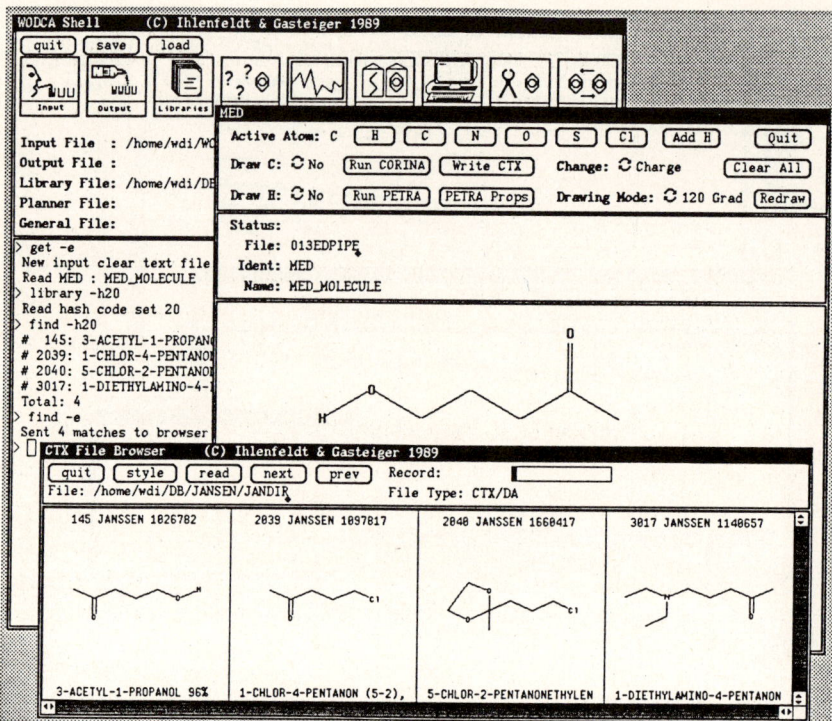

Figure 4: Results of a WODCA hash code transformation search. The hardcopy shows the MED editor, the WODCA main program and the match list display tool. The applied hash code transformation is a substitution pattern recognition.

go beyond a definition as a substructure. Transformations range from simple element normalization, ring system, carbon chain and substitution patterns to sophisticated operations such as oxidations and multistep operations.

A strategic bond severing search enables the user to tag an arbitrary number of bonds in a compound as synthetically important. The user specifies the mode of breakage, limiting factors and a hash code transformation and the computer automatically creates the fragments of the input structure for all combinations (within satisfied limits) of broken strategic bonds. The fragments, where the opened bonds are saturated by fragments depending on the bond breakage mode are subjected in addition to a hash code transformation and matches the transformed fragments against the data base. This search method often gives surprising insights into possible starting materials for a complex synthesis problem. The SESAM synthesis planning program uses this mechanism for the lookup of possible starting materials after setting the bond breakage mode to a synthon scheme.

Command Language

All features of the WODCA system can be invoked also by batch command files. The WODCA main program may in fact run isolated as a batch job without any user interaction. A set

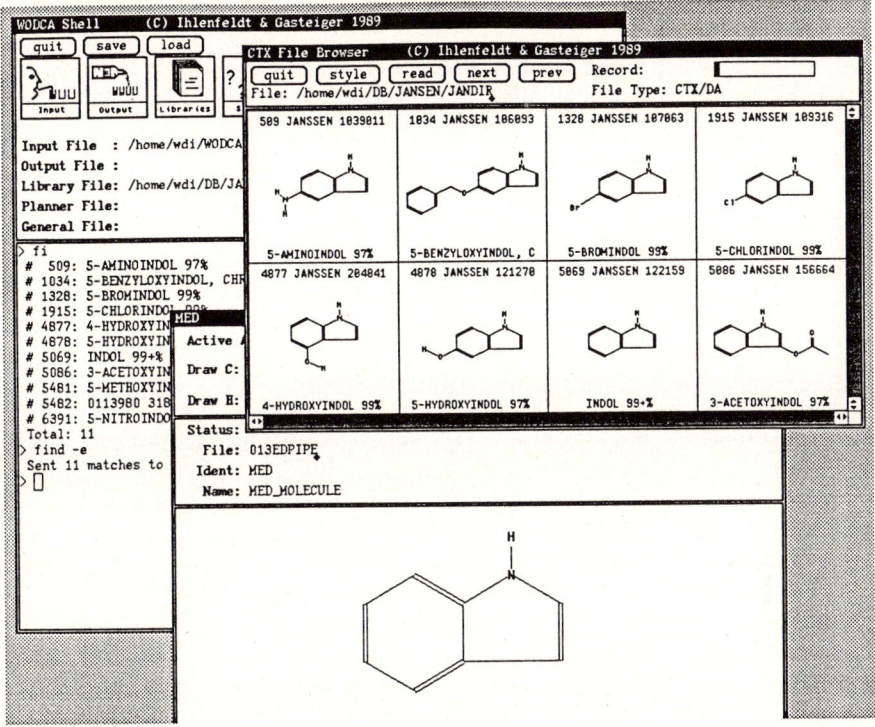

Figure 5: A ring system search for indole derivatives.

of additional commands not applicable in the interactive mode allows conditional reactions to any kind of events. Besides some predefined conditions and user variables the flexibility of the system originates from the FORTH subsystem. Using a FORTH dialect as a programming language, access to virtually every internal state variable is guaranteed. The complexity of evaluations is unlimited. Using FORTH, the FORTH language itself can be expanded to match new requirements. The FORTH subsystem is an integral part of the SESAM synthesis planning system currently under development. The algorithmic knowledge of SESAM is not coded in the program but held externally as FORTH modules.

Molecule Editors

The WODCA main program contains an alphanumeric but mighty molecule editor. However, provisions for the communication with external graphic-oriented and mouse-driven editors have been made. Data may be loaded from these editors at any time and distributed to other sources. The currently supported peripheral editors for the WODCA system are MAGIC, an editor using the PHIGS graphics system, MED, a SunView editor, PROMOL, an experimental editor written in PROLOG and DRMED, an alphanumeric semi-graphics editor running under the control of a SunView shell program.

External Programs

The data in the WODCA memory can be submitted to many programs which calculate properties of molecules. The result files can be read in again. Examples of such programs are:

PETRA	Calculate physicochemical parameters such as atomic charges, polarizabilities, resonance stabilization, heat of formation etc.
EROS 5.X	Reaction and retroreaction prediction [1]
EROS 6	Reaction Prediction [2]
EROS/MS	Mass spectra prediction [3]
CORINA	3D coordinate generation [4]

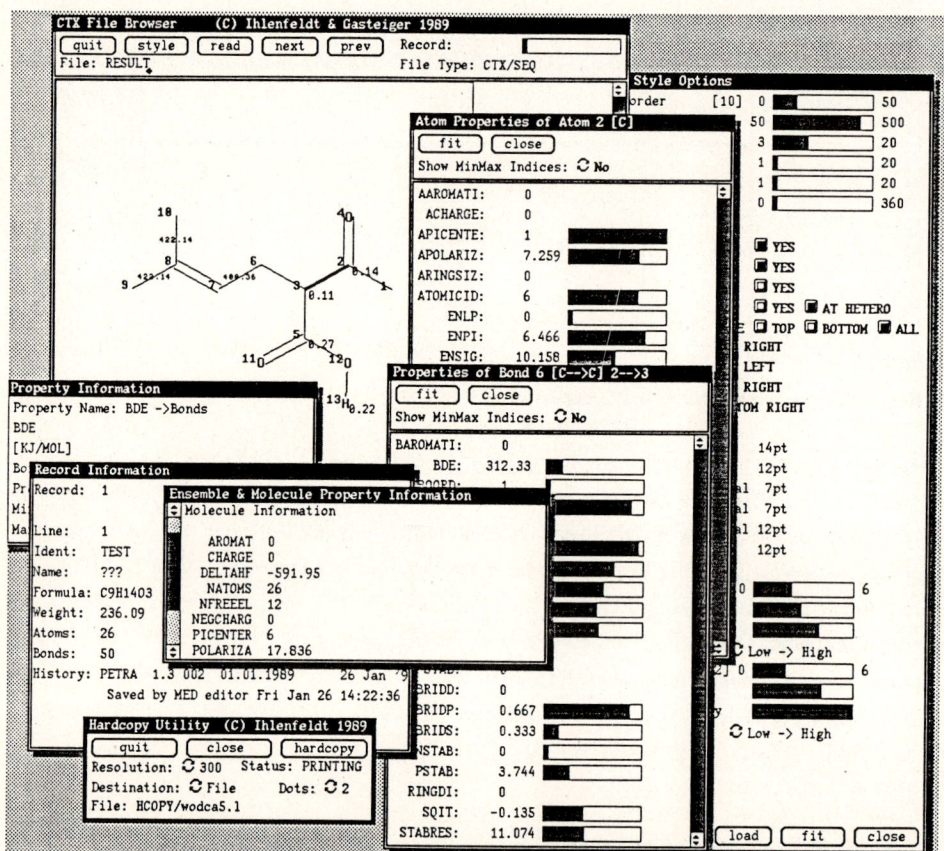

Figure 6: Analysis of the results of a PETRA run using the memory display program with all possible subwindows opened. The graphics display shows atomic numbering, the sigma charge on the most positively charged atoms and some bond dissociation energies. A close-up view on the atomic properties of atom 2 (carbonyl carbon) and on the bond properties of the bond from atom 2 to atom 3 (central atom between the carbonyl groups, fat line) is taken. In addition, the subwindows for general record information, molecule properties and display style options are also opened.

Technical Details

The system runs currently on a SUN 3/80 with 12 MB of chip memory. The operating system is UNIX with SunView as surface layer.

The WODCA program itself consists of about 40.000 lines of FORTRAN 77 subset code. With the exception of the FORTH module (2500 lines C language plus 500 lines FORTH itself to expand the language) all subroutines are plain FORTRAN, although some routines exist for speedup reasons also in a parallel, faster C version. WODCA does not require any graphics output terminal capabilities. These functions are performed by the surrounding peripheral programs.

The peripheral programs with the exception of the MAGIC and DRMED molecule editors are written in C language (each between 1500 and 5000 lines of code). Their input and output operations rely on the SunView window system. They run in parallel to the main process. Communication is basically via standard UNIX named pipes which are accessed in FORTRAN as normal sequential files after they have been created by a small specialized C program. Commands entered in the peripheral programs which require actions from the WODCA main program are reported to the WODCA shell program via a shared IPC (InterProcessCommunication) message queue and transferred by the shell program to the main program as text commands on FORTRAN input unit 5. All menu- and button-events processed by the shell program are sent to the main program in the same fashion as alphanumeric input which could also have been typed in manually in the main program.

With all peripheral programs running (including the SUN operating system and two system shells, but without external programs such as the EROS system) and medium sized graphical displays, the memory requirement is about 12 MB. For most purposes, 8 MB of real memory are sufficient for satisfactory response times. The memory requirement may however be reduced drastically by deselecting some of the peripheral programs which are not needed for a specific application within the configuration tool which starts the WODCA system.

References

[1] Gasteiger J, Hutchings MG, Christoph B, Gann L, Hiller C, Löw P, Marsili M, Saller H, and Yuki K (1987) *Topics Curr Chem*, 137:19–73

[2] Gasteiger J, Röse P, this proceedings

[3] Gasteiger J, Hanebeck W, this proceedings

[4] Hiller C and Gasteiger J (1987) In Gasteiger J, editor, *Software-Entwicklung in der Chemie 1, Proceedings des Workshops "Computer in der Chemie", Hochfilzen, Tirol Nov. 1987*, pages 53–66, Springer-Verlag, Heidelberg

COMPUTER - ASSISTED ANALYSIS OF QUALITATIVE STRUCTURE/ACTIVITY RELATIONS OF ORGANIC MOLECULES

K. Bley and I. Ugi

Lehrstuhl I für Organische Chemie, TU München
Lichtenbergstrasse 4, 8046 Garching

Abstract: The topic is a theoretical approach to the search for activity - carrying substructures. For this the theoretical foundations of structure/activity relations are exploited. A data structure is presented that affords structure/activity correlations, and its realisation in CORREL II, a modified version of CORREL. The upper bound of performance that can be achieved by present day computers and the corresponding presequisites are discussed. Practical examples are given.

1. CONCEPTIONAL FOUNDATIONS

1.1 General Concepts

Some definitions must established first for this approach to the problem of correlating structure and activity of organic molecules. One of these concerns a hierarchical classification of the structure of organic molecules.

1. empirical formula (that we may call the primary structure)
2. chemical constitution (secondary structure)
3. stereochemical structure (tertiary structure)
4. association of molecules to aggregates (quaternary structure)

Thus the structural information that is needed for a structure/property correlation begins with the empirical formula, that indicates the kind and number of atoms in a given molecule (primary structure). The chemical constitution is described by stating for each atom its covalently connected neighbours. The stereochemical structure (tertiary structure) of rigid molecules is given by the relative spatial arrangement of the atoms. The stereochemical features of flexible molecules must be defined in a different, more abstract manner [1]. The quaternary structure of multimolecular aggregates that may include solvent molecules, is determined by the structural features of the individual molecules, together with the interactions of the participating species.

The application of the classical key/lock model to enzymes and their substrates requires detailed knowledge of the quaternary structure of the whole system, including all data concerning electronic and spatial effects and, the solvent environment of the respective molecules.

If all those data were available, a structure/activity correlation would be possible.

However, the required comprehensive data are seldom available, and, at best, only for very few molecules, since such data can only be obtained by laborious investigations.

Also, the number of required variables and parameters will exceed the present computational means.

It is recommendable to reduce the complexity of the problem. When the quaternary and, if needed, even the tertiary structures are neglected, molecular structure is reduced to a list of bonds. Its computational treatment can be conveniently accomplished on the basis of an algebraic model of constitutional chemistry, as formulated by Ugi and Dugundij [2].

The first version of the program system CORREL [3] - [5] for structure/activity correlation is based on a reduction of the problem to constitutional chemistry.

1.2 The Concept of CORREL

CORREL does not aim at predicting properties or activities for individual compounds. CORREL is designed to recognize those substructures that carry a given property, through an analysis of a list of active and non-active compounds. Thus CORREL finds whether or not there exists a constitution/property relation in the given set of data, and if the answer is positive, it proposes a substructure as being responsible, provided a suitable set of data on a not too small number of compounds is available.

- the approach is based on complete connectivity lists
- a complete set of all substructures is generated, for all the compounds belonging to the dataset
- for each substructure its parent structure is noted

The CORREL approach aims at finding from a set of activity data of a list of active and non-active molecules the substructures that are, from a statistical standpoint, significantly responsible for the considered activity. That may include the solution of formidable combinatorial problem [6]. Thus a relationship can be established between structure and activity, provided that the activity is a consequence of the chemical constitution of the considered molecules. Note that this does not encompass total neglect of the tertiary and quaternary structures, because these depend strongly on the respective chemical constitution.

If no structure/activity correlation is obtained by CORREL, this means that the activity is mainly a function of the tertiary and quaternary structural features, and that there is no direct relation between these and the corresponding chemical constitution. This negative information is often valuable.

2. THE DATA STRUCTURE OF CORREL

2.1 Substructure Basis

The essence of CORREL is a substructure library containing, within suitably chosen bounds, all substructures of a given set of molecules. These substructures are connected by a network of pointers that indicate the "father-son-relations" of the substructures. Thereby it is possible to find directly all parent molecules of a given substructure, in order to evaluate this substructure.

This data concept that was presented by J. Friedrich et. al. [3] - [5] is now used in many modern substructure correlation programs, e. g. RESY,KOWIST,HTSS,S4 etc. [7] - [11]. Since each substructure represents also all and any smaller substructures that it contains, a query finds all parent molecules with a given substructure in a single hit.

A numerical value can be assigned to a given substructure, that indicates its probability to be endowed with the considered property. This number can be computed from the chemical constitution of the active and non-active molecules in the list of investigated molecules. If such an evaluation does not pinpoint a subset of distinguished parent structures, this may serve as a hint that the property of interest is not strongly dependend on the constitutional features of the molecules, but is rather determined by other factors, e. g. stereochemical ect.. The compounds with the highest values of assigned numerical probabilities of activity, give leads to new active structures.

2.2 Structure Generation and its Limitations

If all substructures are generated from a given parent structure, combinatorial explosion may occur from a certain molecular size on, i. e. the number of substructures generated exceeds the capacity of present-day computers. The previous versions of CORREL had rather low upper bounds for the size of parent molecules and generated substructures, as well as the cardinality of the lists of compounds. Either only substructures with a graph theoretical radius ≤ 2 were retrieved, or only fairly small molecules could be subjected to complete fragmentation. Besides, the size of the database was also limited at the start of the program by the data access method (hash codes).

Our present approach affords the use of substructures with a graph theoretical radius of ≤ 3, as well as - with sufficient computing power and memory space - taking into account the hydrogen atoms, not only the skeletal atoms of the molecules. This is accomplished through a new data access method without predefined upper bound for the number of structures and substructures to be stored. The volume of data is reduced by selection of the "sphere" centers of the substructures by chemical criteria. A new structured mode of storage and modified accessing procedures have been introduced for substructures, in order to achieve a higher density of data, and thereby an increase in the size of the database.

3. THE REALIZATION OF CORREL

3.1 Implementation

CORREL was first implemented in 1974-1979 as a PL/1 program [2],[3]. The present version of CORREL (called CORREL II) was produced within the framework of a completely new implementation in FORTRAN 77. It can be operated with any FORTRAN-compiler that meets the ANSI 3.9 77 standard. The only extension needed is the possibility to manipulate INTEGER numbers at the bit level (OR, AND, NOT and SHIFT). As a rule, this is available on all compilers, in view of the future standard. Experimental implementations exist in a PC-version (Ryan-McFarland-FORTRAN 2.11), as well as in a CYBER-version (CDC-NOS/VE-FORTRAN).

The system has undergone substantial changes with regard to many essential algorithms. In fact, relative to the original CORREL, CORREL II is a completely new system of programs. CORREL II has a much wider scope and fewer limitations than the previous version.

3.2 Implementational Bounds

The previous approach was confined to substructures with a graph theoretical radius $\leq$ 2, or to molecules with only few atoms. Furthermore, the maximum size of the database was predetermined at the start of the program. Within the framework of the complete re-implementation a new substructure generating fragmentation procedure was installed, and a totally different approach to data access was introduced. CORREL II is capable of processing molecules of practically any size, i. e. within the present bounds of 255 atoms per molecule (based on coding in one byte).

This is achieved by first clipping out "major fragments" of a graph theoretical radius of $\leq$ 3 that are subsequently subjected to total fragmentation. The centers of the abovementioned major fragments are selected according to chemical considerations. Heteroatoms or sp^2-hybridized carbon atoms are favored as centers of fragments.

Database access is now organized in terms of so-called Bayer-Trees. Accordingly, the upper bound for the size of molecules to be processed is only limited by the available memory space, the internal numerical representation, and the allocated computer time. The number of the structures depends on the internal numerical representation (as a rule, 32-bit-INTEGER) and reaches therefore $2^{31}-1$ ($\approx 2*10^9$) sets of data.

Due to data compression at the bit-level, the size of molecules is currently restricted to 255 atoms and a maximum coordination number of 6.

Fragmentation is done by a new fragmentation algorithm which can optionally be switched to smaller spheres. The maximum size of fragments generated is a skeletal graph-theoretical radius of ≤ 3. Moreover, hydrogen atoms can be included in the fragments, if the memory size and computer time limitations permit.

4. EXAMPLE

The processing of a toxicological dataset [12] may serve as an example of structure/property correlation by CORREL II.

The dataset comprises 105 organic molecules of which 21 are endowed with the property "accumulation in the tissues of fish". The remainder of the listed molecules do not exhibit this property. All molecules are fragmented to substructures with a graph theoretical radius ≤ 3. Three atoms is the minimum size of the resulting fragments. Only skeletal atoms and the hydrogens of functional groups are taken into amount.

Fragmentation of the 105 parent molecules according to the procedures in [13] yields ca. 30000 substructures that meet the fragmentation criteria. Evaluation of the fragments according to "accumulation in the tissues of fish" leads to 1,3-dichlorobenzene derivates as the active compounds (see fig. 1). This result is in full agreement with the hitherto known lipophilicity data on chlorinated hydrocarbons and as well as their interpretation [14].

(fig. 1)

5. PERSPECTIVES

In the coming phase of testing CORREL II, the use of larger, more realistic sets of data has priority. Under the given restrictions on memory space (at our insitute ca. 150 MB disc space is available for such studies), roughly 300000 substructures can be stored within CORREL's contiguous network of father - son- relations. This will suffice to assess the present power of CORREL II, and to reveal the future needs of modification and expansion. After some improvements CORREL II will be ready for routine use.

6. REFERENCES

1. Ugi, I. et. al. (1984) "Perspectives in theor. Stereo-chemistry", Springer, Heidelberg
2. Ugi, I., Dugundji, J. (1973) Topics Curr. Chem., **39**, 19 ff.
3. Friedrich, J. (1979) Dissertation, TU München
4. Friedrich, J., Ugi, I. (S1980) J. Chem. Res., **70**
5. Friedrich, J., Ugi, I. (M1980) J. Chem. Res., 1301, 1401, 1501
6. Anal. Chim. Acta, Proccedings of the IX. ICCCRE 1989, in press
7. Bawden, D. (1983) J. Chem. Inf. Comput. Sci., **23**, 14
8. Meyer, E. (1988) E. Sens, Anal. Chim. Acta, **210**, 135
9. Hansch, C., Leo, A., Elkins, D. (1974) J. Chem. Doc., **14**, 57
10. Nagy, M. Z., Kozics, S.,Veszpremi, T., and Bruck, P., (1988) "Chemical Structures" W. A. Warr (ed.), Springer-Verlag
11. Gauglitz, G. (Ed.) (1989), "Software Entwicklung in der Chemie 3", Springer-Verlag Heidelberg
12. Fontain, E. (1983) Diplomarbeit, TU München
13. Bley, K. (1983) Diplomarbeit, TU München
14. H. Bader (1985, 2.Auflage), "Lehrbuch der Pharmakologie und Toxikologie", VCH-Verlag Weinheim

PREDICTION OF THE THRESHOLD SOOT INDEX FOR HYDROCARBON FUELS WITH RANDIC'S TOPOLOGICAL INDICES

Dr. Degen

Daimler Benz AG
D-7000 Stuttgart 60

Abstract : Topological indices and simple structural descriptors were used to calculate the threshold soot indices of 87 hydrocarbons of different types. Regression analysis yielded for the topological indices an only slightly better result than for the simple structure parameters. Better equations could be achieved by dividingthe dataset into two appropriate cluster and correlating each cluster seperately to the TSI - values.

Introduction: In our attempts to improve motor combustion processes , properties of the hydrocarbon fuel play an important role. Since soot formation is one of the most undesired effects in diesel combustion, the fuels propensity to form soot is a matter of considerable concern. An ASTM test (1) takes the maximum smoke-free flame height, called smoke point, of a laminar diffusion flame as reciprocal measure of the tendency of the fuel to produce soot. Similar devices have been used in several other studies (2-5), but since the type of apparatus strongly influences the measured numerical values, these studies could be compared only in there major trends.

Calcot and Manson (6) reviewed the literatur and defined a numeric scale, called threshold soot index (TSI), which eliminates the apparatus factors, so that the fuel datas could be compared quantitatively. The TSI is defined as

$$TSI = a\ (MW/h) + b \quad (\text{Eq 1}),$$

where MW is the fuels molecular weight, and h is the maximum smoke-free flame heigth. The constants a and b scale the measured values from 0 to 100 and eliminate the influence of the experimental setup.

No simple relationship was found between molecular structure and TSI. Qualitatively the TSI increases with the number of carbon atoms, the carbon/hydrogen ratio, the compactness and the degree of branching of the molecular structure. Because TSI are additive, the soot forming propensity may be estimated for fuel mixtures too, if the TSI values for pure hydrocarbons could be computed.

Hanson and Rouvray (7) have used topological indices to predict TSI values for pure hydrocarbons. They found, that amongst other, the so called hydrogen deficiency index (HD), and the connectivity index of Balban are the best descriptors for the TSI. The HD - index is simply the sum of the number of all multiple bonds and rings in the molecule and is therefore an nondiscriminating index. Their best equation was nonlinear. So the influence of hydrogen deficiency and molecular branching could not been separated.

The Data Set : In this study we use TSI values for 87 pure hydrocarbons of different types, given by Olson and coworkers (5). A complete set of Randic connectivity indices (8) was calculated for these hydrocarbons and a set of simple molecular parameters, including the HD-index, was taken for comparison (see Table 1).

The connectivity indices for molecules, containing unsaturated bonds, can be calculated in two ways. Murray (9) suggested, that the molecular graph should be the same for saturated and unsaturated molecules. The multiple bond is counted as one edge and only represented by the endpoint valencies. Randic proposed, that a double bond has to be counted as two edges (Fig. 1) . The endpoint valencies are the same. The major difference in the two ways of computation is, that in Randic's method the unsaturated character of the molecule is represented by indices of the chain typ. Only Randic's connectivitiy indices gave satisfying correlation, so this indices were chosen for this study.

The correlation between the TSI and the topological indices were examined first by factor analysis. Factor analysis gives us a good impression, not only of the relationship between the TSI and the topological parameters, but also of the colliniarity of the used indices. It gives us a rough estimation about the possible correlation coefficient and the possible number of independent regressors too. So factor analysis serves us as a starting point for a systematic multiple linear regression analysis.

Table 1. Definition of the Struktural Descriptors Used in the Regression Analysis

a. simple parameters

1. n - C   : number of carbon atoms
2. n - H   : number of hydrogen atoms
3. n - DB : sum of double bonds and rings
4. CH0   : number of tert. carbon atoms
5. CH1   : number of =CH- groups
6. CH2   : number of -CH - groups
7. CH3   : number of -CH groups

b. topological indices

path type	chaintype
CHI0 =	CHAIN2 =
CHI1 =	CHAIN3 =
CHI2 =	CHAIN4 =
CHI3P =	CHAIN5 =
CHI4P =	CHAIN6 =

cluster type	path - cluster type
CHI3C =	CHI4PC =
CHI4C =	CHI5PC =

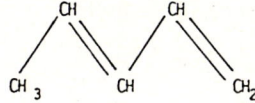

molecular graph according Murray	molecular graph according Randic
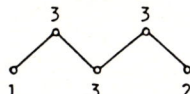	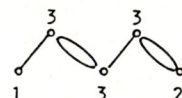
= 1·(1·3)$^{-1/2}$ + 2·(3·3)$^{-1/2}$ + 1·(3·2)$^{-1/2}$	= 1·(1·3)$^{-1/2}$ + 2·(3·3)$^{-1/2}$ + 1·(3·3)$^{-1/2}$ + 2·(3·2)$^{-1/2}$
1.652	2.394

Fig. 1

Results and Discussion : In Fig. 2 the pattern of the factor analysis for the simple structure parameters (number of carbon atoms, of hydrogen atoms, of double bonds, of methyl-groups etc.) is shown. The TSI strongly correlates with the double bond equivalent and the number of tert. carbon atoms. This reflects the high TSI values of aromatic and highly branched molecules. The best linear correlation equation was :

$$TSI = -2.91 + 14.88\ CH0 + 7.69\ CH1 + 0.74\ CH2 - 1.71\ CH3$$
$$(N = 87,\ r = 0.9691,\ F = 312.69,\ RSS = 3454\ )\ (Eq\ 2)$$

The pattern of the factor analysis for the Randic indices is shown in Fig. 3. The TSI is strongly correlated to the indices of chain type and the path-cluster type. These indices also represent the unsaturated character and the amount of branching of the molecule. The best linear equation was found to be:

$$TSI = 2.78 + 24.43\ CHAIN2 - 6.56\ CHAIN3 + 0.98\ CHAIN4$$
$$+ 4.90\ CHAIN5 + 73.25\ CHAIN6$$
$$(N = 87,\ r = 0.9778,\ F = 347.34,\ RSS = 2501\ )\ (Eq\ 3)$$

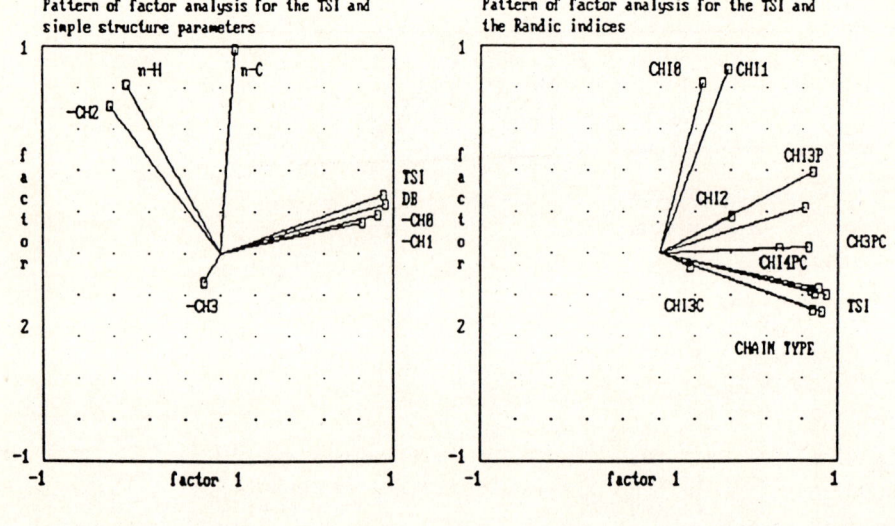

Fig. 2   Fig. 3

This equation is only slightly better than the equation with the simple parameters. Fig. 4 compares the measured TSI-values and the calculated TSI-values. The reason for the remaining scatter is on the one hand the poor quality of the measured data. But on the other hand the distribution of the measured hydrocarbons on the TSI-scale is quite disadvantageously. Most of the alkanes, alkenes and the cycloalkanes have TSI-values between 0 and 20. Only the relative few aromatic hydrocorbons have TSI-values greater then 20.

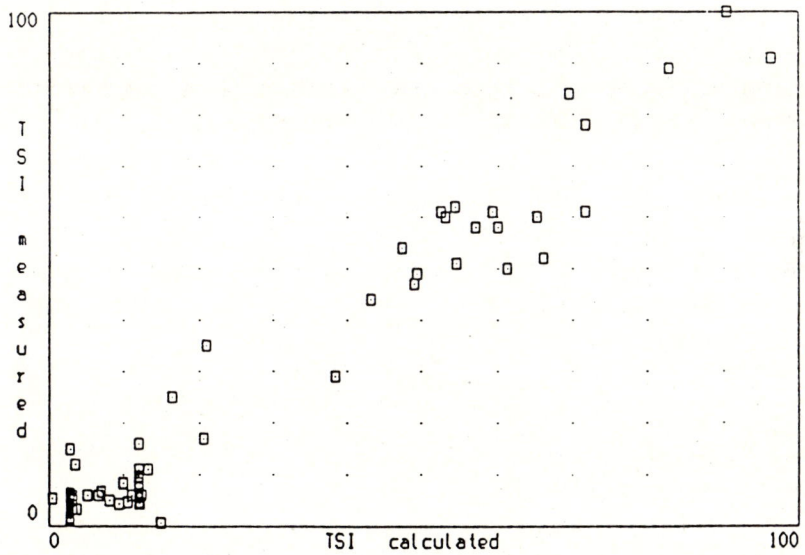

Fig. 4

Correlation equations for hydrocarbons of the same type showed to be much more satisfying. Such equations sometimes are combined using an extra parameter, which has to be chosen carefully for every class of hydrocarbon. The appropriate class has to be selected by the user according his chemical knowledge. This procedure is not suitable for automatic TSI calculation.

Our approach is, to divide the set of measured hydrocarbons into two clusters, which are regressed separately, to find optimal regressors and correlations. The TSI value of a new hydrocarbon can now be calculated only from the molecular graph and without further chemical knowledge, by first determining the appropriate cluster and then using the right correlation equation.

The dataset was clustered first by using the well known KMEANS procedure (10). The used variables were the TSI-values, and the indices CHAIN2 and CHAIN3. Cluster one contains all the aromatic compounds while cluster two contains nearly all nonaromatic hydrocarbons with exception of 1.4 butadiene. So the result seems reasonable.

We then took the two cluster to perform Fisher's discriminating analysis which gives us the linear equation for the decision plan (see Fig. 4 a,b).

$$TSI = 39.48 - 5.65\ CHAIN2 - 7.53\ CHAIN3 \quad (Eq\ 4)$$

For hydrocarbons with unknown TSI the TSI value has to be guessed. This can be done by using the linear equation 5.

$$TSI = 1.17 + 20.91\ CHAIN2 + 21.63\ CHAIN3 \quad (Eq\ 5)$$

Equation 4 and 5 can of cause be combined. Fig. 5 shows that this procedure gives a unique and reasonable classification.

Fig. 6 a,b shows the pattern of the factor analysis for both clusters. Again the TSI correlates best with indices of the chain type. But for the type B (aromatic) hydrocarbons this correlation is not as strong as for the type A hydrocarbons (alkanes, alkenes, cycloalkanes).
There remains however some correlation with the indices from path - cluster type. But this correlation seems to be weaker for both clusters than for the complete set.

Multiple linear regression gives for cluster A :

$$TSI = 0.16 + 11.88\ CHAIN3 + 1.99\ CHI3P + 1.30\ CHI3C$$
$$(r = 0.8613,\ RSS = 249.60,\ n = 64)\quad (Eq\ 6)$$

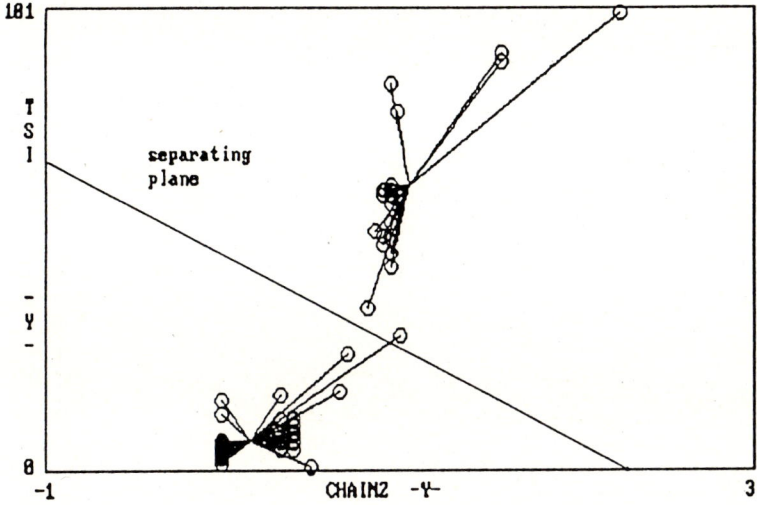

Fig. 4a .

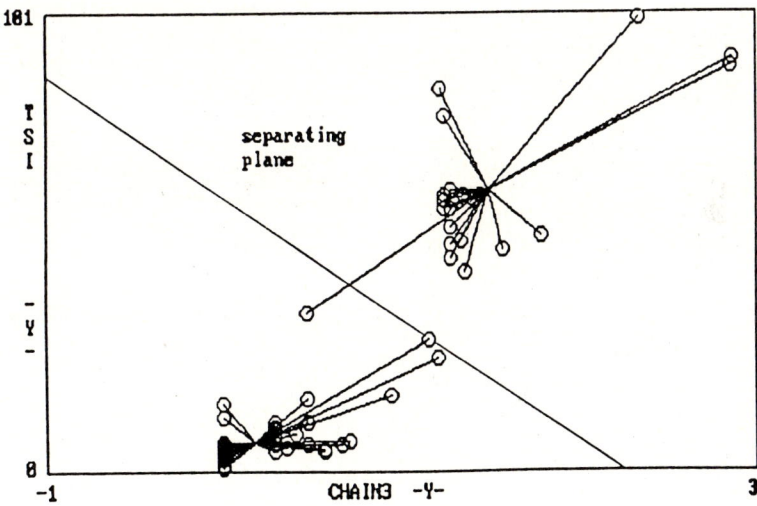

Fig. 4b .

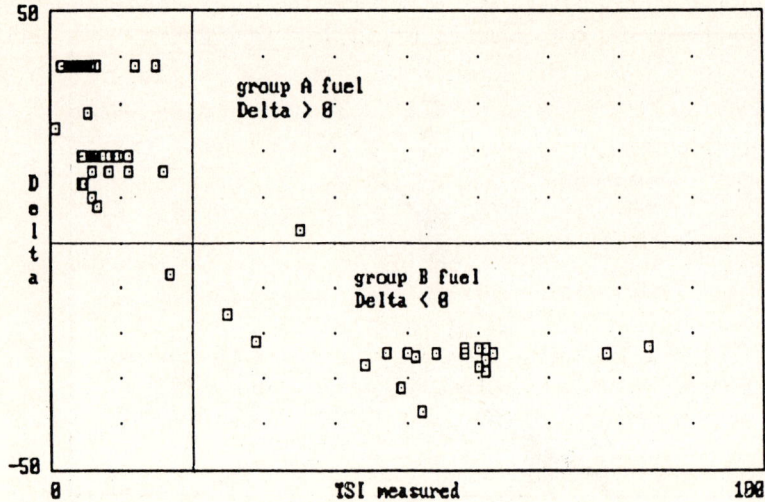

Fig. 5.

and for cluster B :

TSI = -19.27 + 37.22 CHAIN2 + 2.07 CHAIN5
+ 0.24 CHI3P + 12.54 CHI3C
(r = 0.9367 , RSS = 996.34 , n = 23) (Eq 7)

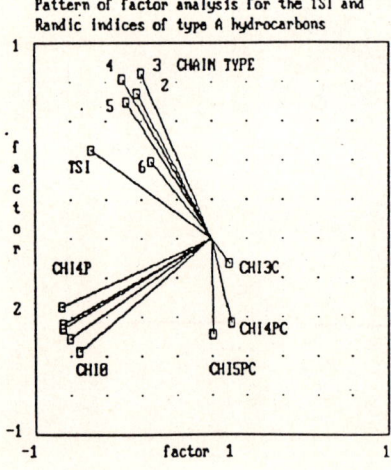

Fig. 6 a.

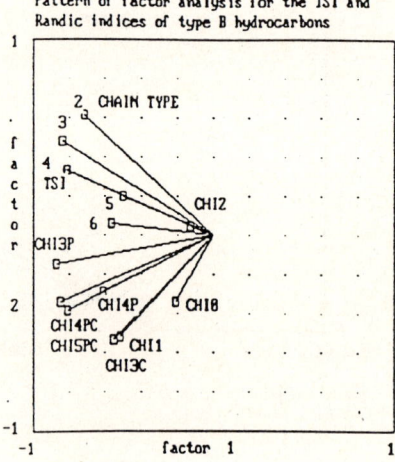

Fig. 6 b.

Although both correlation coefficients are smaller than the correlation coefficient of eq 3 the residual sum of squares (RSS) of the combined equation is smaller than in eq 3 (RSS eq 3 = 2501, RSS comb. eq 6 and 7 = 1245.9). Therefore also the correlation coefficient for the combined equation 4 and 5 is better than for eq 3 (0.9778 eq 3 versus 0.9884 comb. eq 4 and 5) . Fig. 7 shows the calculated and the measured TSI values for both clusters.

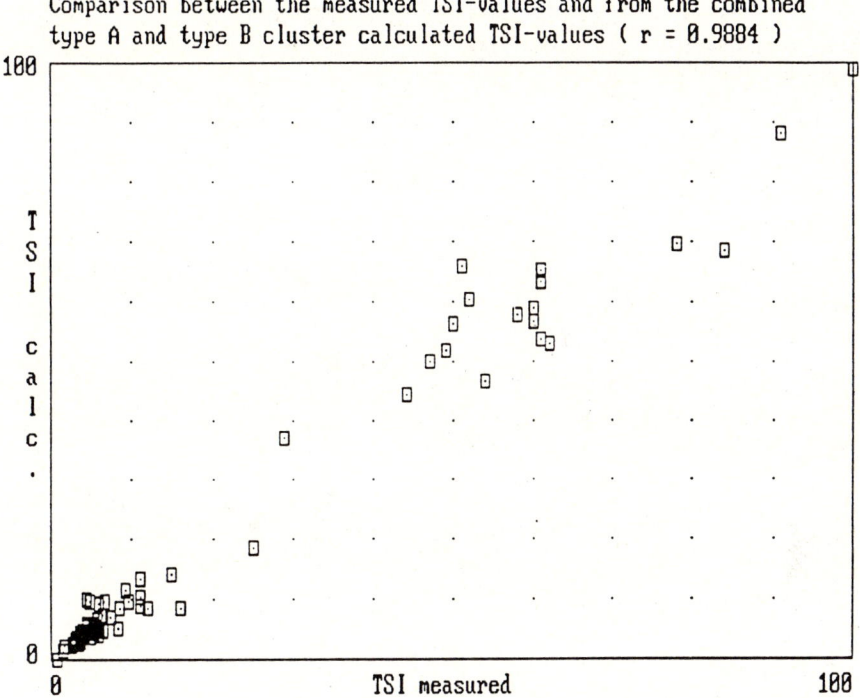

Fig. 7.

References:

1. American Society for the Testing of Materials, D-1322, 1975
2. Minchin S T (1931) J. Inst. Petrol. 17:102
3. Clarke AE, Hunter TG, Garner FH (1946) J.Inst.Petrol.32:627
4. Hunt RA (1953) Ind. Eng. Chem. 45:602
5. Olson DB, Pickens JC, and Gill RJ (1985) Combust. Flame 62:43
6. Calcote HF, and Manos DM (1983) Combust. Flame 49:289
7. Hanson MP, and Rouvray DH (1987) J. Phys. Chem. 91:2981
8. Randic MJ (1975) Am. Chem. Soc. 97:6609
9. Murray WJ, Kier LB, and Hall LH (1975) J. Pharm. Sci. 64:1987
10. Walther H, and Ngler G, Graphen - Algorithmen - Programme Springer Verlag, Wien , New York, 1987

A GRAPH THEORETICAL METHOD TO ESTIMATE SUBSTANCE DATA

R. Brüggemann and B. Münzer

GSF/PUC
Ingolstädter Landstr. 1, D-8042 Neuherberg

Abstract: The lack of data is a major problem for predicting the fate of chemicals in the environment. Data sets taken from handbooks or online databases are often incomplete and inaccurate. We have developed two computer tools to complete such data sets via estimation and to test the quality of the data set. Both programs use methods from graph theory to make optimal use of a complex system of relations while maintaining a very simple user interface.

DTEST (an acronym for DaTa ESTimation) is a computer program to estimate the physico-chemical data needed for environmental fate models using mainly property-property relations. DTEST uses concepts from graph theory to select the most appropriate relations depending on the initial data set.

DXVAL (Data cross VALidation) tests a set of substance data (usually taken from literature) for redundancy and consistency. This is achieved by comparing the given (literature) data with values estimated with the DTEST algorithm. DXVAL is intended to test the quality of a data set rather than just the quality of single data. DXVAL indicates if the data set is not complete enough to allow cross validation, it can detect if a seemingly large data set has been obtained by estimation from just a small set of experimental data, and if there are any inconsistencies in the data set. In any of these cases, DXVAL can then be used for a sensitivity analysis within the data set.

DTEST : A NETWORK OF ESTIMATION FORMULAS

DTEST uses mainly property-property relations to estimate missing data from a given data set. Table 1 shows the quantities which are needed or which can be calculated by DTEST. For an explanation of the symbols and abbreviations, see table 2.

Table 1: Crossbar representation of all relations in DTEST.

	LOGKOW	KOC	BP	VP	HENRY	MOLW	SOLW	THOD	BCF	VFSG	VLEBAS	DW	DG	RELEAS	PROVOL
LOGKOW	.	X	.	.	.	.	X	.	X	.	.	.	.	.	.
KOC	X	.	.	.	.	.	.	.	.	.	.	.	.	.	.
BP	.	.	.	X	X	.	.	.	.	.	.	.	.	.	.
VP	.	.	.	.	X	.	.	.	.	.	.	.	.	.	.
HENRY	.	.	.	.	.	.	.	.	.	.	.	.	.	.	.
MOLW	X	.	.	.	X	.	X	.	.	.	.	.	X	.	.
SOLW	X	X	.	.	X	.	.	.	.	.	.	.	.	.	.
THOD	.	.	.	.	.	.	.	.	.	.	.	.	.	.	.
BCF	.	.	.	.	.	.	.	.	.	.	.	.	.	.	.
VFSG	.	.	.	.	.	.	.	.	.	.	.	X	.	X	.
VLEBAS	.	.	.	.	.	.	.	.	.	.	X	.	X	.	.
DW	.	.	.	.	.	.	.	.	.	.	.	.	.	.	.
DG	.	.	.	.	.	.	.	.	.	.	.	.	.	.	.
SUMFOR	.	.	.	.	.	X	.	X	.	X	.	.	.	.	.
MP	X	.	.	X	.	.	X	.	.	.	.	.	.	.	.
VPT	.	.	.	X	.	.	.	.	.	.	.	.	.	.	.
TVP	.	.	.	X	.	.	.	.	.	.	.	.	.	.	.
BPP	.	.	X	.	.	.	.	.	.	.	.	.	.	.	.
PBP	.	.	X	.	.	.	.	.	.	.	.	.	.	.	.
NRINGS	.	.	.	.	.	.	.	.	.	.	X	.	.	.	.
USEPAT	.	.	.	.	.	.	.	.	.	.	.	.	.	X	.
PROEG	.	.	.	.	.	.	.	.	.	.	.	.	.	.	X
PROUSA	.	.	.	.	.	.	.	.	.	.	.	.	.	.	X
PROWOR	.	.	.	.	.	.	.	.	.	.	.	.	.	.	X

A figure X in the crossbar means that the quantity corresponding to the current row is needed in an estimation formula for the quantity corresponding to the current column. The crossbar inside the program uses numbers representing the different estimation formulas instead of the plain figure X in table 1.

Table 2 is the form we use to gather and enter data for a substance. It is included here because it explains all the abbreviations used in this text.

Table 2: The data sheet for DTEST and the fate models in E4CHEM

NAME _____

SUMFOR _____
       (chemical sum formula)

HENRY   _____ (--) Henry - coefficient
KOC   _____ (cm^3 H$_2$O/g C) partition coefficient organic carbon / water
LOGKOW _____ (--) logarithm of n-octanol - water - partition coefficient

If the above values are unknown :
BP   _____ (Kelvin) boiling point
VP   _____ (Pascal) vapour pressure at 293 Kelvin
SOLW _____ (kg/m^3 H$_2$O) water solubility
MP   _____ (Kelvin) melting point

If BP is unknown :
BPP _____ (Kelvin) boiling point at a pressure of PBP Pascal
PBP _____ (Pascal) pressure used when measuring BPP

If VP is unknown :
VPT _____ (Pascal) vapour pressure at the temperature TVP
TVP _____ (Kelvin) temperature used when measuring VPT

MOLW   _____ (g/Mol) molecular weight
NRINGS _____ (--) Number of aromatic and heterocyclic rings
VFSG   _____ (cm^3/Mol) molecular volume
VLEBAS _____ (cm^3/Mol) molecular volume according to LeBas

THOD _____ (kg O$_2$/kg) theoretical oxygen demand
BCF   _____ (--) bioconcentration factor

PKA   _____ (--) pKa value
ACID _____ (--) acidity (1: acid; -1: alkaline; 0: non-dissociating)

DG _____ (m^2/d) gas diffusion constant
DW _____ (m^2/d) water diffusion constant

RPHOTO _____ (1/d) photodegradation rate by OH - radicals
RSED   _____ (1/d) degradation rate in sediment
RSOIL   _____ (1/d) degradation rate in soil
RWATER _____ (1/d) degradation rate in water

PROVOL _____ (t/year) production volume in West Germany
PROEG   _____ (t/year) production volume in the European Community
PROUSA _____ (t/year) production volume in the USA
PROWOR _____ (t/year) world-wide production volume

RELEAS _____ (%) release into air
        _____ (%) release into water
        _____ (%) release into soil

Here is a translation of the crossbar in table 1 to a more readable form :

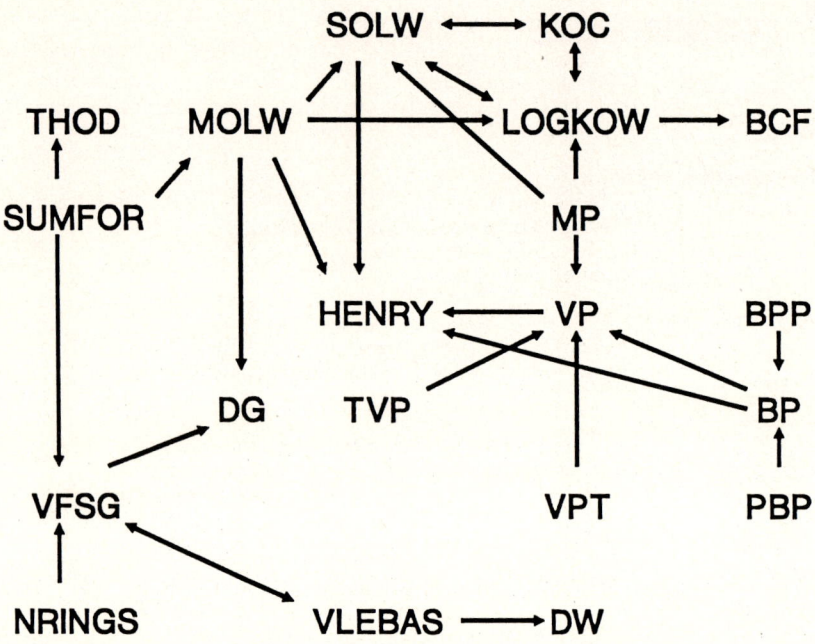

<u>Figure 1:</u> Data interdependencies in DTEST. An arrow always points from a property needed in an estimation formula to the property estimated with this formula.

This picture shows no information on which arrows belong to which formula. Furthermore, the priority of the formulas is missing. Showing all these in just one graph would make the picture virtually unreadable.

The next figure shows the complete set of all estimation paths that eventually lead to one substance property. The dimensionless Henry coefficient is chosen because of its central role in our fate models (together with K_{OC} and LOGKOW).

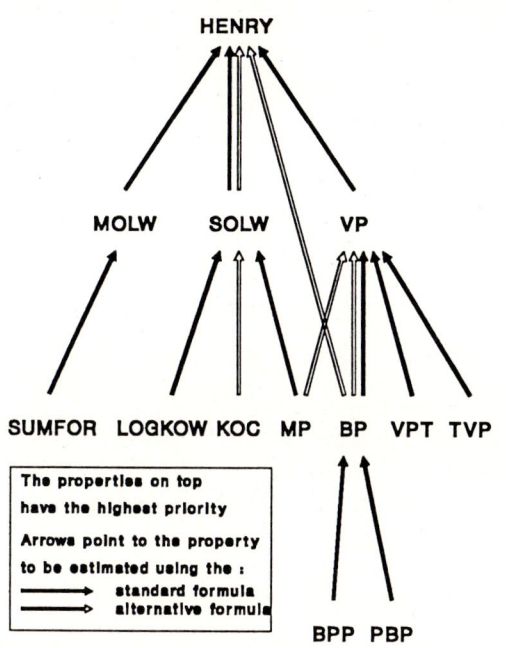

This is the estimation tree for the Henry coefficient in DTEST. Each set of arrows pointing to a property name represents an estimation formula. If there are not enough data to use the first estimation formula (filled arrows), the second best formula (hollow arrows) will be chosen. If this also fails, DTEST proceeds with the next lower level in the graph to fill in the gaps.

Figure 2: The estimation graph for the dimensionless Henry coefficient in DTEST

This picture shows there may be several ways to estimate the Henry coefficient with a given set of data. Usually every one will yield a different result. This means that an algorithm has to choose the appropriate combination of formulas such that no data that can be estimated are overlooked and for each value the most reliable estimation path is chosen. Usually the shortest possible path is chosen to minimize error propagation. In DTEST, this is achieved by an ordering of estimation formulas which leads to a sophisticated hierarchy. The above figure gives an impression of how this hierarchy looks like.

These complicated interiors of DTEST are hidden by a very simple shell : The user just enters his data in a syntax like "HENRY = 0.0625", and triggers the estimation with the command "DTEST". The estimation is executed automatically, the estimated values are displayed on the screen and kept for later use.

EXAMPLE : ESTIMATING WATER SOLUBILITIES WITH DTEST

The principles of the selection of the appropriate estimation equations are given elsewhere (1), therefore a brief example suffices here.

A well established method to estimate the solubility in water is given by Mackay (2):

$$SOLW = MOLW * 1.8 * F / K_{OW}$$

$$F = Min (1, exp(6.8*(1-MP/293)))$$

where F is a correction factor introduced by Yalkowsky (3,4). Slightly better results will be produced with the following relation, recommended in "The Handbook of chemical property estimation methods" (5):

$$SOLW = MOLW * 9.506 * F / K_{OW}^{1.339}$$

DTEST uses the second formula in the recommended range, i.e for values of LOGKOW between 0.3 and 4.7. Outside of this interval Mackay's formula is used.

The next table shows a compilation, done by data given in (6).

Table 3: solubility (g/l) and log Octanol-water distribution coefficient of 13 selected compounds

	SOLW	LOGKOW
2-Chloroaniline	3.8	1.90
4-Chloroaniline	3.9	1.83
Chlorobenzene	0.471	2.84
1-Chloro-4-nitrobenzene	0.225	2.39
1,2-Dichlorobenzene	0.156	3.38
1,3-Dichlorobenzene	0.111	3.60
1,4-Dichlorobenzene	0.078	3.52
Hexachlorobenzene	$6.2*10^{-6}$	5.31
1,2,4-Trichlorobenzene	0.049	4.02
1,3,5-Trichlorobenzene	0.006	4.49
4-Chlorophenol	27.	2.39
Ethylbenzene	0.16	3.15
Di(2-ethylhexyl)Pthalate	$0.3*10^{-3}$	5.11

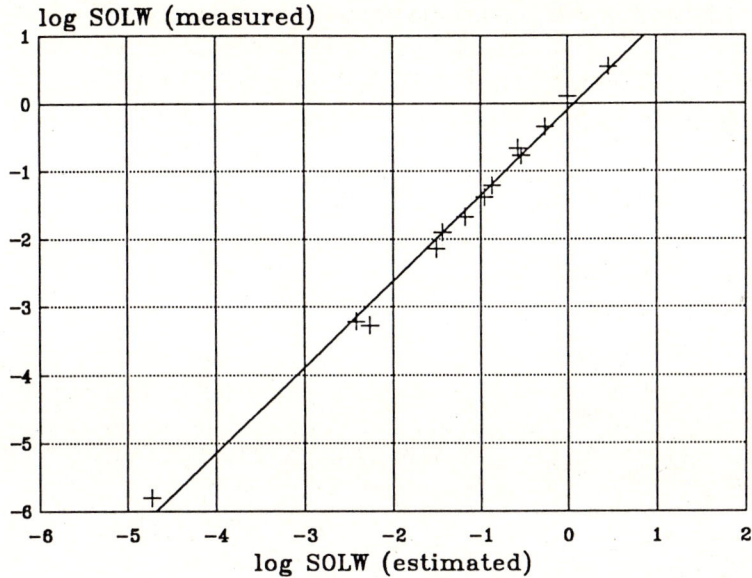

Figure 3: Measured water solubility versus estimated values

Two things can be noted in this graph. The correlation between measured and estimated values is quite good; there is not much deviation from the regression line. The regression hits the origin quite well. On the other hand, the slope of the regression line is about 5/4, while the ideal value should be 1. We could obviously use a "custom-made" estimation formula for this (rather homogeneous) set of substances with a better fit to the experimental values. But this formula would produce unpredictable errors for other substance classes. The advantage to have a quite general valid equations like those above is always preferred over an equation for which validity for different classes of chemicals is not established.

DATA CROSS VALIDATION : DXVAL

If a data set has been compiled from handbooks, databases etc., it is difficult to find out the quality of the data. Consulting original literature is usually too tedious, time-consuming and expensive for routine applications. However, there are some criteria which can be derived from the data set without further information, and which can raise or lower one's confidence in the data. The program DXVAL (Data cross VALidation) combines these criteria in a simple, but effective way.

The first level of DXVAL makes an overall validation of the data set. For every entry, DXVAL uses the DTEST algorithm to estimate the value from the remaining data. This estimated value is then compared with the original value. If there is a big discrepancy, there may be an error in the original data. Of course, the estimation formula could also be inaccurate for this substance. On the other hand, it is rather improbable that an estimated value exactly matches the actual value. If this is the case, the original value has probably not been measured but estimated with the standard formulas also used in DTEST.

The second level of DXVAL is used to cross validate one property with the rest of the data set. First, the selected property is compared to its estimated value, just as in the first level. Then, each of the other properties is deleted in turn from the data set, and again an attempt is made to estimate the selected property. If this estimation yields a different result than the estimation with the complete data set, it will be displayed. If deleting a property will make an estimation impossible, this will also be noted.

The following example demonstrates how DXVAL helps pin-pointing the "weak spots" in the data set.

EXAMPLE : USING DXVAL ON P-DICHLOROBENZENE

Table 5: These are the data we have found for p-dichlorobenzene :

```
NAME    = P - Dichlorobenzene
SUMFOR  = C 6 H 4 CL 2        (chemical sum formula)
MOLW    = 147.0   g/mol       (molecular weight)
SOLW    = 0.087   g/l         (water solubility)
BP      = 447.1   K           (boiling point)
VP      = 234.6   Pa          (vapour pressure)
KOC     = 1361    cm³ H₂O/g   (partition coefficient org carbon - water)
LOGKOW  = 3.52                (partition coefficient n-octanol - water)
HENRY   = 0.0625              (dimensionless Henry coefficient)
MP      = 326.2   K           (melting point)
```

To start the first level of DXVAL, the user just has toe type "DXVAL" to get the following results :

Table 5: This is the result of running DXVAL (first level) with these data :

```
MOLW    = 147.0              g/mol
SOLW    = 0.087  0.0125      g/l              (strong aberration)
VP      = 234.6  76.93       Pa               (strong aberration)
KOC     = 1361.              cm³ H₂O/g        (obviously estimated)
LOGKOW  = 3.52   3.148
HENRY   = 0.0625  0.1628                      (strong aberration)
```

The first number is the original and the second the estimated value. If both are equal, the second value is omitted. It is no surprise that the molecular weight was computed correctly. The original value for KOC however was obviously already estimated; it is highly improbable that the estimate should be exactly the same as an experimental result.

The estimate of the water solubility is not too good. Therefore the validity of the estimation formula should be more closely examined for this substance. The vapour pressure and the Henry coefficient also show strong aberrations. Let us give a closer look to the Henry coefficient using the second level of DXVAL :

Table 6: Cross validation of HENRY for p - Dichlorobenzene (second level)

```
  HENRY  = 0.0625    0.1628
 -SOLW   : HENRY  = 1.133
 -VP     : HENRY  = 0.05336
```

"-SOLW" indicates that the next value for HENRY has been estimated without using SOLW. We only get a different estimate for the Henry coefficient if the water solubility or the vapour pressure are deleted from the data set. Deletion of any other data has no effect on the estimate.

If the water solubility is deleted from the data set, it has to be estimated to estimate the Henry coefficient. In this case, the gap between original value and estimate gets much wider. This is another reason to distrust the estimate for the water solubility in this data set.

If the vapour pressure is deleted, the estimate for the Henry coefficient matches its literature value quite well. This indicates that the reported value for the vapour pressure might be false. DXVAL can not squeeze any more information on the vapour pressure from the given data set :

Table 7 : Cross validation of the vapour pressure for p - Dichlorobenzene

```
VP     = 234.6  76.93     Pa
-BP    : VP     could not be estimated
-MP    : VP     could not be estimated
```

At this point, we had two options to continue our task (which was running a simulation model for p-dichlorobenzene). With no further information on the data, we could start our simulation with the original and then again with the estimated value for the vapour pressure. If the results do not depend sensitively on the vapour pressure, the case need not be pursued any further.

Actually we did some more literature research and some more estimations with CHEMEST (7) (a more sophisticated estimation program). In both cases we found the same disagreement about the value of the vapour pressure that has been reflected by DXVAL.

ENVIRONMENTAL FATE MODELS

DTEST has been developed as part of E4CHEM, a toolbox to assess the fate of chemicals in the environment. In this task, mathematical models can help save time, money and to understand the underlying principles. To run such models a set of substance properties is needed, which is seldom completely measured or even qualitatively known. The simplest model in this toolbox is EXTND (8) (EXposition TeNDency). This model predicts the general tendency of the substance to accumulate in air, water or soil. If the Henry coefficient and LOGKOW are given (estimated by DTEST), EXTND can be run on a piece of paper by just putting markers on an "exposition map".

The following figure shows the result for five compounds. It is easily seen that electron withdrawing substituents and more polar groups lead to a more distinct loading of aquatic ecosystems.

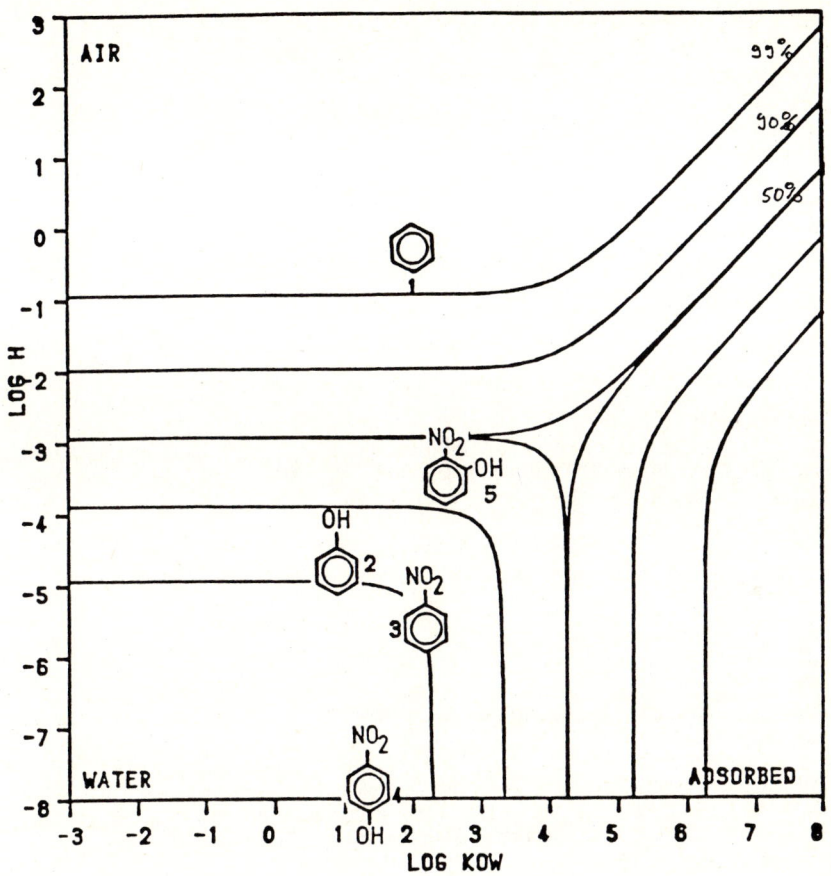

COMPARTMENT VOLUMES (M³)
  AIR   6.0•10⁹
  WATER   7000000
  ADSORBED 30000

  1 71-43-2   BENZENE
  2 108-95-2   PHENOL
  3 98-95-3   NITROBENZENE
  4 100-02-7   4-NITROPHENOL
  5 88-75-5   2-NITROPHENOL

<u>Figure 4:</u> Exposition map: The lines refer to constant mol-percentage in air, water, and soil. Corresponding to the 90% or 99% lines there are "regions", where a substance is a potential hazard for aquatic systems (left lower corner), for the air (left upper corner) and soil (right lower corner).

SUMMARY

DTEST is a program which calculates substance properties. An important facility of DTEST is, to choose automatically suitable paths to estimate missing properties. These involves more than one step with consequences of error propagation. The module DXVAL of DTEST does the job of the cross validation and shows discrepancies which arise from the bias in the data or in the used estimation formulas.

Once the literature data are written on a piece of paper, it takes about half an hour to estimate and cross validate the data set; with a little practise, it will take even less time.

1 Brüggemann R, Münzer B (1988) In: Jochum C, Hicks MG, Sunkel J (eds) Physical property prediction in organic chemistry :303. Springer, Berlin Heidelberg New York
2 Miller MM, Wasik SP, Huang GL, Shiu WY, Mackay D (1985) Env Sci Technol 19:522
3 Yalkowski SH (1979) Ind Eng Chem Foundam 18:108
4 Mackay D, Bobra A, Shiu WY, Yalkowski SH (1980) Chemosphere 9:701
5 Lyman WJ, Reehl WF, Rosenblatt DH (1982) Handbook of chemical property estimation methods, McGraw-Hill, New York
6 Howard PH (1989) Handbook of environmental fate and exposure data for environmental chemicals Vol I, Lewis Publishers,
7 Lyman WJ, Potts RG, Magil GC (1983) CHEMEST : User's Guide; a program for chemical property estimation, A.D. Little Inc, Cambridge
8 Brüggemann R, Münzer B (1987) In: Umweltmodelle und rechnergestützte Entscheidungshilfen für die vergleichende Bewertung und Prioritätensetzung bei Umweltchemikalien, FE-Bericht 106 04 016/Anhang I.5, Umweltforschungsplan des Bundesministers für Umwelt, Naturschutz und Reaktorsicherheit, Bonn

PIMO, a program visualizing HMO-results by producing transferable graphics output

Beate Heinrich, Erich Schreiner, Dieter Spielbauer,
Markus Wolperdinger und Hans-Ulrich Wagner

Ludwig-Maximilians-Universität, Institut für Organische Chemie

Karlstraße 23, D-8000 München 2

Abstract: The program PIMO presented here uses either simple topological input matrices or CHEMTEXT[R]-connectivity tables to produce graphics output of the results of HMO- or Ω-calculations.

The CIP project in the chemistry department of the university has been installed in 1988. There were several courses in BASIC, FORTRAN an PASCAL. In a continuation course in PASCAL there was the exercise to develop a molecular orbital program. The HMO model, as given by E.Heilbronner and H.Bock (1), is very useful in learning both programming and MO theory. We followed the example shown by H.D.Breuer (2) at the CIC workshop 1988 in Tübingen.

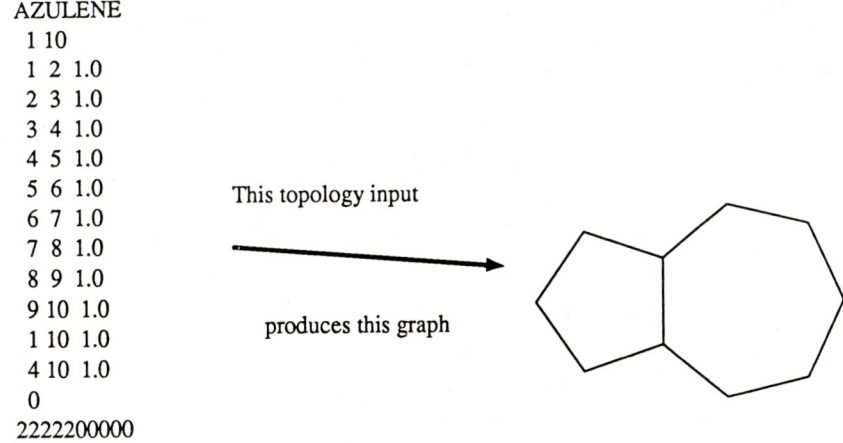

Fig.1. Input and graph output of the program PIMO

The program PIMO is written in TURBO PASCAL (3). Several units coming from the computer journal series CHIP-Spezial (4) have been included. The program PIMO generates from a simple HMO topology matrix a standard geometry of the π-system, using standard bond lengths and bond angles. It is programmed only for planar molecules. Especially several condensed rings are calculated nicely (Fig.1).

This standard geometry is used to produce a graphical output of the π-system in consideration and to visualize the molecular orbitals which are calculated by the program. The total π-density and the π-charge can be analyzed graphically, too (Fig.2). It is easy to include heteroatom parametrization.

The program PIMO performs modern menu techniques to be easily used by students in the appropriate courses in theoretical chemistry. The graphical output is either on screen (pixel oriented) or it is stored in HPGL (5) plotter files (vector oriented). The HPGL files can be transferred to CHEMTEXT (6) documents. Fig. 3 shows an example in which six output files from PIMO have been included in to a CHEMTEXT document. This example is especially valuable in preparing text book documents.

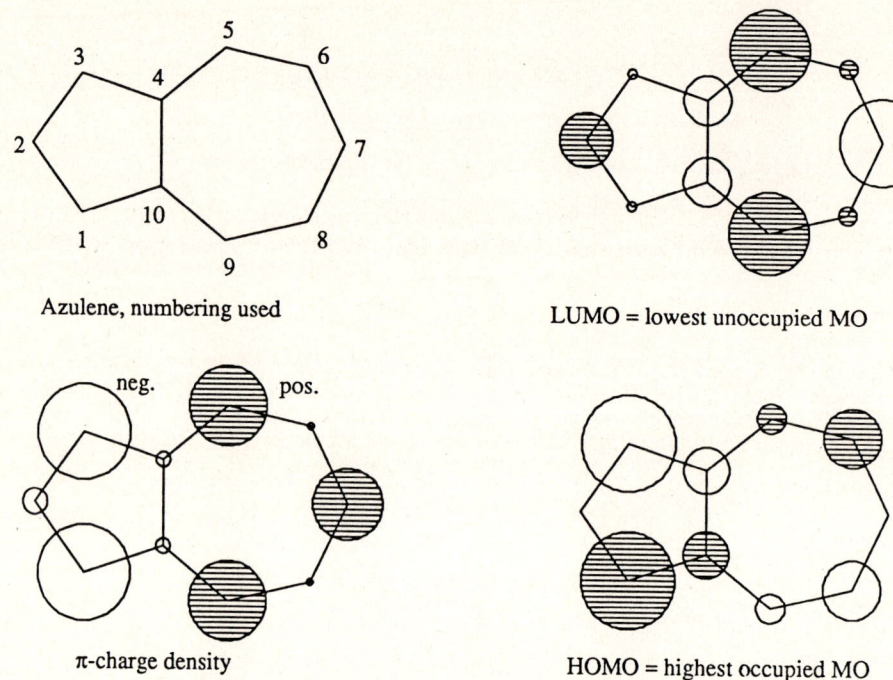

Fig.2 MO-diagrams for azulene. Calculated and drawn using the program PIMO. The graphs are transferred as HPGL files to CHEMTEXT, reduced, labeled, and given the desired layout.

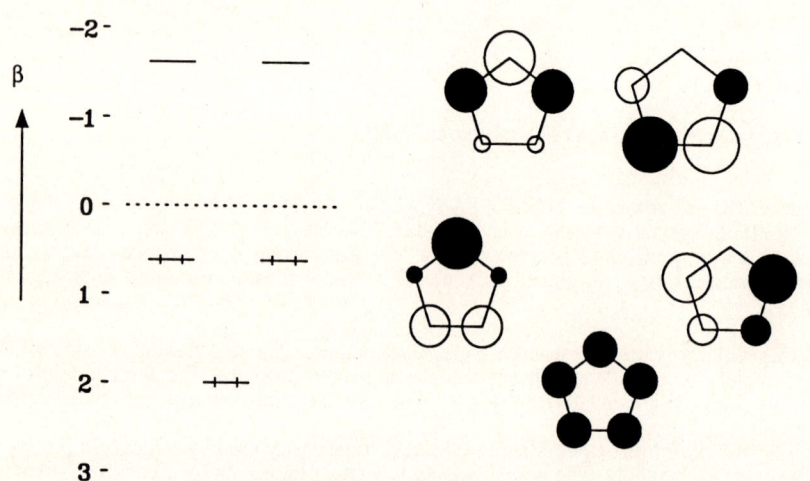

Fig. 3 HMO-diagram for cyclopentadienylanion

There are options to calculate coefficients for first and second order perturbation theory evaluations, different polarisabilities and localization indexes. The MO-method can be changed to the Ω-**method**.

Literature and References:

1 E.Heilbronner and H.Bock, Das HMO-Modell und seine Anwendungen, Band 1 bis 3, Verlag Chemie Weinheim 1968.

2 CHIP-Spezial, Turbo-Pascal Programmier-Praxis, Hefte 1-11, Vogel-Verlag, Würzburg 1985-1988.

3 H.D.Breuer, Ein Programm zum Einsatz im computerunterstützten Unterricht, in G.Gauglitz (Ed.), Software-Entwicklung in der Chemie 3, Springer-Verlag Berlin Heidelberg 1989, Seite 409.

4 Turbo Pascal 4.0, Heimsoeth & Borland, München 1987.

5 HPGL = Hewlett-Packard Graphics Language. The two letter graphics instructions's mnemonic is suggestive of its role. Hewlett-Packard Graphics Plottermanual,Hewlett-Packard Company, San Diego, CA 92172-1899.

6 CHEMTEXT is a modul of the series CPSS (Chemists Personal Software Series) by Molecular Design Limited; MDL-Europe, CH 4123 Allschwil, Mühlebachweg 9, Schweiz.

The Calculation of Electrolyte Solution Properties with the Help of the ELDAR Data and Method Bank Exemplified by Electrolyte Conductance

J. Barthel, H. Popp, and G. Schmeer

Institut für Physikalische und Theoretische Chemie der
Universität Regensburg, D-8400 Regensburg, West Germany

Abstract: Chemical models based on mean force potentials are generally used to calculate properties of electrolyte solutions from measurements with the help of the Gauss-Marquardt method for least squares fits. For these calculation processes ELDAR offers a data bank managing fact knowledge, a method bank managing algorithmic knowledge and a communication manager supplying the input of the modules. The interaction of data and method bank for the automatic calculation of electrolyte solution properties and the production of basic data under the control of the communication manager is exemplified for electrolyte conductance.

1. ELECTROLYTE SOLUTIONS

The special feature of electrolyte solutions is the existence of interactions of the type ion-ion, ion-solvent and ion induced solvent-solvent interactions. These interactions have their origin in the electric charges of the ions. They are observed throughout the whole concentration range showing non-ideal excess properties of the solution even at very low concentrations, in contrast to liquid mixtures of uncharged compounds. This feature requires particular methods for the treatment of electrolyte solutions in thermodynamics and transport processes.
Electrolytes, Y,

$$Y = (X_1^{z_1})_{\nu_1}(X_2^{z_2})_{\nu_2}(X_3^{z_3})_{\nu_3}\ldots \qquad (1)$$

dissociate completely or partially under the influence of a solvent to produce electrolyte solutions; $X_i^{z_i}$ are the constitutive ions of valence z_i, present with stoichiometric number ν_i in an electroneutral element of the electrolyte.

The theories of electrolyte solutions are based on two-particle correlation functions g_{ij} of ions i and j and their mean force potentials, W_{ij}, derived from statistical mechanics at McMillan-Mayer level. At this level the ions are charged spheres of radius a_i imbedded in the bulk solvent which is considered as a dielectric continuum of relative permittivity ε. The ions may be surrounded by solvation shells made up by orientated solvent molecules. Pair correlation functions and mean force potentials are connected by the relationship

$$W_{ij} = -kT \ln g_{ij} \qquad (2)$$

Different levels of approximation of the mean force potential, which includes Coulombic, W_{ij}^{el}, and non-Coulombic, W_{ij}^*, interaction energies

$$W_{ij} = W_{ij}^{el} + W_{ij}^* \qquad (i,j = +,-) \qquad (3)$$

may be considered, depending on the nature of the ions and solvents and on electrolyte concentration. The non-Coulombic potential, W_{ij}^*, encompasses ion-dipole, ion-induced dipole, and „chemical" contributions, such as H-bonding. The chemical model of electrolyte solutions at low to moderate concentrations [1] considers it as a step potential. Then the mean force potential is given by the relationships

$$W_{ij} = \begin{cases} & W_{ij}^{el} & W_{ij}^* & \text{range} \\ & \infty & r \leq a \\ \frac{e^2 z_i z_j}{4\pi\varepsilon_0 \varepsilon} \frac{1}{r(1+\kappa r)} + \text{const.} & a \leq r \leq R \\ \frac{e^2 z_i z_j}{4\pi\varepsilon_0 \varepsilon} \frac{\exp[\kappa(R-r)]}{r(1+\kappa R)} + 0 & R \leq r \end{cases} \quad (4\text{-a,b,c})$$

where R is the mutual ion distance up to which paired states of ions are considered to be ion pairs, a is the closest possible distance of the ions i and j in the solution, κ is the reciprocal radius of the ion cloud around a central ion.

Methods for the calculation of pair-correlation functions and various physical and chemical properties of the solution derived from pair-correlation functions and mean force potentials are implemented in ELDAR. At moderate to high concentrations the pair-correlation functions are obtained with the help of integral equation methods such as MSA (mean spherical approximation), HNC (hypernetted chain), or their appropriate combinations [2].

With an actual content of 350.000 data of physical and chemical properties of 46.000 chemical systems, 11.200 literature references and 40 documented modules for the calculation of solution properties the ELDAR data and method bank actually is the largest data base and management system for electrolyte solutions. The chemical systems (solvents and electrolyte solutions) result from combinations of 2000 electrolytes with 700 solvents. Beyond that ELDAR contains 37.000 thesaurus terms (descriptors and component names) with their semantic references.

2. ELECTROLYTE CONDUCTANCE OF 1,1 ELECTROLYTES

Among the properties of liquid mixtures and solutions electrolyte conductance is a unique phenomenon of electrolyte solutions. Its variable in dilute solutions is the equivalent conductance, Λ, given by the relationship

$$\Lambda = \frac{\sigma}{c} \quad (5)$$

where σ is the specific conductance and c the molar concentration. At each concentration the electrolyte conductance is the sum of the ion conductances of cation, λ_+, and anion, λ_-.

$$\Lambda = \lambda_+ + \lambda_- \quad (6)$$

The theory of electrolyte conductance yields the conductance equation

$$\Lambda = \alpha\{\Lambda^\infty - S\sqrt{\alpha c} + E(\alpha c)\ln(\alpha c) + J_1(R) \cdot (\alpha c) - J_2(R) \cdot (\alpha c)^{\frac{3}{2}}\} \quad (7)$$

for symmetrical electrolytes at low to moderate concentrations. Eq.(7) makes use of the concept of ion association which considers paired states of oppositely charged ions at distances $a \leq r \leq R$ as non-conducting ion pairs. The portion of equally charged ions at distances $a \leq r \leq R$ is negligible at low concentrations. In Eq.(7) Λ and Λ^∞ are the equivalent conductances at concentration c and infinite dilution, respectively, and α is the fraction of the conducting ions. The coefficient S is the coefficient of Onsager's limiting law. E, $J_1(R)$, and $J_2(R)$ are the parameters of the extended theory. Below distance R the ions behave unaffected by the external field and hence α is connected to the thermodynamic association constant, K_A, as obtained from the investigation of every thermodynamic property of the solution. The association constant K_A can be interpreted as the equilibrium constant of the ion-pair formation

$$C^+ + A^- \rightleftharpoons C^+A^{-\ 0} \quad (8)$$

where C^+ and A^- are the free (conducting) ions and $[C^+A^-]^0$ is the ion pair when C^+ and A^- have distances $a \leq r \leq R$. $y'_\pm$ is the activity coefficient of the chemical model [3].

$$K_A = \frac{1-\alpha}{\alpha^2 c} \cdot \frac{1}{{y'_\pm}^2} \qquad\qquad y'_\pm = \exp\left[-\frac{\kappa q}{1+\kappa R}\right]; \qquad\qquad (9\text{-a,b})$$

$$\kappa = \sqrt{\frac{1000 N_A e^2}{\varepsilon_0 \varepsilon kT} \cdot \sum_j (\alpha_j c_j z_j^2)}; \qquad\qquad q = \frac{e^2 |z_+ z_-|}{8\pi \varepsilon_0 \varepsilon kT} \qquad\qquad (9\text{-c,d})$$

Ion conductances, λ_i, and their limiting values, λ_i^∞, at zero electrolyte concentration are available via transference numbers, t_i and t_i^∞

$$\lambda_i = \Lambda t_i; \qquad\qquad \lambda_i^\infty = \Lambda^\infty t_i^\infty \qquad\qquad (10\text{-b,c})$$

The transference number can be determined with high accuracy from independent measurements. A table of λ_i^∞-values, established for a variety of ions in a given solvent permits the estimation of Λ^∞ via Eq.(6) for an electrolyte whose conductance never has been measured. If the association constant, and hence also the distance parameter R, of this electrolyte are available from thermodynamic measurements, its conductance can be completely calculated with the help of Eq.(7). Examples are given in Ref.[4]. Even if λ_i^∞-values are unknown for the investigated solution, they can be estimated with the help of anchor values and solvent viscosities [5] or with lesser accuracy with the help of Walden's rule. The development of a rule bank guiding the ELDAR user to simulated reliable values is the actual task of the ELDAR development group.

3. STATISTICAL FUNDAMENTALS

A system of modules for linear and non-linear optimization and statistical data analysis is implemented in the method bank of ELDAR for the calculation of the properties of electrolyte solutions and the establishment of the rules of the rule bank. It contains modules for the interpolation with B-splines, for least-squares fit, and for statistical analysis.

3.1 Parameter Estimation by the Least-Squares Method

In case of electrolyte conductance the input data are the equivalent conductances $y_1, y_2, \cdots, y_n$ measured at concentrations $x_1, x_2, \cdots, x_n$. For associated electrolytes the conductance equation, Eq.(7), with the parameters Λ^∞, J_1, J_2, and K_A is used in connection with Eqs.(9). The most commonly used technique of parameter estimation is the fitting of the equation by the least squares method. In contrast to the maximum-likelihood method, in which values of parameters are selected to yield the highest probability of the measured data for a given *a priori* probability distribution, the least squares method is free of assumptions on the distribution of the variables. If the data differ in accuracy, non-negative weighting factors w_i are attached to the residuals ϵ_i^2. The weigthed least squares method applied to the property equation $g(\vec{\theta}, x_i)$ (here conductance equation) with the vector of parameters $\vec{\theta}$, has its solutions at the stationary point $\vec{\theta}^*$ of the objective function $\phi(\vec{\theta})$

$$\phi(\vec{\theta}) = \sum_{i=1}^n \frac{1}{w_i}[y_i - g(\vec{\theta}, x_i)]^2 = \sum_{i=1}^n \frac{1}{w_i}\epsilon_i^2 \rightarrow min(\vec{\theta}) \qquad\qquad (11)$$

If the Hessian matrix (matrix of the second partial derivatives of $\phi(\vec{\theta})$ with respect to $\vec{\theta}$) is positive definite at the stationary point $\vec{\theta}^*$, then $\phi(\vec{\theta})$ has a local minimum at this point. If the Hessian matrix is positive definite for all values of $\vec{\theta}$, then the stationary point $\vec{\theta}^*$ is the global minimum.

The partial derivatives of polynomials are *linear in* $\vec{\theta}$; they yield normal equations which are easy to solve. In contrast, the conductivity equation produces non-linear normal equations which must be solved by iteration, starting with favourably chosen initial values for the parameters, $\vec{\theta}^0$.

The least squares algorithms are characterized by their approximation of the hyperspace $\phi(\vec{\theta})$ and the iteration of the parameters from $\vec{\theta}^{(i-1)}$ to $\vec{\theta}^{(i)}$.

The Gauß-Newton method [6] develops the Taylor series of $\phi(\vec{\theta})$ at $\vec{\theta}^{(l)}, l = 0, 1, \cdots$ and truncates the series after the linear term. If this method converges, it converges quickly. However, it has some weak points [7]
- if the initial values $\vec{\theta}^0$ are chosen far away from the final solution, then properly speaking, linearization of $\phi(\vec{\theta}^0)$ is not admissible; the parameter corrections $\vec{\Delta\theta}$ may yield a local minimum or even lead outside the limits of convergence,
- convergence is not monotonous; in the sequence of iterations also higher values of the objective function $\phi(\vec{\theta})$ may occur,
- the inverse of the normal matrix N must not exist; N may be a singular or numerically unstable matrix.

The Gauß-Marquardt strategy [8] tries to diminish the problem related to the inversion of the normal matrix N by the addition of a positive number λ to every diagonal element of N rendering the modified matrix positive definite and easy to invert. A standardized comparison of five nonlinear fit algorithms favours the Gauss-Marquardt algorithm with regard to computer time and convergence [9].

3.2 Improvements of Least Squares Fits

The Gauss-Marquardt method can be improved by
- appropriate guess of the initial values $\vec{\theta}^0$,
- effective numerical differentiation methods if an explicit analytical derivative of $\phi(\vec{\theta})$ does not exist,
- the control of the parameter corrections $\vec{\Delta\theta}$ at the first few iteration steps, and
- interpolation or extrapolation of $\vec{\Delta\theta}$ at exponential overflow.

The choice of the initial values is decisive for the success in locating the minimum. In ELDAR stored information on parameters of previously solved problems can be used, or the initial values are computed by solving the function $g(\vec{\theta}, \vec{x})$ for a number of data points equal to the number of parameters, or part of the parameters is estimated with the help of a simplified version of the model equation. In the case of electrolyte conductance the limiting law is used with the conductances at the lowest measured concentrations to yield starting values of Λ^∞ and K_A.

In cases where analytical derivatives are not accessible, numerical differentiation is useful with difference quotients of adaptative interval lengths [10].

If the starting values of the parameters are located far away from the final solution, the first correction steps may be so large that vector $\vec{\theta}$ exceeds the convergence limits. Then the correction intervals $\vec{\Delta\theta}$ of the first 6 iteration steps may be shortened according to the relation [11]

$$\vec{\Delta\theta} = \frac{\vec{\Delta\theta}}{2(6-i)}$$

Because every investigated function $g(\vec{\theta}, x)$ is continuous and continuously differentiable with respect to its independent variables x, the parameters $\vec{\theta}$ of the function are generally determined by the zeroes of the function and its derivatives. The parameters are adjusted the better the more zeroes are in the interval covered by measurements or near to the interval limits. For some ill-behaving functions without zeroes the parameters can be still determined with the help of the asymptotes of the functions at the limits of their domains of definition, far away from the intervals of measurements. In this case, occuring for non-linear exponential functions of the temperature dependence of transport processes, the domain of convergence of the least squares fit is very small and exponential overflow may occur during the iteration process. An algorithm for interpolation and extrapolation [12] improves this situation.

An universal module of ELDAR, LSQFIT, for linear and non-linear parameter estimation by the Gauss-Newton or Gauss-Marquardt method, which takes into account all improvements of the solution strategy mentioned above is implemented and documented in the relation MODULE; an extract of the documentation of this module is given in Table 1.

Table 1

Documentation of module LSQFIT

```
NAME       :LSQFIT
TITLE      :LEAST SQUARES FIT FOR VARIOUS MODELS BY GAUSS-NEWTON
            OR GAUSS-MARQUARDT METHOD
CATEGORY   :STATISTICS//DATA ANALYSIS//FITTING//FUNCTION // OPTI-
            MIZATION METHOD//STATISTICAL ESTIMATION//FIT
DESCRIPTORS:LEAST SQUARES FIT//GAUSS-NEWTON FIT METHOD//MARQUARDT
            FIT METHOD//PARAMETER ESTIMATION//NON-LINEAR REGRESSION
            //MULTIPLE REGRESSION
SOURCE     :Brand,S.; Datenanalyse,BI,Mannheim,1975,p165-218;
            Schmeer,G.; Der Einfluss der Umgebung des Reaktions-
            zentrums..,Habilitationsschrift, Regensburg 1981,p.194-205
DATE-VERS. :870214(1.00)
EXTERNALS  :COVMAT(MATCON)//MATIGJ(MATIGJ)//MATEQR(MATEQR)
CONNECTION :A model specific module with 19 entries is neccessary.
            It is specified in the parameter list.
PROG.LNG.  :FORTRAN 77
MEMORY     :32K SOURCE; 31K OBJECT
ACCURACY   :REAL*8
FILES      :
STANDARD   :FORTRAN 77
THEORY     :Least squares method (LSQ) is a....
REMARKS    :The achievement of the global minimum......
```

3.3 Test for Outliers

If data tupels from different publications are used for optimization, an *a priori* outlier test can be carried out for the values of the property variables belonging to the same value of the independent state variable. An *a posteriori* outlier test is installed in ELDAR for the distribution of the residuals of the fitted equation. The module STAHYP contains a maximum likelihood-quotient test for the largest value in the sequence of the absolute residuals [13]. STAHYP probes the discordancy of this value in a normal sample with unknown mean μ and variance σ^2. A null hypothesis assumes that the measured data are normally distributed with unknown μ and σ^2. The test statistics, $t = \frac{x_n - \bar{x}}{s}$, for the largest deviation is compared for a given significance level α to the α-quantil of the distribution of the test statistics t [13]. If t does not correspond to the respective α-quantil, the null hypothesis is rejected and the corresponding value designated as an outlier.

3.4 B-Spline Fit for Interpolation

The module APRBSP of ELDAR is used for the interpolation and graphical representation of data. It approximates cubic splines satisfying local concavity or convexity constraints to a given set of data points and offers a B-spline representation of cubic splines [14].

4. DATA BANK - METHOD BANK INTERACTION

4.1 Features of the Data and Method Bank

A data, method and rule bank, and their interaction controlled by a communication manager make up the knowledge base system ELDAR [15].

The fact knowledge is mapped into Codd's relational model with an extension to repeating attributes [16] in it's 1st normal form, which reduces the number of basic relations from 29 (without repeating attributes) to the following 9 (with repeating attributes) relations [15]: Literature and measured data are mapped into the relations LITERATURE, CHEMICAL SYSTEM and DATA, semantic connections of terms are represented by a THESARUS relation, the documentation of modules, of input/output parameters of modules and of rules are mapped into the relations MODULE, EFFECT, PARAMETER and RULE. The calculated coefficients of the property equations with full information on their calculation given by the primary keys of the chemical system, the data and modules used with their input parameters and also the parameters of the statistical analysis are mapped into the relation BASIC-DATA. The definition of these basic relations with their attributes is given in Ref.[15].

The data bank of ELDAR [17] actually contains in its more than 350.000 data tupels about 6000 data blocks (one data block consists of 1 to 500 data tupels) on temperature- and concentration- dependent electrolyte conductance, covering 95 % of the known measured data. The ELDAR method bank contains
- a collection of about 40 modules designed by principles of normalization given in Ref.[18]; they are classified as modules of numerical mathematics, statistical analysis, physical properties of electrolyte solutions based on actual theories, and modules for different purposes, such as the conversion of different concentration variables into SI units, etc.,
- the documentations of all modules,
- a management system (compiler, linker, library manager, editor, graphic tools, etc.),
- a communication manager connecting user, data bank, and method bank.

In the following paragraph a typical example is given to illustrate the functions of ELDAR: The calculation of the coefficients $\Lambda^\infty, J_1(R), J_2(R)$ of Eq.(7) and K_A (via α) of Eq.(9). This example stresses the role of the communication manager for the automation of calculations beginning with the allocation of appropriate data from the data bank, the establishment of the sequence of modules required by the problem, the organization of the calculations, and coming to an end with the storage, output, print or plot of the result. It must be stressed that the user is informed on the screen about the procession of the automatic program and may choose alternatives at various decisive stages.

4.2 Calculation of Electrolyte Solution Properties

The calculation of electrolyte solution properties is a many-sided process requiring data of various other properties in addition to that which is investigated. In ELDAR the calculation process is organized recursively. It is controlled by the communication manager (CM), which selects the needed modules and supplies their input demands.

The equivalent conductance of potassium iodide in propanol at 25^0 C has been chosen to exemplify the calculation of electrolyte solution properties with the assistance of the CM. The parameters of the conductance equation are the limiting equivalent conductance, Λ^∞, the association constant, K_A, and the distance parameters of the system solvent molecule - electrolyte, a and R, which are included in the terms J_1 and J_2 of Eq.(7).

1. The user requests the measured data from the data bank:
   FIND FROM DATA WHERE NAME=1-PROPANOL AND NAME=KIl1 AND VARIABLE = CONDUCTIVITY AND VARIABLE=TEMPERATURE=25 CL AND COMPONENT NO.=2
   He asks the thesaurus for the stored descriptors and component names, if he is unsure of the correct spelling of the component name:
   FIND FROM THESAURUS WHERE NAME=KI
   He gets the information that the terms KIl1, KI, POTASSIUM IODIDE, and KALIUM IODIDE are synonyms in ELDAR.
2. The user selects the data for the calculation from the last hits of his query which also can be displayed with their bibliography on the screen. He selects individual values from the table on screen or specifies the wanted data by the hit number with the command:
   WORK DATA 1 .
3. This command also activates the communication manager
   - The CM converts all data to SI units independent of their units in the original literature. For some conversions such as molality, m, to molarity, c, the CM needs the density of the solution, ρ, at given temperature. In ELDAR the density can be calculated via a polynomial of two variables which is obtainable from the method bank

$$\rho = P(m,T) = \sum_{j=0}^{k} m^j \sum_{i=0}^{n} a_{j(n+1)+i} T^i \qquad (12)$$

   The CM procures the polynomial coefficients, $a_{j(n+1)+i}$, from the BASIC DATA relation of the data bank via an automatic query. If the search is sucessful it calculates the densities at the wanted concentrations and temperatures. If the query fails, the CM asks the user for the density data.
   - It determines the variables of state and their values (here: p = 101325 Pa, T = 298.15 K).
   - It constructs the calculation process of data analysis using the information which it has extracted from the query, its dialogue knowledge, and the retrieved data.
   - If the CM is not able to recognize completely the calculation problem, it asks the user for the lacking information.
4. The CM searches for the modules needed in the calculation process by the query:
   FIND FROM PARAMETER WHERE INPUT TERM=EQUIVALENT CONDUCTANCE
   If the query yields several effects, corresponding to several modifications of the property equation, the CM offers all of them to the user to select the wanted one. Alternatively at this stage the user may build up his own query for special modules or effects if he is unsatisfied with the proposal of the CM e.g.:
   FIND FROM PARAMETER WHERE EFFECT= ASSOCIATION CONSTANT AND EFFECT=LIMITING CONDUCTANCE

The selected parameter documentation is activated by the command: WORK PARAMETER

5. The CM interprets the parameter documentation, Table 2, reading the attribute TYPE and searching for the input parameters which are of the type of I or IC. It tries to supply the input demand from the attribute DOMAIN of the PARAMETER relation, or from the relations DATA or BASIC DATA, or with the help of calculations, or from the user, proceeding in the following order

5.1 PARAMETER relation. If the attribute DOMAIN in Table 2 contains predefined

Table 2
Extract of the Parameter Documentation for
the Electrolyte Conductance Equation

```
MODUL_NAM:LSQFIT      NUMBER: 9
PARAMET| DATA TYPE | DOMAIN        |TYPE|   PARAMETER TERM       |UNIT
------------------------------------------------------------------------
CONEQU  |SUBR()    |;CONEQU        |IC  |CHEMICAL MODEL//TRANSPO|
        |          |               |    |RT PROPERTY            |
LM1     |I4        |1,2;2          |IC  |GAUSS-NEWTON METHOD//MA|
        |          |               |    |RQUARDT METHOD         |
ISTYP   |I4        |4              |IC  |START VALUE,AUTOMATIC  |
IPSI(1) |I4        |1              |IC  |IONIC CONDUCTANCE      |
IPSI(2) |I4        |1              |IC  |FUNCTIONAL DEPENDENCY  |
IPSI(3) |I4        |INTEGER+       |I   |CHARGE NUMBER//ELECTROL|
        |          |               |    |YTE                    |
IWEIGH  |I4        |1              |IC  |NON-ERROR TYPE         |
ADJ     |A(4)B     |T,F;TFTT       |IC  |INDICATOR OF FIXED PARA|
        |          |               |    |METERS                 |
Y       |A(NGES)R8 |REAL           |I   |BASE POINTS,ABSCISSE AN|ML/M3
        |          |               |    |D ORDINATE//CONCENTRATI|M2/OM
        |          |               |    |ON//CONDUCTIVITY       |
SIGMAY  |A(NGES)R8 |REAL+0         |W   |ERROR                  |
PSI(1)  |R8        |1.,2.          |IC  |CHEN EFFECT WITHOUT//CH|
        |          |               |    |EN EFFECT              |
PSI(2)  |R8        |REAL+          |I   |ION DISTANCE           |M
PSI(3)  |R8        |REAL+          |I   |DISTANCE PARAMETER//ION|M
        |          |               |    |PAIR                   |
PSI(4)  |R8        |REAL+          |I   |TEMPERATURE            |K
PSI(5)  |R8        |REAL+          |I   |PERMITTIVITY//SOLVENT  |
PSI(6)  |R8        |REAL+          |I   |VISCOSITY//SOLVENT     |PAS
NAK     |A(20)R8   |REAL           |I   |NATURAL CONSTANT       |
CONSTG  |A(20)R8   |REAL           |O   |CONSTANTS              |
PAR     |A(4)R8    |REAL           |W   |START VALUE PARAMETER  |
A       |A(4)R8    |REAL           |O   |LIMITING CONDUCTIVITY//|M2/OM
        |          |               |    |J1-TERM//J2-TERM//ASSOC|
        |          |               |    |IATION CONSTANT        |M3/ML
ITER    |I4        |1,..,50        |O   |NUMBER OF STEPS,CALCULA|
        |          |               |    |TED                    |
IERROR  |A(10)I4   |1,6,31,32,     |O   |ERROR LABEL            |
        |          |80,81,82,83    |    |                       |
FL1     |R8        |REAL           |O   |WEIGHTED SUM OF SQUARED|
        |          |               |    | RESIDUALS             |
EVR     |A(NGES)R8 |REAL           |O   |BASE POINTS,FITTED     |M2/OM
BE      |A(20)R8   |REAL           |O   |EIGENVALUES            |
BE1     |A(M)R8    |REAL           |O   |ACTIVITY COEFFICIENT   |
```

values the CM takes these default values, such as CONEQU for the module name, 2 (Marquardt method) for LM1, and 4 (automatic estimation of start values) for ISTYP.

5.2 Dialogue knowledge. It collects all facts which are used during the dialogue session as module inputs and provides them on request.

5.3 BASIC DATA relation. It manages the parameters of the respective property equations such as the polynomial coefficients of temperature dependence of permittivity from which the permittivity at given temperature is calculated.

5.4 DATA relation. If basic data do not exist the input data must be calculated by the method bank, here for instance the viscosity of 1-propanol at 298.15 K. In this special case the CM asks the data bank for temperature dependent viscosity data of 1-propanol:
FIND FROM DATA WHERE NAME=1-PROPANOL AND VARIABLE=VISCOSITY AND VARIABLE=TEMPERATURE AND COMPONENT NO=1
Then it collects the hits by:
WORK DATA
Finally it writes the viscosity data on a communication file and selects the appropriate module for interpolation using heuristic knowledge about the needed module or asking the data bank by the MODULE relation. The search for the module and the parameters is performed automatically using the relations MODULE or PARAMETER:
FIND FROM PARAMETER WHERE OUTPUT PARAMETER=VISCOSITY
If the query yields more than one effect the CM asks the user for his choice; then the CM activates the selected parameter documentation of the module, e.g. the Vogel-Fulcher-Tammann equation for the temperature dependence of the viscosity [19].

$$\eta(T) = AT^n exp[-\frac{B}{T-T_0}] \qquad (13)$$

According to the parameter list of the module which represents Eq.(13) the CM supplies the input by repetition of steps 5.1 to 5.4 for the module viscosity, and builds up the queries automatically. It starts the calculation process with the needed input values and receives the output values of this module from the method bank.

5.5 User's decisions. In case of the conductance equations an always required user's decision concerns the version of the conductance equation (with or without Chen effect) and the determination of the adjustable parameters (two, three or four parameter fits [20]). Other optional decisions can be made in connection with the model parameters (distance or potential parameters).

6. The CM writes the complete module input and its interpretation key in the communication file [15], as shown in Table 3.

Table 3

Example of a Communication File

```
 8 1 4 6 1 1 1 1   1    1   22 01 LSQFIT(CONEQU)*
PROPANOL//K1I1
PSY00   101325.    PA  TSY00@000RUUUNA$UU2.9815E+02K         E20
XSY002000RLUULA$UU     MOL/M3 ICD00@000RLUPUD$UU    M2/OHM
TFTT
CONEQU 2 4 1 1 1 2.
 1 1 0
2. 3.53E-10 10.43E-10 2.9815E+02 20.436 1.967E-3
```

```
1.756666E-01   2.375264E-03   7.125793E-07        03HA142
4.840817E-01   2.188309E-03   6.564927E-07        03HA142
6.721828E-01   2.105644E-03   6.316931E-07        03HA142
9.173911E-01   2.022641E-03   6.067924E-07        03HA142
1.258155E+00   1.927986E-03   5.783958E-07        03HA142
1.625222E+00   1.847984E-03   5.543952E-07        03HA142
2.580019E+00   1.683475E-03   5.050424E-07        03HA142
4.012091E+00   1.521727E-03   4.565181E-07        03HA142
03HA142( 8)
Interpretation key........
```

7. The CM activates the method bank with the communication file and executes the outlier test (STAHYP) of the weighting factors and of the property variables.
8. The method bank estimates the initial values and calculates the parameters with the help of the least squares method (LSQFIT). The intermediate values of the parameters and the χ^2-values are shown on screen at every iteration step. During the calculation process of the equivalent conductance the module ACOEFF calculates the activity coefficient, Eq.(9b), in an iterative manner at every concentration.
9. The CM displays the final values of the parameters and their statistical significance, writes them on a file, see Table 4, or plots the fitting curve, see Figure 1.

Table 4

Results of the Estimation Process

```
            **** LEAST SQUARE FIT **** with LSQFIT(CONEQU)
COMPONENT(S): PROPANOL//K1I1
VARIABLES   : Concentration [MOL/M3]   Conductance [M2/OHM MOL]
SOURCE      : 03HA142( 8)

Data tupel  : 8; Degree of freedom= 5; Number of Outliers= 0;
Iterations  : 8; Sum of Weighted Residuals=  .2271499853D-10

Sigma FIT   : .2131431381D-05

          *** PARAMETERS OF THE EQUATION ***

LAMBDAoo =  .2600495115D-02  +/-   .3286004872D-05   M2/OHM MOL
J1-TERM  =  .2281534694D-03  +/-   .0000000000D+00
J2-TERM  =  .3439335012D-04  +/-   .1503915422D-05
ASS.CON. =  .3169473485D+00  +/-   .4969376867D-02   M3/MOL

Concentration   Conductance        Conductance-calc.   Difference
  [Mol/M3]                [M2/Ohm Mol]
-------------------------------------------------------------------
 .17566660      .23752640E-02      .23738700E-02      .13937240E-05
 .48408170      .21883090E-02      .21889090E-02     -.60000460E-06
 .67218280      .21056440E-02      .21084930E-02     -.28493810E-05
 .91739110      .20226410E-02      .20230270E-02     -.38556750E-06
1.2581550       .19279860E-02      .19276290E-02      .35727860E-06
1.6252220       .18479840E-02      .18446590E-02      .33254040E-05
2.5800190       .16834750E-02      .16844080E-02     -.93341810E-06
4.0120910       .15217270E-02      .15220230E-02     -.29639340E-06
```

Concentration [Mol/m3]	ALPHA	YPSILON+-'
.17566660	.95856839	.89992154
.48408170	.90848482	.85011175
.67218280	.88481691	.83100514
.91739110	.85873004	.81170355
1.2581550	.82865551	.79104362
1.6252220	.80180085	.77363211
2.5800190	.74845784	.74102531
4.0120910	.69303168	.70894859

```
z = 1
T = 298.15 K              p = 101325.0 Pa
eta = 0.1967 mPas         DK = 20.44
a = 3.53E-10 m            R = 10.43E-10 m
q = 1.3712E-009 m         chen = 2.0
S1 = 1.72508              S2 = 53.7231
S = S1*LAMBDAoo+S2           = 98.5837
E1 = 13.0087              E2 = 59.3284
E = E1*LAMBDAoo-chen*E2      = 219.63096
```
Ionic Distance from J1 = 10.4300 (Angstroms); Adjustment = F
Ionic Distance from J2 = 12.4329 (Angstroms); Adjustment = T
Electrostatic Part of Association Constant = .2513174D+03 L/MOL

*** STATISTICS ***

Weight-matrix of parameters
```
   .23768058D+01
   .00000000D+00    .10000000D+01
  -.24110370D+05    .00000000D+00    .49785643D+09
   .33201729D+03    .00000000D+00   -.46152805D+07    .54357711D+05
```

Eigen-values of the matrix
```
   .52140857D+01    .20084386D-08    .86410411D-04    .10000000D+01
```

Correlation-matrix of parameters
```
   .10000000D+01
   .00000000D+00    .10000000D+01
  -.70089823D+00    .00000000D+00    .10000000D+01
   .92370497D+00    .00000000D+00   -.88718760D+00    .10000000D+01
```

10. The CM maps the result of the calculation process into the BASIC DATA relation and writes it on a buffer file ready for storage in the data base. Table 5 shows an example of the BASIC DATA relation.

The information on the literature and data of ELDAR is available online in the DECHEMA data base DETHERM; FIZ (Fachinformationszentrum) Chemie, Berlin, manages a PC-Version of the ELDAR data and method bank.

Acknowledgements: The authors are grateful to the Bundesministerium für Forschung und Technologie for financial support.

Table 5

Tupel of the BASIC DATA Relation

```
PRIMARY KEY: ICDAL
SYSTEM-TAG : E20
COMPONENTS : 1-PROPANOL//POTASSIUM IODIDE
DATA SOURCE: 03HA142(8)
DEG.FREEDOM: 5
SIGN CHANGE: 3
ERROR TYPE : 0
MEAN WEIGHT: 1.0
VARIANCE   : 0.2131E-5
USED MODULE: CONEQU 2 4 1 1 1 1 TFTT
INT. CONST.: 1 1 1
REAL CONST.: 2. 3.53E-10 10.43E-10 298.15 20.436 1.967E-3
STATE VAR. : PRESSURE     101325 PA
             TEMPERATURE 298.15 K
STATE RANGE: CONCENTRATION,LOWER 0.17566 MOL/M3
             CONCENTRATION,UPPER 4.01209 MOL/M3
VAR. DOMAIN: CONDUCTIVITY,LOWER  .152E-2 M2/OHM
             CONDUCTIVITY,UPPER  .260E-2 M2/OHM
COEFFICIENT: LIMITING CONDUCTIVITY .002600 M2/OHM .3286E-5
             J1-TERM               .2281E-3       .0
             J2-TERM               .3439E-4       .1503D-5
             ASSOCIATION CONSTANT  0.3169 M3/MOL  .49E-2
```

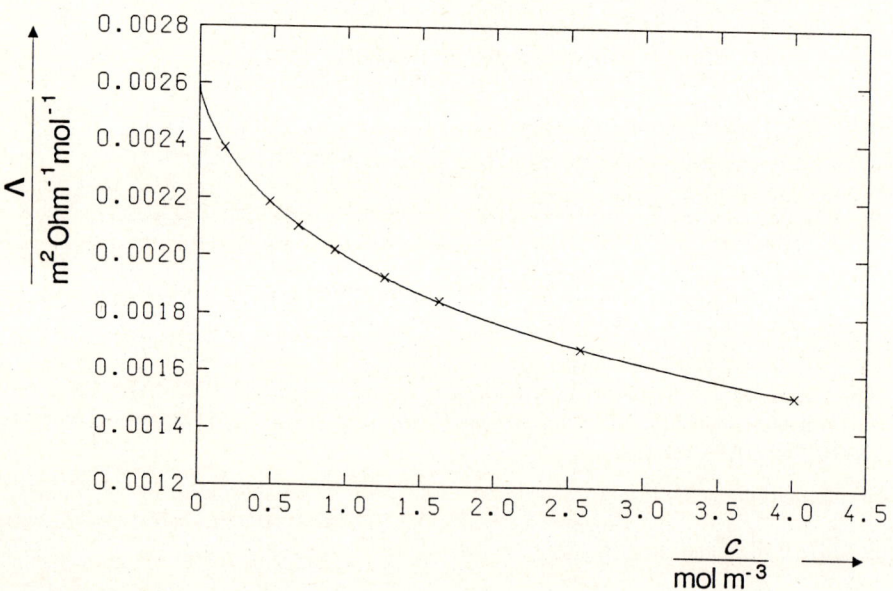

Figure 1. Concentration dependence of the equivalent conductance of KI in 1-propanol at 25^0C. The full line represents the computed equivalent conductance according to Eq.(7) with the coefficients of Table 4. The crosses (x) show the experimental data.

References:

1. Barthel J (1979) The Temperature Dependence of the Properties of Electrolyte Solutions. I. A Semi-Phenomenological Approach to an Electrolyte Theory Including Short Range Forces. Ber Bunsenges Phys Chem 83:252
2. Kunz W, Barthel J (1990) Vapor Pressure Measurements on Nonaqueous Solutions - Part 6. Some Remarks on Integral Equations. J Sol Chem 19: (in press)
3. Barthel J, Buchner R, Wittmann H (1984) Leitfähigkeit und dielektrische Eigenschaften wäßriger $CdSO_4$-Lösungen. Z Phys Chem NF 139:23
4. Barthel J (1985) The Chemical Model of Electrolyte Solutions and its Use for Calculating Solution Properties. In Glaeser PS (ed) The Role of Data in Scientific Progress. Elsevier Publishers, 337
5. Krumgalz BS, Barthel J (1984) Conductivity Studies of Electrolyte Solutions in Dimethylformamide at various Temperatures. Z Phys Chem NF 142:167
6. Brandt S (1975) Datenanalyse. BI, Mannheim
7. Kutas T (1984) A New Method for Solving Nonlinear Least Squares Problem. In: Havranek T, Sidak Z, Novak M (ed) COMPSTAT. Physica Verlag, Heidelberg 415
8. Marquardt DW (1963) An Algorithm for Least Squares Estimation of Nonlinear Parameters. J Soc Indust Appl Math 11:431
9. LeRoux S, Messean A, Vila JP (1982) Standardized Comparsion of Nonlinear Model Fitting Algorithms. In: Caussiunus H, Ettinger P, Tomassone R (ed) COMPSAT. Physica Verlag, Heidelberg, 306
10. Engeln-Muellges G, Reutter F (1984) Formelsammlung zur Numerischen Mathematik, BI, Mannheim
11. Schmeer G (1981) Der Einfluß der Umgebung des Reaktionszentrums auf Hydrolysereaktionen der Ester und Amide aliphatischer Carbonsäuren, Habilitationsschrift, Regensburg
12. Bard Y (1974) Nonlinear Parameter Estimation. Academic Press, New York
13. Barnett V, Lewis T (1978) Outliers in Statistical Data. Wiley & Sons, Chichester
14. Dierckx P (1980) An algorithm of cubic fitting with convexity constraints. Computing 24:349
15. Barthel J, Popp H (1989) Die Methodenbank des wissensbasierten Systems ELDAR zur (statistischen) Analyse von Elektrolytlösungen auf einem Mikrocomputer. Report BMFT-FB : 106 3211 0
16. Popp H (1984) Mensch - Mikrocomputer Kommunikationssystem. Management Expertensystem in der Chemischen Industrie auf der Basis eines universellen Daten- und Prozeduralmodells auf einem Mikrocomputerverbundsystem. Dissertation, Regensburg
17. Barthel J, Popp H (1986) ELDAR, eine benutzerfreundliche relationale Elektrolytdatenbank zur Erzeugung von Stoffdatentensoren auf einem Mikrocomputerverbundsystem. Forschungsbericht BMFT: 106 3208 9
18. Barthel J, Popp H, Schmeer G (1988) Die ELDAR-Methodenbank für Elektrolytlösungen. In: Gasteiger J (ed) Workshop „Computer in der Chemie" Softwareentwicklung in der Chemie 2. Springer, Berlin
19. Melsheimer J, Langner K (1980) On the difficulties of evaluating viscosity/temperature data. Colloid & Polymer Sci 258:160
20. Barthel J, Gores H, Schmeer G, Wachter R (1983) Nonaqueous Electrolyte Solutions in Chemistry and Modern Technology. In: Boschke FL (ed) Topics in Current Chemistry. Springer, Berlin 111:33

TRIANGULATION OF MOLECULAR SURFACES

W. Heiden, M. Schlenkrich, C.-D. Zachmann, J. Brickmann

Institut für Physikalische Chemie I
Technische Hochschule Darmstadt
Petersenstr. 20, D-6100 Darmstadt

Abstract: An algorithm for the generation of a solid molecular surface from a given number of surface points in 3D space is presented. This surface is represented by a triangle mesh. Such a mesh of triangles is not uniquely defined by the position vectors of the surface points, and there is no straight forward algorithm for triangulation. In this paper a hierarchy of algorithms is described which successfully works for the automatic triangulation of molecular surfaces of all sizes. Some examples are given for the application of this solid surface approach in molecular modelling of proteins.

I. Introduction

The graphical representation of molecular surfaces and surface properties becomes increasingly important in molecular science. This is a consequence of the dramatic increase of computer power, which allows the calculation of a variety of molecular quantities. The flood of numerical data resulting from such calculations can be qualitatively studied with a conveniently chosen graphical representation. Characteristic patterns can easily be recognized just by inspection. A typical example is the representation of the electrostatic potential of a charged probe migrating along the van der Waals surface of a molecule. The conventional way of representing these data by molecular graphics techniques is the colour coding of each surface point, and the display of the point set on the screen. This method, however, has all the disadvatages of the old vector graphics procedure, i.e. with increasing complexity of the molecular scenario there are increasingly many dots and lines interfering on the screen. Displaying the molecular surface in raster graphics technique, and coding surface properties according to a colour scale helps to represent complex information much more clearly.

In this paper we present a program system for the generation of a triangle mesh from a dotted surface. The solid surface can be displayed on the basis of sequences of triangles as a smooth manifold by using the standard Gouraud-shading algorithm [1] which is implemented in most modern display hardware. Interactive handling of such surfaces becomes possible with the hardware capacity of modern workstations. In section II the basic algorithm of the triangle mesh generation is described while section III deals with the handling of topologically complex surface regions. In section IV some examples for the application of the technique to the treatment of proteins are given. In the last section some conclusions are drawn.

II. Basic Triangulation Algorithm

In the algorithm presented in this work the three dimensional coordinates of surface points are sufficient as basic input for triangulation. The normal vectors on these points can be used as additional information if these data are available. The program generates a list of triangles, representing the closed surface of the object. The smoothness of the surface is determined by the density of input points. All parameters for the triangulation algorithm are generated automatically from the given point set. The general strategy of the program is a multiple run of a basic triangulation routine with varying control parameters. A neighbour list for all points is generated before the actual triangulation procedure starts. The program then starts from an initial triangle sequentially adding new triangles with one side in common to a previous one. For many objects this basic scheme generates the triangle mesh for the complete surface, and no sophisticated handling of complex regions is necessary. The basic algorithm can be described as follows:

For every point a list of nearest neighbours - including their distance from the reference point - is generated. The execution time of this calculation scales linearly with the number of points, due to the usage of the cell index method [2]. The only input parameter necessary for the neighbour list is the maximum distance to those points which are included in the list. This distance can be generated automatically from the density of the points in three dimensional space when one restricts the number of neighbours to 20 - 60. The generation of a neighbour list for about 30000 points needs only 10 minutes on a Silicon Graphics workstation IRIS 4D.

For each triangle a normal vector is calculated. The direction depends on the order of the three points forming this triangle. The direction of the normal vector on the initial triangle defines the inner and the outer space of the molecule. The direction of the normals of adjacent triangles is related to the initial one. In order to set the initial normal vector correctly, the first triangle starts at a point which is maximal along one of the coordinates. The two closest neighbours are chosen to complete the initial triangle. The normal vector of this triangle is chosen such that its projection on the coordinate is positive. The initial triangle is now framed by three "outer" lines. Outer lines are stored in a connection stack.

Starting from one outer line, the basic algorithm searches for the nearest point not excluded by certain restrictions. Such restrictions are necessary in order to prevent the formation of triangles which cannot be part of a smooth surface (for example intersecting triangles, see Ref.3). The major restriction is given by a maximum angle between the normal vector of the new triangle and the normal vectors on the adjacent points. If no normal vectors are available from the input file, the vectors of the points are calculated as an average of the adjoining triangle normals. With every new point a new triangle is generated. Two new outer lines are added to the connection stack while one old outer line is removed, i.e. this line cannot be incorporated in another triangle. This procedure is repeated until no more outer lines are present in the connection stack.

Topologically complex regions are defined as areas in the 3D point set for which no new triangle can be formed within the basic algorithm. In order to exclude such a region from further treatment within the first cycle of triangulation, all triangles adjacent to points of a border line (i.e. a line from which no new triangle point can be found) are removed. During this process outer lines are deleted, while inner lines become restrained lines. During the further triangulation, restrained lines are treated like outer lines but with the condition that from such a line no search for a new triangle may be started.

Another criterion for the detection of error-causing regions within the basic algorithm is an overflow of the number of connections from one point (vertex number). This number normally ranges between 4 and 11. Moreover, triangles representing a planar intersection with an existing surface are excluded in the basic routine. All triangles in contact with points identified as sources of error are deleted analogously to the procedure mentioned above.

If there are still restrained lines existing when the basic triangulation is finished additional steps have to be performed in order to generate a closed surface.

III. Handling of Topologically Difficult Regions

After running through the basic algorithm cycles the surface triangulation is finished with the exception of those regions where this algorithm failed. These regions form gaps in the surface which are framed by restrained lines.
In a first attempt all restrained lines are changed to outer lines and the basic algorithm is started again. In this step many gaps can be closed, since the number of possible errors is reduced considerably by the incorporation of most points in already completed surface regions.
If there are still gaps in the surface, a more sophisticated second attempt is necessary. The gaps are now treated independently. For each of them restrained lines are changed to outer lines and the basic triangulation algorithm is used. The parameters for this algorithm (maximum distance for the neighbour list and maximum angle between point normal and triangle normal) are increased. If this attempt is also not successful, the procedure is carried out a second time with reversed order of outer lines used as a basis for a new triangle.

A last ultimate triangulation step is performed if all previous attempts to generate a closed surface have failed: all points of restrained lines of each gap are connected in a specific order [3].
Two major effects in the dotted surface cause problems in triangulation. For some regions the point density may be too small in order to define a surface. For these regions a

recalculation of a sufficient number of surface points is necessary. There are also problems when the point density over the total surface is strongly varying and does not allow triangulation of the whole object without a change of parameters. Both problematic cases can be generally circumvented by a proper choice of surface points.

IV. Applications in Protein Design

There are two major applications derived from the triangulated surfaces. With modern graphic workstations like the Silicon Graphics IRIS 4D it is possible to display interactively solid objects by using the triangulation list. Using shading and shadow algorithms [4,5] a "realistically" looking picture of topologically complex objects like proteins can be drawn (see figure 1). Qualities such as the electrostatic potential on this surface can be shown by colour coding.

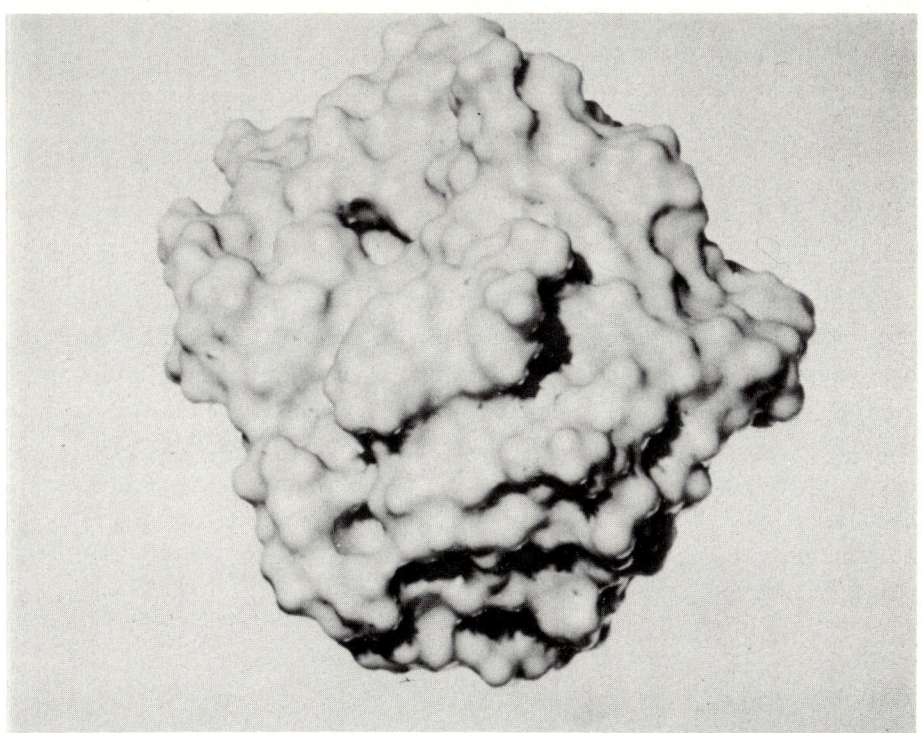

Figure 1 :
Bovine β-trypsin, solid model of a triangulated van-der-Waals surface.

Triangulation also opens new possibilities of evaluating complex properties of the object. One application is the calculation of surface integrals, but also all volume integrals can be obtained from Ostrogradski's formula [6]. This presents an elegant way of calculating surface area and volume of molecules.

The main advantage of the triangulated surface is the well defined neighbourhood of each point. Taking advantage of this special feature, another algorithm calculating the two main surface curvatures at each point has been developed (For details see Ref.7). Knowing the main curvatures, convex and concave surfaces can be easily distinguished. This approach opens the possibility for an automatic localization of pockets on protein surfaces. This might be an essential tool in the automation of detection of receptor sites. Also for molecular modelling it is helpful to display the curvatures as colour coded information on the solid surface. This allows the fast localization of cavities in the protein.

With today's powerful graphic workstations, interactive docking of small molecules into pockets on protein surfaces can be done very effectively using the solid surface representation of triangle meshes.

V. Conclusion

A program for an effective triangulation of surfaces from any given number of surface points has been presented. It has also been shown that the representation of objects by solid surfaces has essential advantages compared to the conventional display of dotted surfaces. These advantages become even greater when more powerful graphic workstations with high interaction rate are used. This additional tool allows the presentation of qualities necessary for molecular mechanics on the complex surface of molecules. Topological quantities can be most easily analyzed using this type of representation.

Acknowledgement

The image shown in this paper is generated using the package MOLCAD which has been developed in our group. In this context we like to thank Martin Knoblauch and Michael Waldherr-Teschner for their cooperation. This work is supported by the Bundesminister für Forschung und Technologie, Bonn, the Fonds der Chemischen Industrie, Frankfurt, and Silicon Graphics Inc., Mountain View (USA) and München.

References

1 Gouraud H (1971) In: IEEE Transactions on Computers C20(6):623
2 Quentrec B, Brot C (1975) J.Comput.Phys 13:430
3 Heiden W, Schlenkrich M, Brickmann J J Comp Aided Mol Design
  to be published
4 Brickmann J (1983) J Mol Graphics 1:62
5 Schlenkrich M, Heiden W., Brickmann in preparation
6 Huron M, Claverie P (1972) J Phys Chem 76:2123
7 Zachmann C-D, Schlenkrich M, Heiden W, Brickmann J
  in preparation

MOBY: MOLECULAR MODELING ON THE PC

Udo Höweler

Institut für Organische Chemie der WWU
Wilhelm-Klemm-Str. 4, D-4400 Münster

Abstract: The molecular Modeling programme MOBY is introduced that makes the standard features of high level modeling tools available to PC users, without the limitations on speed and the restrictions on the molecular systems found for other PC programmes. Several extensions to the standard features like a detailed energy analysis and quantum chemical methods have been implemented. Some examples for the applicability of MOBY are shown on the poster.

Molecular modeling methods are of increasing importance in chemistry, pharmacology, and biology. Current applications cover the areas of structure determination for small molecules, receptor-substrate interactions, drug design, structure-reactivity relationships, and protein engineering.

Some features provided by molecular modeling programmes are: i) display of X-Ray structures; ii) generation of molecular structures from scratch; iii) optimisation of molecular structures based on a molecular mechanics force field; iv) preparation of input data for quantum chemical and force field programmes; v) visualisation of the results; vi) modification of known experimental structures with monitoring of the geometrical and energetical changes; vii) comparison of structures; viii) analysis of relative stabilities of conformers.

Most of these topics depend strongly on fast graphics facilities while quite an amount of computer power is needed to perform the force field calculations. Thus, molecular modeling programmes are usually implemented on graphic workstations. But, most of the time to solve a structural problem is usually needed to generate a molecule, comprehend its structure and its reactivity and to analyse the results of calculations. Thus much of the computing power of the workstations is wasted.

On the other hand, the availability, the enhanced graphics and the computing power of the most recent series of PC's makes them the ideal hardware to generate structures, to prepare the input for calculations and to analyse the results of calculations that themselves may be run on more powerful mashines. Today only two programmes are widely used for molecular modeling on the PC: ALCHEMY and PCMODEL. Both are versions of modeling programmes running on workstations and are tailored to the memory size and the restricted capabilities of the PC and its operating system. They both provide only a subset of the features of the original programmes.

The molecular modeling programme MOBY is intended to close the gap between the high level modeling tools running on expensive hardware and the low level systems that do not perform satisfactorily. It provides the general modeling features enhanced by options for the analysis of the results, that are even not available in high level programmes, without the limitations on speed and the restrictions on the magnitude of molecular systems found for the PC programmes.

Prominent features of MOBY are:
- display and manipulation of up to 2000 centers
- input of data in "every" type of format
- input of Brookhaven Protein Data Bank files
- interactive generation of 3D structures
- AMBER force field originally parametrised for proteins and nucleic acids and extended by some generalised center types for organic molecules
- energy calculations and minimisation of up to 150 centers interacting with ALL of the 2000 centers
- partial optimisation of user defined substructures
- modification of internal and cartesian coordinates of user defined substructures (with energy monitoring)
- quantum chemical calculations using the MNDO and AM1 Hamiltonian for substructures with up to 60 atomic orbitals
- color coding of centers according to their properties like partial charges and energy contributions.

A few application are shown on the poster starting with the structure of the plastocyanine PCY1 extracted from the Brookhaven Protein Data Bank. PCY1 is a protein of the photosynthesis system and consists of about 890 centers in the united atom representation. The full structure as well as the backbone representation are shown. Some amino acids of importance for the electron transfer properties are shown as spacefilling dot surfaces surrounded by the stick model of the remaining protein centers.

A fraction of the protein consisting of two strands of a β-sheet region are used to depict some further analytical features of MOBY. One picture shows the electrostatic potential for a positive point charge of unity in a distance of 200 pm to the van der Waals surfaces of the centers within the substructure. This potential is shown for the isolated substructure and for this substructure within the protein environment.

In order to study the attractive forces between the two strands, the distances from H-atoms to acceptor atoms within 220 pm are depicted and color coded according to their actual value in increments of 10 pm within the range from 220 pm to 180 pm.

Every NH fragment forms a hydrogen bond to the nearest CO group. The strengths of these bonds can be estimated by color coding the centers according to the hydrogen bond term specific to the underlying AMBER force field.

The geometries of the β-sheet region prior to and after force field optimisation for the isolated substructure are compared in order to show the influence of the surrounding protein centers on the conformation of the amino acids within the strands. Color coding of the centers according to their contributions to the force field energy reflects the relaxation of the geometry during the optimisation.

At last the superposition of snap shots of a molecular dynamics simulation at 300 K for a few H_2O molecules in the vicinity of a hydrophilic salt bridge at the outer surface of the protein are shown. The geometries are color coded according to the time elapsed since the beginning of the simulation. The simulation monitors the opening of a hydrophilic pocket built by the salt bridge (LYS and GLU residues) and a nearby SER side chain, allowing one H_2O molecule to move into this pocket.

Although only a small subset of the features of MOBY is presented, the wide range of applications that can be covered even with a modeling tool running on a PC is clearly seen.

ROCOCO: REFERENCE ONLINE LIBRARY FOR THE COMPUTER AIDED CONSTRUCTION OF MOLECULAR GEOMETRIES
COMPUTER AIDED CONSTRUCTION OF REALISTIC MOLECULAR MODELS USING A KNOWLEDGE BASE (I)

Benno Krieg und Peter Keller

Institut für Organische Chemie der Freien Universität Berlin
IBM Deutschland Kompetenzzentrum Chemie

Abstract: This article describes a general concept for the construction of molecular models which accurately mirror the structure of the real molecule without large scale optimization. Imitating the procedure of an experienced chemist the computer programme system analyses the topological data of a molecule - presented as a 2D structural formula - and determines the wanted geometric parameters using an internal data base, which contains the values of the respective fragments of the molecule. The results can be displayed in tables or printed onto a plotter. In the current stage of development the programme together with the data base is a "comfortable, electronically readable handbook" for bond lengths and bond angles of aliphatic and heteroaliphatic compounds as well as strain-free aliphatic cyclic and aliphatic heterocyclic compounds containing the elements C, H, O, N, F, Cl, Br and I.

THE SITUATION

The aim, to develop a 3D molecular model from a 2D structural formula, has not been completely satisfactorily solved, although computer aided methods have made great advances. Most of the developed methods work as follows:
1. structure building from standard fragments
2. energy minimization through stepwise variation of molecular parameter

An unfavourable selection of the starting structure(s) leads to very time-consuming calculations and moreover, may lead to localized minima.

KNOWLEDGE PROCESSING - A NEW WAY TO MOLECULAR MODELS

The geometry of a specific molecule is the result of optimizations made by nature. Our knowledge about the geometry is constantly improving (X-ray analysis, NMR-investigations etc.). Through construction of these models the obtained data should be available in the form of a knowledge base. Indications of this procedure can be found in the recently published programmes WIZARD (1) and AIMB (2). A general concept for the use of data bases and rules for stored knowledge from several experimental and theoretical structure determinations for use in the investigation of bond lengths and bond angles does not, to our knowledge, to date exist. We have attempted to follow such a course, through the development and application of the programme ROCOCO (Reference Online Library for the COmputer aided COnstruction of Molecular Geometries).

ROCOCO

The information about the bond partners, the bond order of the atoms and their charge, contained in the structural formula of the molecular model to be constructed, is transfered to the programme by way of a conection table (mol file). From this file the components of the molecule in the form of bond centred and atomic centered fragments are determined and coded, then, these fragments are searched for in the internal data base of the programme. The obtained data base values and standard values for all bond lengths and bond angles are stored in a new file (add file). This file can be displayed in tables or printed onto a plotter.

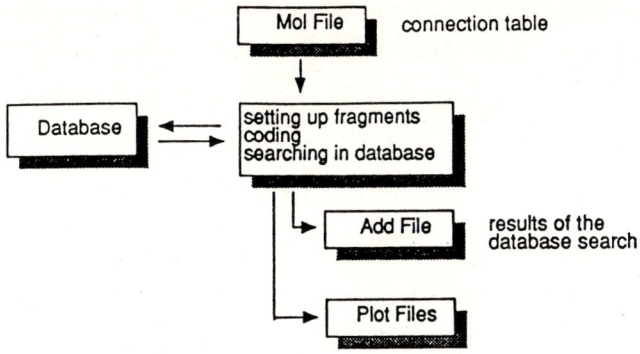

Fig. 1. Schematic map of ROCOCO.

Input of the 2D Molecular Structure

The programme system ROCOCO receives the 2D molecular structure in the form of a so called mol file. For its generation a suitable mol file editor is required, similar to the one implemented in the programmes CHEMBASE (R) and CHEMTEXT (R) (both from the firm Molecular Design Limited). (We are currently working on our own comfortable 2D structure editor.)

Currently, the following inputs are permitted:
- aliphatic compounds as well as strain-free aliphatic cyclic compounds and aliphatic heterocyclic compounds.

- up to 20 non-hydrogen atoms of the elements C, O (bond order 1 - 3), N (bond order 3 and 4), F, Cl, Br and I.

The Knowledge Base

The internal data base of ROCOCO contains bond lengths and bond angles of exactly specified molecular fragments. The data are divided into 7 different categories, corresponding to their origin: X-ray analysis, neutron diffraction, electron diffraction, micro wave spectroscopy, infrared spectroscopy, quantum chemical calculations, and force field calculations (molecular mechanics).

Searching terms; determination and coding: For the coding of fragments (searching terms) linear codes were developed:
- atoms (a_code and ext_a_code)
- bonds (b_code and bs_code)
- angles (w_code and ws_code)

In the following paragraphs they are briefly described.

Coding of atoms (a_code and ext_a_code): The coding of an atom occurs currently in this way:

eehc (a_code)
    ee = symbol of the element
    h = degree of hybridization (relating only to the p-orbitals, for hydrogen arbitrarily set to zero)
    c = charge (0 = uncharged, 3 = singular positive charge, 5 = singular negative charge)

Furthermore, an extended atomic code (ext_a_code) for the elements C, O and N is generated. It is rather similar to the above described a_code, except that it contains, in the 2nd position, information about the number of the H-atoms bound to this atom.

enhc (ext_a_code)
    n = number of the H-atoms on element e

Examples:
C_30: a_code of an C-atom, sp^3-hybridized, uncharged
C330: ext_a_code of a CH_3-group
O_30: a_code of an O-atom, sp^3-hybridized, uncharged
O130: ext_a_code of an OH-group
Cl30: a_code of an Cl-atom

If the element is marked with only a single letter then, in the 2nd position, there will be an underline ("_").

Coding of the bond centered fragments (b_code and bs_code):

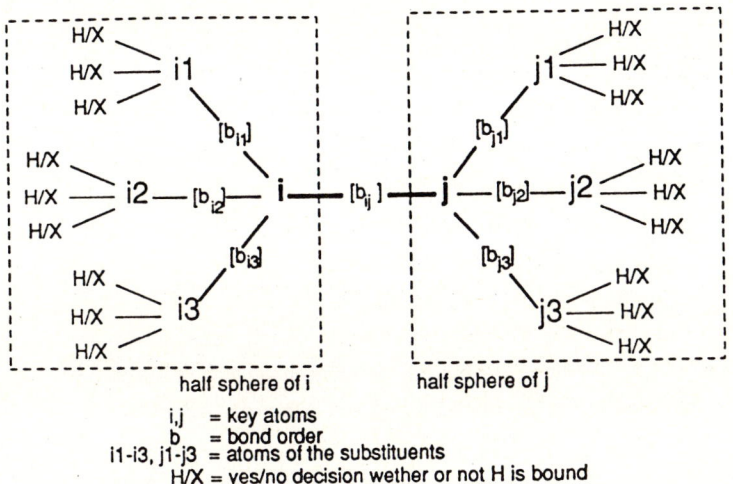

i,j = key atoms
b = bond order
i1-i3, j1-j3 = atoms of the substituents
H/X = yes/no decision wether or not H is bound

Fig. 2. Schematic picture of a bond between the key atoms i and j with regard to their environment.

Rules are required for the atoms of the bond and for the substituents of each half-sphere to decide their order of priority. For the atoms i and j the following applies:
(I.1) The priority decreases with descending order number.
(I.2) The priority decreases with descending p-contribution to the degree of hybridization (sp^3-sp^2-sp^1).
(I.3) The priority decreases with descending (numeric) value of the charge code.

For i1 to i3 and j1 to j3 the following applies:
(II.1) The priority decreases with descending bond order (of the bond to i or j, respectively) (triple - double - single bond).
(II.2) Like (I.1)
(II.3) Like (I.2)
(II.4) Like (I.3)
(II.5) The priority decreases with descending numer of bound H-atoms.

The following rules are provided to decide the priority of the half-spheres.

(III.1) The half-sphere which has a key atom with a higher priority according to the rules for (I) has the higher priority.

(III.2) If no differences are to be found in (III.1), you proceed as described in the following.
For each half-sphere, the substituents i1 to i3 and j1 to j3 are ordered according to rule (II), enabling comparisons between them.

(III.2.1) You compare the substituents with the highest priority from both half-spheres. The order of priority of the half spheres is determined according to (II).

(III.2.2) If no difference is found, you compare the substituent atoms with intermediate priority from both half-spheres. Then the assignment is again according to (II).

(III.2.3) If no difference is found again you arrange the substituent atoms with the lowest priority according to (III).

If, after employment of all these rules in the programme, no differences are found the two half-spheres will be assumed identical.

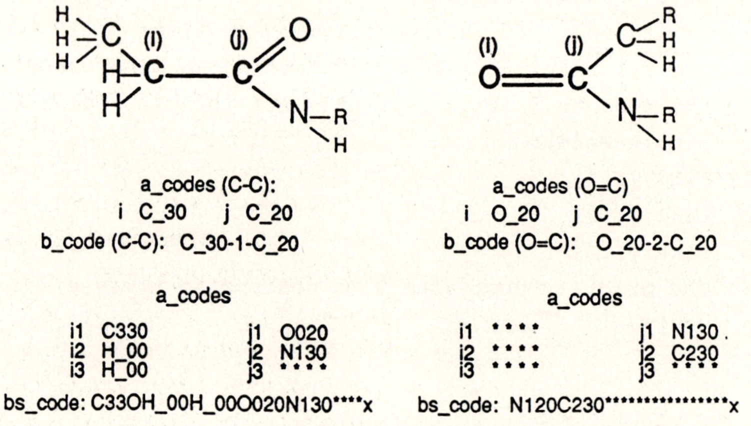

a_codes (C-C):
  i C_30   j C_20
b_code (C-C): C_30-1-C_20

a_codes (O=C):
  i O_20   j C_20
b_code (O=C): O_20-2-C_20

a_codes
i1 C330   j1 O020
i2 H_00   j2 N130
i3 H_00   j3 ****
bs_code: C33OH_00H_00O020N130****x

a_codes
i1 ****   j1 N130
i2 ****   j2 C230
i3 ****   j3 ****
bs_code: N120C230****************x

Fig. 3. Examples of coding of bond centered fragments.

Coding of the atomic centered fragments (w_code and ws_code):
The angle code applies to a specific angle of an atom (internal sphere). It is composed of the a_code of the atom and the bond orders of the angle defining bonds.
e1n1h1c1e2n2h2c2(e3e3h3c3e4e4h4c4)******x (ws_code)
  n = number of the bound H-atoms
  * = vacant position

You have to decide between angle-defined substituent atoms $i1_{in}$, $i2_{in}$ (internal sphere) and the others $i1_{out}$, $i2_{out}$ (external sphere). In ws_code the first 8 symbols are reserved for the

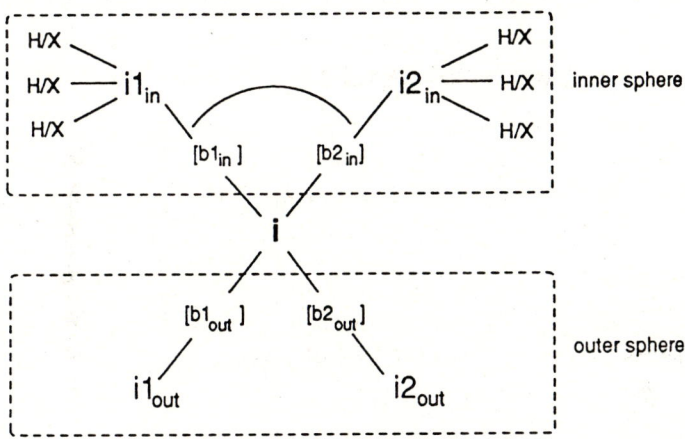

Fig. 4. Schematic picture for the coding of an atomic centered fragment (examples see fig. 5).

substituent atoms of the internal sphere ordered according to the results from rule (II) for the two ext_a_codes. There then follows an open round bracket to indicate the beginning of the external sphere. Both the substituents of the external sphere are indicated by the following a_codes (symbol 10 to 17). Here you have to dispense with the ext_a_codes because these two substituents have

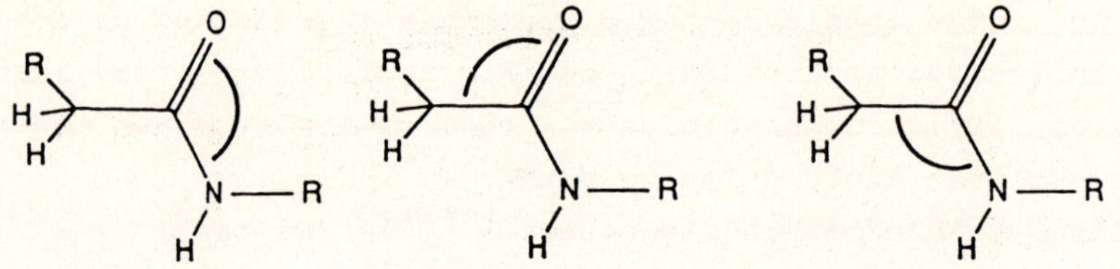

w_code: C_20-12
ws_code:
O020N130(C_30****)******x

w_code: C_20-12
ws_code:
O020C230(N_30****)******x

w_code: C_20-11
ws_code:
N130C230(O_20****)******x

Fig. 5. Examples of coding of atomic centered fragments.

a reduced influence on the value of the angle. The next symbol is a closing round bracket. If the external sphere contains none or just a single substituent, the vacant spaces up to and including position 24 will be filled up with the symbol "*". The symbol "x" in position 25 indicates completion.

RESULTS

The following example, using ß-amino-propionic acid (protonized), demonstrates how ROCOCO works. After providing the connection table (mol file) via a 2D structure editor, ROCOCO_1 is started and in the main menu of the view port the option MOL FILE INPUT is activated. A list of the existing mol files appears. Shortly after the choice of the file name (mouse click) under which the connection table is stored, the information appears that the loading is finished and you begin SEARCH IN DATABASE. Now ROCOCO_1 executes the following tasks (see fig. 6 and 7, too):

- generation of the atomic codes (a_code and ext_a_code)
- generation of the bond codes (b_codes and bs_codes)
- generation of the angle codes (w_codes and ws_codes)
- search for the corresponding values in the data base and in the list of the standard values
- construction of the results file (add file)

The contents of this file can be displayed on the screen with the option VIEW ADD FILE and printed as a hardcopy, if required. With

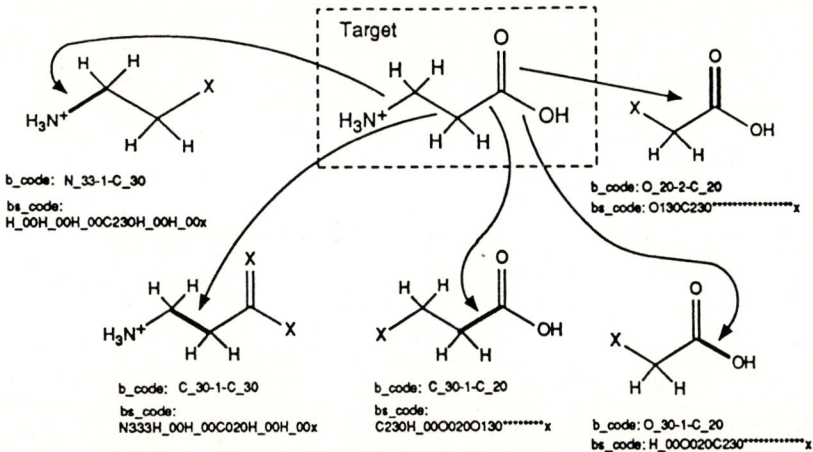

Fig. 6. Generation of bond codes (example: protonized β-amino-propionic acid).

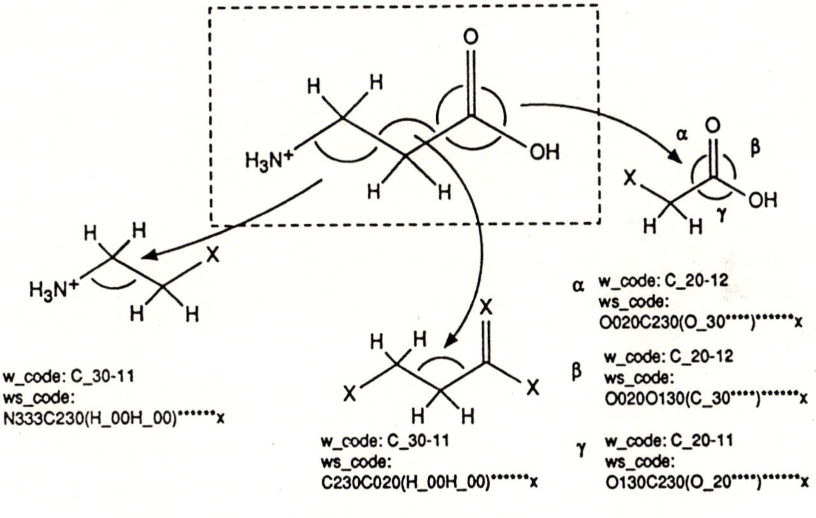

Fig. 7. Generation of angle codes (example: protonized β-amino-propionic acid).

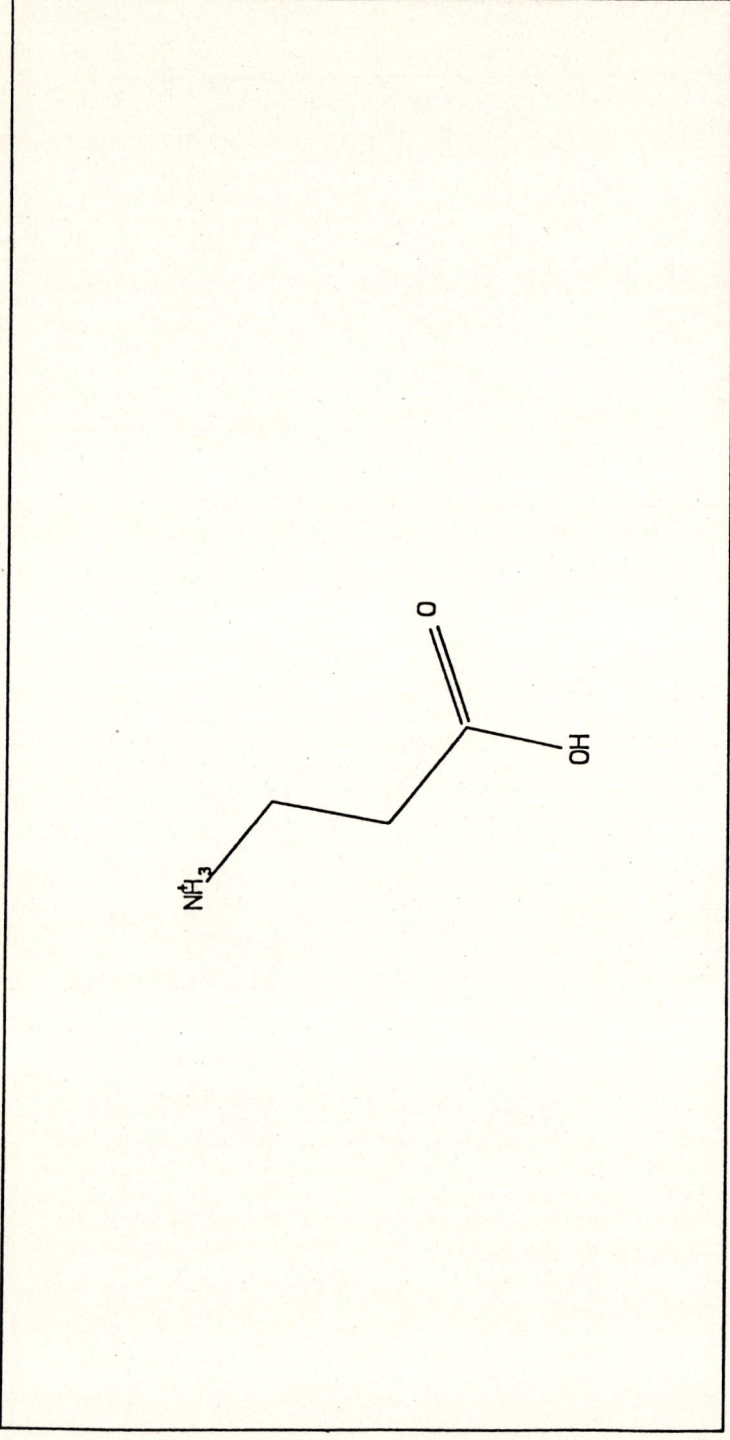

Filename : EXAMPLE2.PLT
Date   : 23.10.1989
Comment   :

Fig. 8. Structure formula of protonized ß-aminopropionic acid (laser plot).

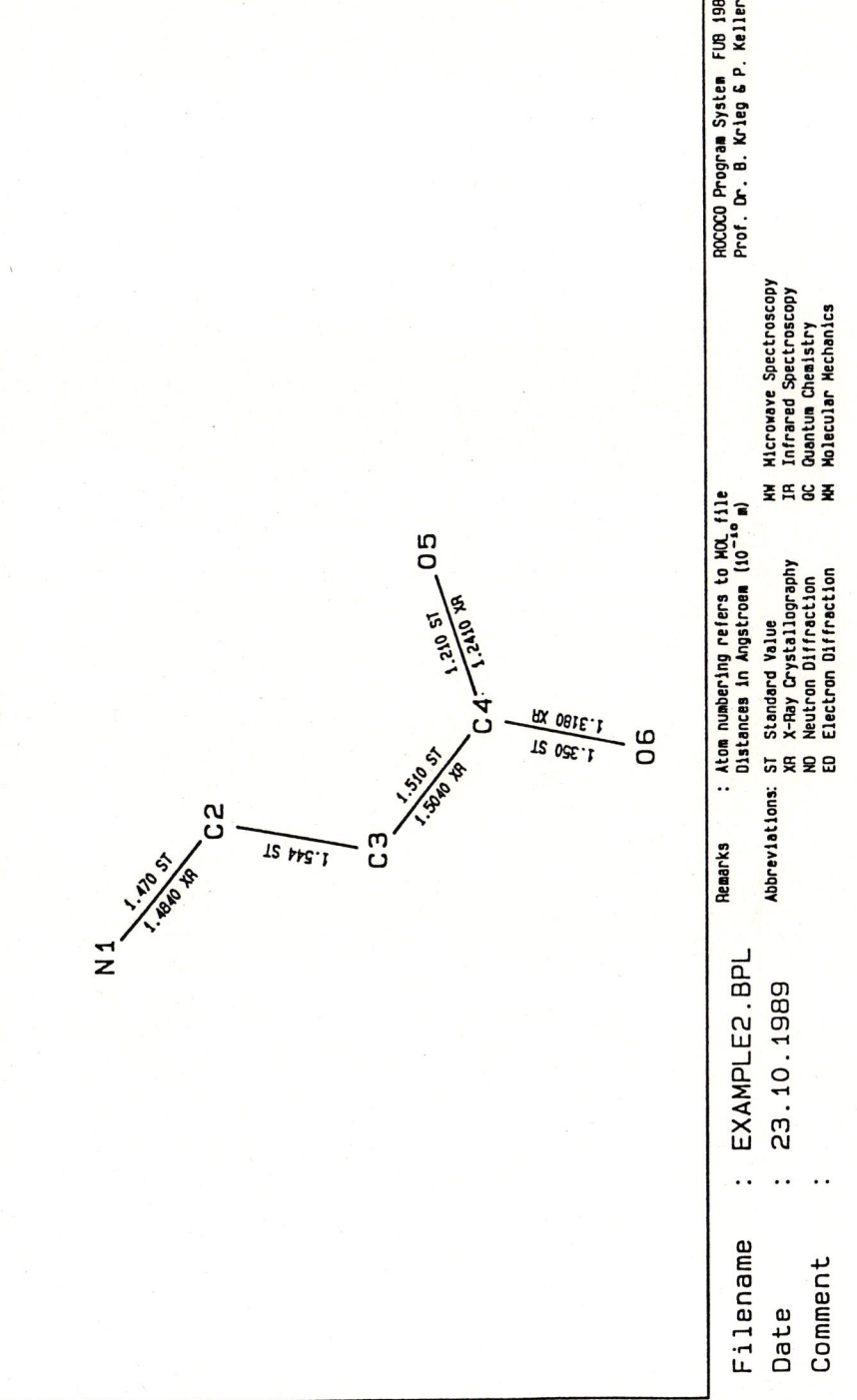

Fig. 9. Atomic distances of protonized β-aminopropionic acid (laser plot).

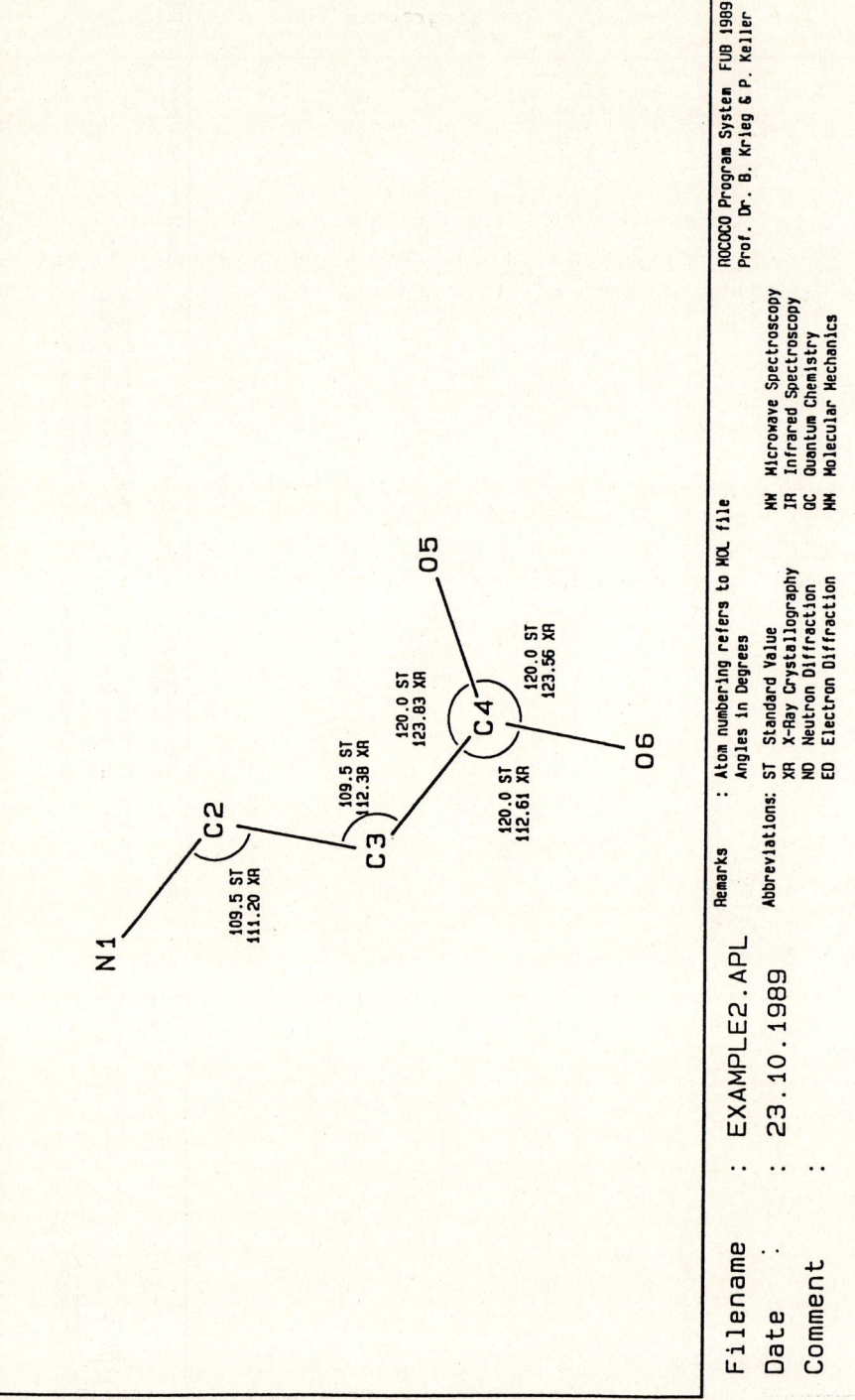

Fig. 10. Bond angles of protonized β-aminopropionic acid (laser plot).

the option 2D PLOT FILES the results can be printed onto a plotter or, by initiating a laser plot programme, onto a laser printer (see fig. 8-10).

EXPERIMENTAL DATA

ROCOCO was programmed with the Turbo Pascal Compiler 5.0 and the Database Toolbox 4.0 (both from Heimsoeth & Borland). It is compatible with IBM PS/2 models and with IBM AT (VGA- or EGA-graphic).

LITERATURE

1 Dolata DP, Carter RE (1987) J Chem Inf Comput Sci 27:36
2 Wipke WT, Hahn MA (1988) Tetrahedron Computer Methodology
   1 2:141

Der Firma IBM Deutschland danken wir für finanzielle Unterstützung.

MMGEO
A VERSATILE TOOL FOR MMX USERS

N. Reuter

Universität Freiburg,
Institut für Organische Chemie und Biochemie,
Albertstraße 21, D 7800 Freiburg/Breisgau

Abstract: MMGEO is a versatile tool for MMX users, it is a so called software interface between MM2/MMX and other programs like SCHAKAL, CHEMTEXT and CDCHEM. It has many other build-in functions for calculating atomic distances, bond angles and dihedral angles.

HISTORY OF MMGEO

Molecular mechanics, so called force field calculations, get more and more importance in our days. New and better programs appear in the scene. One of the most wide spread programs in molecular mechanics is Allingers MM2 [1]; he developed MM1 first in the early seventieth, changed in the middle of the seventieth to MM2 and ended so far with MM3 in 1988. Because of its wide spread and its conversion from a mainframe version [2] to a one that can be run on every personal computer (IBM clone) [3], the author wrote a kind of software interface to link MM2 data to some other programs, like CHEMTEXT [4], SCHAKAL [5] and CDCHEM [6].

The author learned handling with force field data in a seminar at the university of Freiburg with MM2-82 running on an UNIVAC mainframe, data must be entered directly with an editor or via an implemented Saunders-algorithm in the MM2 I/O-program STR [7], the results of such calculations could only be seen from the various printouts. On this UNIVAC MM2 was linked to SCHAKAL with a procedure called PIT [8], so it was simple to get pictures from the MM2 data files. In 1989 the university changed the mainframe, an IBM-3090 was installed. On the IBM force field calculations can be done, but no I/O-program exists, no images can be generated.

The second possibility to do force field calculations was the PC version of MM2 MMX87 and its input/output program PCMIO. These programs run on every personal computer equipped with a numeric coprocessor and a graphics board. But handling data with the FORTRAN written PCMIO could be very boring, because all entries or angle-/distance-calculations had to be confirm by one or more pressed keys. The input format was restricted and after each calculation option the graphic image was slowly recalculated an drawn. This anger lead the author to write a new program called MMGEO [9], in which such options (atomic distances, dihedral angles and bond angles calculations) can be done easily. The input format of the commands had not (or only little) to be restricted, calculations had to be done quickly and simple images had to be generated.

To start a command line version was best, a parser interpreted the first implemented commands and their parameters. During the phase of development many suggestions from colleques were additionally implemented.

Now MMGEO reached the version number 2.61 (November 1989) and has the following available commands:

COMMAND SUMMARY OF MMGEO

Calculation Commands:

DIS atom#1 atom#2 - atomic distances between atoms #1 and #2.

ANG atom#1 atom#2 atom#3 - angle between three atoms #1 to #3.

DIH atom#1 atom#2 atom#3 atom#4 - dihedral angle made by four atoms #1 to #4, if it is a H-C-C-H chain a Karplus-1H-coupling constant range is shown.

Graphic Commands:

PIC parameter - draws a wire model on the screen, only connected atoms are drawn. Parameters are ALL, all atoms (connected and attached) will be drawn, and NUM, all atoms are numbered.

BAL parameter - draws a very simple 'ball model', radii are not Van der Waals radii, z-coordinates are sorted, i. e. atoms in front hide atoms in rear. Parameter can be ALL, all atoms are drawn.

ROT X Y Z - Rotates the molecule versus x-, y- and z-axes, calls PIC to show molecule.

SET param#1 param#2 - can be used to configure MMGEO to a specific graphic adapter. First order parameters are CGA, EGA, HGC and VGA. If you use the second order parameter SAVE (SET EGA SAVE) a configuration file will be written on disk, this leads MMGEO not to search the highest graphic adapter build in your system. If SET without parameter is entered, MMGEO searches the best graphic adapter build in your computer system (VGA > EGA > HGC > CGA).

SIN degrees - Single step rotation mode, molecule can be rotated by pressing x, y, z, or X, Y, Z clock- or counter clockwise, if degrees is not specified 5 degrees are optional.

Interface Commands:

CDC - generates a CDCHEM compatible data file.

MOL - generates a CHEMTEXT compatible molfile, all bonds are single bonds, no stereo chemistry is build in.

SCH - generates a SCHAKAL compatible data file, lone pairs are transferred to the atomic type LP.

Divers Commands:

ATT - shows the attached atom list of the current data file.

CAT - cataloging function to see what files are on your disk.

CLS - clears the screen.

CON - shows the connected atom list of the current data file.

COP - shows the copyright notice.

DAT - shows date and time of day and the elapsed time using MMGEO.

DIR - same as CAT.

DOS - DOS shell, return with EXIT from DOS.

END - ends program.

EXI - ends program.

HEL - help system, HEL shows all available commands. More information on an item can be get by typing HEL command, for example HEL SCHAKAL shows the SCHAKAL command help screen.

LOA - loads a data file.

NAM - shows the name of the MMX data file (if entered).

<u>QUI</u> - ends the program.

<u>TIM</u> - same as DAT.

<u>VER</u> - shows version number of MMGEO.

<u>VID</u> - shows the current video mode.

Supplement:

MMGEO reads MMX and MM2 data files, unkown formats are rejected. The program can also be aborted by pressing the function key <F3> during every action in MMGEO.

MMGEO is fully written in Turbo-BASIC [10], the source code includes approximately 1800 lines of code. Graphic adapter detection is made by interrupt programming.

How to get MMGEO

MMGEO is a ShareWare product, send a formatted diskette and a stamped envelope to the author to get a version. If you find it useful you are pleased to send the author a registration fee of DM 30.- or US$ 20.-, it is also possible to have a cheap annual update service. MMGEO is available on 5.25" or 3.5" diskettes.

Future Developments

Some new implementations will be made soon. Some colleques want to have the Saunders algorithm to be implemented, because PCMIO then is no more necessary for quick structure input. A next improvement is an automatically rotating molecule mode to simplify the single step rotation mode. The most fundamental improvements will appear in the version 3.00 and higher, MMGEO will be fully mouse driven with pull-down menues. Atoms in calculation commands will only be clicked to enter.

Hardware Requirements:

MMGEO runs on every IBM personal computer or 100% compatible computer, minimun RAM is 384 KB, one diskette drive or hard disk, any of the following graphic adapters: CGA, EGA, VGA or hercules graphic card.

References:

1) N L Allinger, Adv in Phys Org Chem, 13, 1 (1976)
2) MM2, QCPE 395, University of Indiana, USA
3) MMX87, Serena Software, Bloomington, Indiana, USA
4) CHEMTEXT, Molecular Design Ltd, Fallon Drive, San Leandro, USA
5) SCHAKAL-88, E Keller, Kristallogr Inst Univ Freiburg
6) CDCHEM, R Kaufmann, c't Magazin für Computertechnik, Heinz Heise Verlag, Heft 7, (1989)
7) STR, QCPE, University of Indiana, USA
8) PIT, H D Beckhaus, Inst für org Chemie und Biochemie, Universität Freiburg
9) MMGEO, N Reuter, Inst für org Chemie und Biochemie, Universität Freiburg
10) TURBO-BASIC 1.00e, Borland International, München.

JCAMP - DX, A STANDARD ?

A.N.Davies, H.Hillig, M.Linscheid

ISAS
Institut für Spektrochemie und angewandte Spektroskopie
Postfach 10-13-52
4600 Dortmund 1
Fed. Rep. Germany

Abstract

The implementation of the JCAMP-DX standard as transport format for Infrared Spectra has been a great step forward in the direction of laboratory communication. The standard has however been implemented by different manufacturers with varying degrees of success. Despite the explicit and clear language used in the publication of the standard, the various data system manufacturers have generally so far failed to produce software capable of complying with the regulations laid down by McDonald and Wilks. The types of errors found in the present generation of JCAMP-DX files and handling software will be discussed with a view to assisting people writing there own JCAMP-DX software. A programme capable of checking the transport file format, diagnosing errors, and where possible correcting errors will be presented. Future developments in the field of JCAMP-DX type standard formats, with particular reference to the structure transport format JCAMP-CS will also be presented.

Background

With the implementation of dedicated computers in infrared spectroscopy there should exist the possibility of making exact spectral data universally available and exchangeable. Unfortunately the widely varying internal digital storage formats is a serious problem. The Joint Committee on Atomic and Molecular Physical Data (JCAMP) with the support of several spectrometer manufacturers took on this problem and produced **JCAMP-DX** [1] . JCAMP-DX is a standard transport format for the exchange of spectra between data systems, irrespective of which internal data format the data systems themselves use.

Following the aims of the developers the format should
- be capable of dealing with all spectroscopy types
- enclose information about sample identity etc.
- be compact
- be computer and human readable
- from the majority computers and telecommunications systems accepted
- flexible and expandable

JCAMP-DX allows the coding of, as well as data curves, line spectra and data tables, sample information and spectral measurement parameters.

The Programme JCAMP-CHECK

As part of the Bundesministerium für Forschung und Technologie (BMFT) project "Spektrendatenbanken Verbundsystem" (SDVS) the Institute for Spectrochemistry and Applied Spectroscopy (ISAS) in Dortmund has been given the function of spectral valuation centre for optical and nuclear magnetic resonance spectra.

Spectral valuation means in particular: the reading in of spectra to the system, checking associated data for completeness and absence of contradictions, carrying out a consistence check between the spectra and structure, assignment of a quality index value.

```
1,  ##TITLE=
2,  ##JCAMP-DX=
3,  ##DATA TYPE=
4,  ##ORIGIN=
5,  ##OWNER=
6,  ##XUNITS=
7,  ##YUNITS=
8,  ##XFACTOR=
9,  ##YFACTOR=
10, ##FIRSTX=
11, ##LASTX=
12, ##NPOINTS=
13, ##FIRSTY=
14, ##XYDATA=
15, ##END=
```

Figure 1. Required JCAMP-DX 'Labelled-Data-Records' (LDR's) for Infrared-Spectra.

Inside ISAS the development of the appropriate software tools has begun. The first major problem to be overcome is the translation of the various formats used to submit data into a general working format. The transfer format JCAMP-DX was opted for as the obvious choice for a working format available to all. Comparison of the various JCAMP-DX spectral formats

---

[1]   McDonald RS, Wilks jr PA (1988) Appl Spectrosc 42(1):151

available from different manufacturers showed however clear deviations from the standard. It was therfore necessary to produce a programme which could handle the diverse JCAMP-DX implimentations and produce an 'ISAS' JCAMP-DX Version 4.24 file which does not diverge from the standard and therfore should be capable of being handled by all the available software.

The programme packet JCAMP_CHECK is a system which checks the JCAMP-format coded spectrum for inconsistencies or missing information such as the prerequisite Labelled-Data-Records (LDR's see figure 1.), also whether or not FIRSTX and FIRSTY in the header zone correspond to the values given in the data section. Spectra files simple errors are corrected automatically and marked as having been corrected, while more significant errors cause the spectra to be side-lined to await more detailed correction. Any error discovery causes the generation of an error file containing the location and type of fault discovered. The correct spectral files are stored and available for further work such as structure/spectra correlation and database building. Figure 2 shows a flow diagram for the programme JCAMP-CHECK.

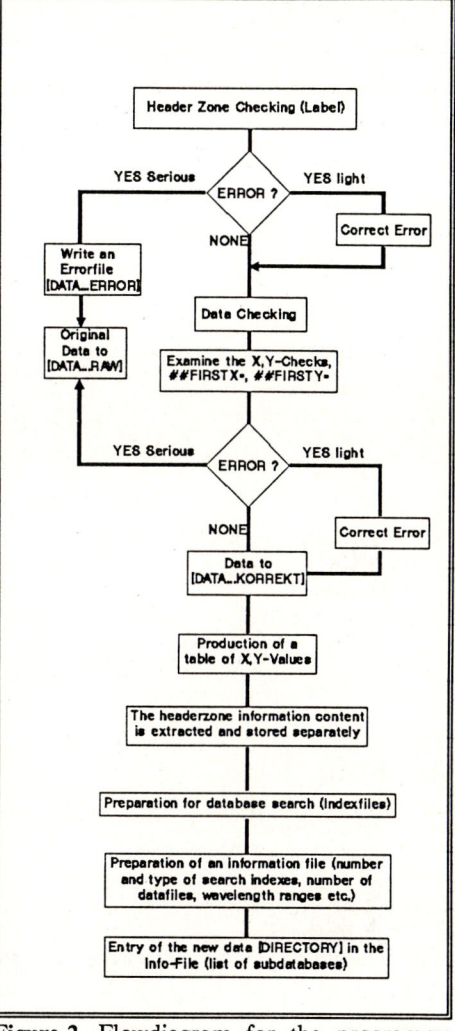

Figure 2. Flowdiagram for the programme JCAMP_CHECK.

Typical Format Errors

There are several areas in which the commercial implimentation of the JCAMP-DX standard is particularly poor. Suprisingly the production of the numerical spectral information is generally achieved without significant difficulties although often the dynamic range of the Y-axis scaling is poorly chosen leading to a loss in resolution. The software should be capable of varying the ##YFACTOR= LDR value to always maintain the maximum resolution availible in this direction on original data station. The major problem in the implimentation of the software to produce JCAMP-DX files seems to be in the simple task of deciding on the information content of the header zone.

Header Zone Errors

Although the precise nature of the required LDR's and how they should be written out is given in the standard, at the time of writing this article not one single commercially available JCAMP-DX data file investigated did not contain errors.

Some examples of the problems we have run into in the last twelve months will be given. Figures 3, 4 and 5 are examples of typical JCAMP-DX header zones as written by currently available commercial software. All three of these examples fails to include the ##OWNER= and ##ORIGIN= information, although one version at least provides the labels even if nothing is written into the field. The ##OWNER and ##ORIGIN information is extremely important when data transfer between institutions or database building is considered as the data source and all copyright information should be stored here. One particular manufacturers software package writes the manufacturers name in under both ##OWNER= and ##ORIGIN=.

```
##TITLE=TEST SPECTRUM
##JCAMP-DX=4.23
##DATA TYPE= INFRARED SPECTRUM
##XUNITS=1/CM
##YUNITS=TRANSMITTANCE
##XFACTOR=0.122074
##YFACTOR= 3.125E-05
##FIRSTX= 300
##LASTX= 4000
##NPOINTS= 3701
##FIRSTY= 133
##XYDATA=(X++(Y..Y))

MISSING:-
   ##ORIGIN=
   ##OWNER=
```

Figure 3. Header zone produced by a commercial JCAMP-DX software package which omits the compulsory LDR's ##OWNER and ##ORIGIN and incorrectly assigns ##FIRSTY.

The LDR ##YUNITS= has also caused some problems, not in the writing of this label itself but in the meaning of the word TRANSMITTANCE. To assist programmers this has been defined within the standard as I_T/I_0 and thereby Y axis values should normally lie somewhere between 0 for a completely absorbing sample and 1 for a completely transparent sample. The Y values in figures 3 and 4 as given by ##FIRSTY=, ##MINY=, and ##MAXY= lie clearly outside the expected range. The reasons for these two abberations are different. The software used to generate the JCAMP-DX file shown in figure 3

```
##TITLE= Test Spectrum
##JCAMP-DX= 4.24
##DATA TYPE=INFRARED SPECTRUM
##ORIGIN=
##OWNER=
##DATE= 89/09/06
##TIME=  09:51.32
##SPECTROMETER/DATA SYSTEM=
      SYSTEM B
##DATA  PROCESSING=
SMOOTHING=none;
##XUNITS= 1/CM
##YUNITS= TRANSMITTANCE
##XFACTOR=1.00
##FIRSTX=1800.00
##LASTX=700.00
##DELTAX=-1.00
##NPOINTS=1101
##YFACTOR=0.0001
##MINY=4.6473
##MAXY=96.2525
##FIRSTY=59.5714
##XYDATA=  (X++(Y..Y))
```

Figure 4. Header zone produced by a commercial JCAMP-DX software package showing several formatting errors.

has taken the actual first Y value and multiplied it by the ##YFACTOR= value before storing it in ##FIRSTY= although the standard requires the actual value here. The JCAMP_CHECK programme can locate and correct this problem producing an error file detailing the type of error and it's location in the original data file. In figure 4 however, the improper values arise from the use of % Transmission values instead of transmittance and hence any check between the actual first Y value and that given in the header zone will not yield an error and correction of such files is a very dangerous matter as the cause of the extra large values could also come from a improperly measured spectrum such as sample in the reference beam of a double beam instrument or a sample which in which light is emitted at certain frequencies rather than being absorbed for example.

```
##TITLE=TEST SPECTRUM
##JCAMP-DX=4.24
##DATA TYPE=INFRARED SPECTRUM
##= SYSTEM C
##XUNITS=1/CM
##YUNITS=ABSORBANCE
##RESOLUTION=    1.92858
##FIRSTX=   4.0008448256E+3
##LASTX=  5.9978922183E+2
##DELTAX=  -9.642913535E-1
##MAXY=  5.0000000000E+0
##MINY= -3.299713134E-4
##XFACTOR= 4.8828125000E-4
##YFACTOR= 2.4414062500E-4
##NPOINTS= 3528
##XYDATA=(X++(Y..Y)

MISSING:-

##ORIGIN=
##OWNER=
##FIRSTY=
```

Figure 5. Header zone produced by a commercial JCAMP-DX software package lacking the compulsory ##ORIGIN, ##OWNER, and ##FIRSTY LDR's.

Errors also occur in the DX files where the software doesn't differentiate between a 'string' value and a 'text' value. As defined in the standard - "STRING data-sets contain alphanumeric fields intended to be parsed by a computer and read by a human" whereas "TEXT data-sets contain descriptive information for humans, not normally intended to be parsed by computer, i.e., title, comments, origin, etc.". This difference is simply explained by reference to figure 4 where the ##TIME= is given by system B as 09:51.32. The ##TIME= reserved label is defined "(STRING). Time when spectrum was measured, in the form HH:MM:SS", so the datafile in figure 4 errs in the use of a fullstop to separate the minutes and seconds where a colon is expected. If ##TIME= had been defined as a text variable this would be irrelevant as any set of characters which a human would recognise as refering to a time would suffice.

A dataset has been received at ISAS with the LDR ##YUNITS= ABSORBANCE where the spectra were actually in transmission units. As yet we have not been able to identify the software which produced these files.

These types of formatting errors are serious as far as we know do not cause problems as long as spectra are exchanged between data systems from the same manufacturer running the same software versions. It is worth noting that in this respect the implementation is better than some recent versions of the EPA Mass Spectroscopy standard software where certain data systems couldn't even read EPA files produced by themselves without loosing all peaks over 10% total ion count. Formatting errors and deviations from the standard start to be a real problem when DX files from one manufacturer are required on a datastation of a second manufacturer.

Breaking out of the circle

Two examples of problems arising when spectra are transferred between different systems will be given. The data systems and manufacturers have been disguised so as not to give a biased impression, the problems described are general to the implementation and the intention at this time is not to single out certain manufacturers for criticism of their particular software.

```
##TITLE= TEST SPECTRUM
##JCAMP-DX= 4.10
##DATA TYPE= INFRARED SPECTRUM
##ORIGIN= MANUFACTURER W
##OWNER= MANUFACTURER W
##DATE= 88/03/03
##TIME= 16:18:56
##CREATED= Thu Mar 03 16:18:56 1988
##SPECTROMETER/DATA SYSTEM=
     SYSTEM A
##XUNITS= 1/CM
##YUNITS= ABSORBANCE
##RESOLUTION= 4.0
##FIRSTX= 3.97335161e+02
##LASTX= 4.00228378e+03
##FIRSTY= 8.422651887e-02
##MAXY= 7.909523807e-01
##MINY= 1.899838820e-02
##XFACTOR= 1.92881146e+00
##YFACTOR= 1.862645149e-09
##NPOINTS= 1870
##XYDATA= (X++(Y..Y))
```

Figure 6. A commercial header file from an early unpublished JCAMP-DX version.

Figure 6 shows the header from a spectra produced on system A by manufacturer W's software. Figure 7 the same spectrum header file after the spectrum had been read into system B from manufacturer X and then saved as a JCAMP-DX file from this system. No data manipulation was instigated only single read and write action so the information content should be retained. Conversion to the internal format of the second system could lead to certain changes in such things as the difference between the wavenumber position of the points on the X-axis if, as is the case here, the data interval for one system in the X direction is fractional laser frequency and the internal data interval on the second system is

```
##TITLE=  TEST SPECTRUM
##JCAMP-DX= 4.24
##DATA TYPE= INFRARED SPECTRUM
##ORIGIN=
##OWNER=
##DATE= 89/11/14
##TIME=  17:08.58
##SPECTROMETER/DATA SYSTEM= SYSTEM B
##DATA PROCESSING= SMOOTHING= none;
##XUNITS= 1/CM
##YUNITS= ABSORBANCE
##XFACTOR=1.00
##FIRSTX=4002.00
##LASTX=397.00
##DELTAX=-2.00
##NPOINTS=1803
##YFACTOR=0.0001
##MINY=2.7207
##MAXY=79.0874
##FIRSTY=5.8426
##XYDATA= (X++(Y..Y))
```

Figure 7. Header zone of the file created from a second datasystem after reading in the data in figure 6. Serious loss of information has resulted.

a rational interval but these changes should only bring a slight loss of resolution.

The original datafile (Fig.6.) is based on an early (and unpublished) JCAMP version and thereby includes a now unrecognized LDR ##CREATED=, it also displays the ##ORIGIN=manufacturer W, ##OWNER=manufacturer W, problem mentioned earlier. This file is however basically correct and shouldn't provide any problems for the receiving software. Figure 7 shows however what poor programming can do to a simple standard.

Probably the most serious problem brought to light here is the loss of the ##OWNER= and ##ORIGIN= information which for our purposes means the loss of copyright notification. This loss was foreseen by McDonald and Wilks and warned against. Even if the particular internal format used by a data system doesn't have fields available for this type of information there is often a comment string associated with a particular spectrum where this data could be stored. It is also an extremely simple task to read in this header information into a separate text file which could then be searched for when a DX file is to be produced from the basically unaltered spectrum. A flow diagram for such a system is given in figure 8.

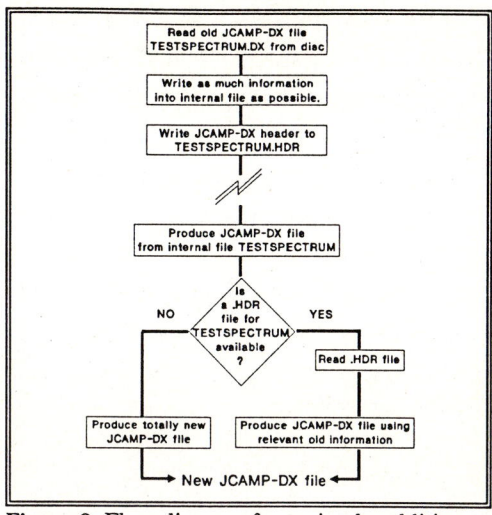

Figure 8. Flow diagram for a simple addition to JCAMP-DX software to overcome the loss of, for example, owner and origin copyright information.

This type of header zone storage would solve many of the other problems shown here such as the alteration of the ##DATE= and ##TIME= data sets despite the fact the information refers to the date and time that the spectra was measured and not that when the spectra was converted to JCAMP-DX format ! Figure 7 also contains a new LDR ##DATA PROCESSING= SMOOTHING= none; which as this information was not included in the original JCAMP-DX file is a rather rash assumption and shouldn't appear in the new DX file.

Closer examination of figure 7 will also reveal an error in the ##LASTX= value which, as it is given as an odd number, is unreachable from ##FIRSTX= in steps of ##DELTAX= !

We have encountered one very serious problem which actually causes changes in the peak positions observed in a test spectrum. As part of the test procedure a peak table was generated on the various data systems as a quick check on whether any spectral information had been lost during the data transfer. This type of checking has been successful and showed up the problem which will be described below.

Having read a particular spectrum in JCAMP-DX format into SYSTEM B it was discovered that the peak table did not correspond with that provided by the system that had measured the spectrum. This JCAMP-DX file was then read into two other systems and the peak tables generated on the other systems confirmed the positions given in the original DX file. Further investigation of the dubious peak table showed a steady shift in the reported peak position to higher wavenumber, the lower the wavenumber of the peak. Figure 9 is a graphical

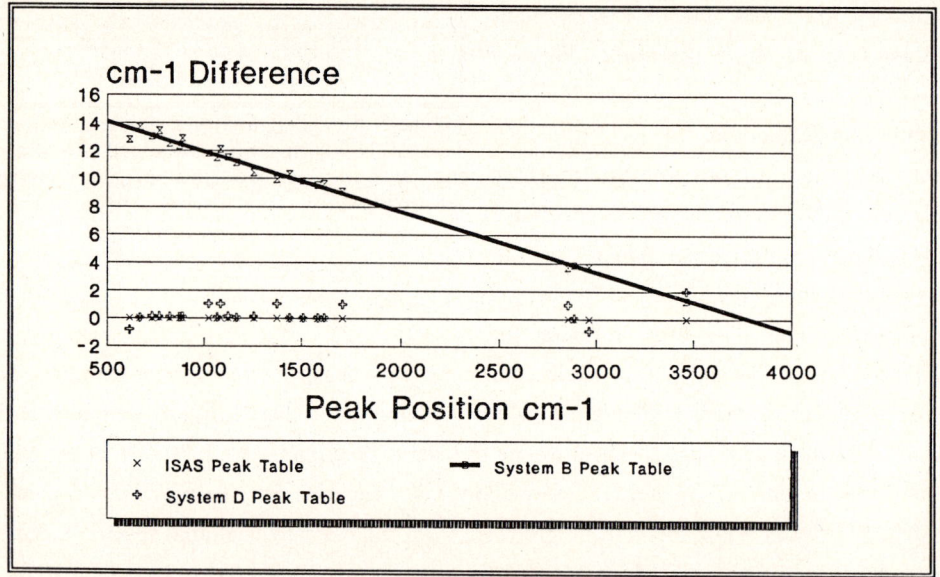

Figure 9. Graph of the difference in the peak positions of the test spectrum when measured on three systems different from that which measured the original spectrum.

representation of these changes in peak position. The x-axis in figure 9 is the peak table values from the original system C JCAMP-DX file and the three data sets shown are the ISAS Database in-house VAX based software which agrees with the original peak positions, system D which agrees but shows numerical rounding errors and system B which has a systematic incremental error.

This particular error was tracked down to problems in the JCAMP-DX software in system B. This system uses a logical x-axis increment for internal spectral storage and so when present with a DX file with a fractional laser wavelength incremental x-axis an internal conversion is required. This type of conversion was successfully carried out by the software on system B in the case of figures 6 and 7 where the ##DELTAX= was not provided and hence needed to be calculated internally. In this case however the source JCAMP-DX file (figure 10.) provides not only ##FIRSTX and ##LASTX but a ##DELTAX as well. System B reads the ##DELTAX value to 2 significant figures and this is used to scale the following ##NPOINTS along the x-axis starting at ##FIRSTX. The in-built x-checking was not carried out by the system B software and the fact that the last x value calculated did not correspond to the to that of the data table or the header zone of the original DX file went unnoticed.

```
##TITLE=TEST 5
##JCAMP-DX=4.24
##DATA TYPE=INFRARED SPECTRUM
##= SYSTEM C
##XUNITS=1/CM
##YUNITS=ABSORBANCE
##RESOLUTION=    1.92858
##FIRSTX=  4.0008448256E+3
##LASTX=  5.9978922183E+2
##DELTAX=  -9.642913535E-1
##MAXY= 6.2156796455E-1
##MINY= -2.633333206E-4
##XFACTOR= 4.8828125000E-4
##YFACTOR= 3.0517578125E-5
##NPOINTS= 3528
##XYDATA= (X++(Y..Y))
```

Figure 10. Original JCAMP-DX spectrum header zone showing the fractional laser wavelength ##DELTAX=.

Conclusions

It has clearly been demonstrated that although the JCAMP-DX standard is widely accepted as a transport format for infrared spectra the poor quality of many of the software implementations is at present a hinderance to wider error-free data exchange.

```
##TITLE= TEST 5
##JCAMP-DX= 4.24
##DATA TYPE=INFRARED SPECTRUM
##ORIGIN=
##OWNER=
##DATE= 89/09/27
##TIME= 14:12.20
##SPECTROMETER/DATA SYSTEM= SYSTEM B
##DATA PROCESSING= SMOOTHING=  none;
##XUNITS= 1/CM
##YUNITS= TRANSMITTANCE
##XFACTOR=0.01
##FIRSTX=4000.84
##LASTX=614.92
##DELTAX=-0.96
##NPOINTS=3528
##YFACTOR=0.0001
##MINY=0.2390
##MAXY=1.0006
##FIRSTY=0.9978
##XYDATA= (X++(Y..Y))
```

Figure 11. JCAMP-DX header zone produced from the same spectrum as that in figure 10 showing the new ##DELTAX value.

Any form of standardization is however better than none at all and with a little care most of the problems described in this paper can be overcome. The main message must be however, not to trust your software as an error free black box but to check exactly what it is doing and that it is doing it properly.

The acceptance of JCAMP-DX by the manufacturers and their involvement in it's development is a great step forward which must be warmly welcomed but

> "There must be a beginning to any great matter, but the continuing until thoroughly finished yields the true glory" [2].

The Future

A standard exchange format for chemical structure information in computer readable form - JCAMP-CS - written along the lines of the JCAMP-DX standard by Gasteiger et.al. has recently been submitted for publication. The implementation of this format as a standard for chemical structure exchange and the correction of the errors in the current breed of JCAMP-DX software releases will make reliable information exchange in digital format between scientists at long last a reality and open up many new routes for quality control and, from a recent example in the literature, pre-purchase comparison of instruments ! [3]

Acknowledgements

The authors would like to thank the Bundesministerium für Forschung und Technologie for financial support and encouragement (project 08 G32 21), and the Ministerium für Wissenschaft und Forschung des Landes Nordrhein-Westfalen.

---

[2] Sir Francis Drake to Sir Francis Walsingham 17th May 1587 (1898) Navy Records Soc XI:134

[3] Ciurczak EW (1989) Two Valuable Tools for Instrumental Evaluation: JCAMP-DX and Price Tags. Spectrosc Int 1(7):12,14

Automatic Interpretation of 2D-NMR-Spectra [1]

Peter Haas and Wolfgang Robien

Institute for Organic Chemistry
University of Vienna
Währingerstraße 38
A-1090 Wien

Abstract: During the past decade two-dimensional NMR-spectroscopy [2] has developed into one of the most powerful tools for organic structure determination. On spectrometers equipped with a sample changer 2D-NMR-spectroscopy can be completely done under computer control producing a vaste amount of data, thus shifting the bottle-neck from accumulation to data interpretation. These sophisticated experimental techniques can only be fully utilized by appropriate computer software supporting the spectroscopist during the whole process of data interpretation. Our recently published approach for automatic interpretation of NMR-spectra [3] has been expanded and is now able to use distance-constraints derived from different types of 2D-NMR experiments. The main advantage of this algorithm is its capability to work only with one-dimensional data thus leading to the possibility to interpret incomplete 2D-NMR data describing only a part of the structure. Some examples selected from natural product chemistry will be given proofing the power of this concept.

INTRODUCTION:

Structure elucidation of organic compounds is usually done by applying sophisticated spectroscopic methods. Within these techniques NMR-spectroscopy plays an important role, because of its high information content and the large number of powerful pulse-sequences available, even during routine operation with a sample changer. Carbon-13 NMR-spectroscopy allows immediate insight into the skeleton of organic molecules, therefore contributing a large amount of information about functional groups and molecular symmetry. The interpretation of one-dimensional ^{13}C-NMR-spectra utilizes mainly the chemical shift values reflecting the electronic structure of the corresponding carbon atoms and their multiplicity giving information about directly attached hydrogens. The carbon-carbon coupling is usually not visible because of the low natural abundance of this isotope (ca. 1.1%). The insensitive INADEQUATE-method allows selection of C-C pairs showing directly the carbon-carbon connectivity pattern of the compound under investigation. Other routinely used techniques like HH-COSY and COLOC gave also insight into different types of coupling networks with the disadvantage of ambiguity concerning the length of the coupling path between two coupled nuclei. The systematic solution to this problem can only be done by calculation of all possible isomers using the whole set of known constraints derived from the experimental data. This is a very tedious task which can be only performed by an appropriate software-package.

The chemical shift values of the ^{13}C-isotope are usually in the range between 0ppm and 225ppm showing mainly sharp lines under BB-decoupled conditions. A very good correlation between chemical shift values and structural features is given, thus allowing efficient storage of the spectra in computerized data bases [4-6]. This structure-spectrum correlation can be used for interpretation of one-dimensional ^{13}C-NMR-spectra as described earlier [3]. The algorithm is based on an automatically derived datafile from our database containing all sequences of 3 connected atoms and their spectral properties, which are used for creation of structural fragments fitting the input conditions. These fragments are now combined to complete, chemically useful structures. The 2D-information (if available) is used at the earliest possible step during combination of these fragments. This concept is therefore also valid for an incomplete set of 2D-information describing only a part of the molecule. The success of the structure elucidation process depends heavily on the contents of the database itself, the algorithm will therefore work only with the information deduced from a large data collection containing most of fragments occuring during solving of "real-world structure elucidation problems".

METHOD:

The knowledge-base used [4] for our investigation contains some 37.000 ^{13}C-NMR-spectra taken from the literature. From these structures all fragments containing sequences of 3 connected atoms are generated leading to a total number of about 500.000 fragments. They are stored with their corresponding spectral pattern after elimination of duplicates creating about 25.000 unique entries in this file. The properties of each atom is encoded into a 4Byte-word, thus compressing the spectral and structural information derived from 37.000 spectra into a 300kByte-file. The input data consist of the molecular formula, the ^{13}C-NMR spectrum and optional information from ^{1}H-NMR, ^{13}C-X couplings and 2D-NMR data. At the moment the following 2D-NMR techniques have been implemented: HH-COSY, HC-COSY, COLOC and INADEQUATE. From the fragment file all sequences are selected, which meet the requirements of the input data. These fragments are combined to α-environments (central atom and directly attached neighbors), which are used to generate all chemically correct structures fulfilling also the distance-constraints as given by the 2D-NMR data or the coupling information. In such cases, where fragments are missing, but the 2D-information is available, the algorithm will automatically insert the missing α-environments, because of its predefined priority-scheme. The 2D-information is only used to cut the combination tree at the earliest possible step, therefore the algorithm will also work without distance-constraints, leading to a larger number of combinations and usually creating more candidate structures. After creation of the list of all possible isomers a ranking procedure sorts them according to a two-parameter function, using the difference between the experimental spectrum and the calculated one and according to the representation of the structure by the reference data collection. Especially the second parameter is very useful for creation of a correct hitlist in such cases, where some of the created structures are in the database.

RESULTS:

Interpretation of 1D-NMR Data:

The first example uses only the molecular formula and the 1D-^{13}C-NMR-spectrum including multiplicity as given in lit. [7]. A total number of 47 proposals are created. The ranking algorithm puts the correct solution at position #1. No one of the 47 structures is contained in the database, proofing the excellent power of our method even using only molecular formula and peaklist without any distance-constraints. The 8 most probable solutions are shown in figure 1 with the correct solution in position #1.

Interpretation of 2D-NMR Data:

The example choosen was 9ß-hydroxycostus-acid using the data as given in lit. [8]. The derived set of constraints leads to an unique solution representing the correct structure. The complete set of input data and the correct solution structure is given in figures 2 and 3.

CONCLUSION:

The algorithm described has proven its excellent power during a six-month period solving about 200 examples. In combination with our ranking procedure more than 95% of the problems encountered could be solved and the correct structure was placed in position #1 even when using only one-dimensional ^{13}C-NMR data and creating a list of candidate structures with about 300 entries. Further expansions and improvements including predefined partial structures and skeletons known to the chemist are planned. The algorithm will be prepared for parallel computation using an IBM-3090/400 computer system.

ACKNOWLEDGEMENT:

The authors want to express their gratitude to the Vienna University Computing Center for access to the IBM-3090/400E-VF computer. This work was supported by IBM within the European Academic Supercomputing Initiative (EASI).

1 Part 7 of the series: "Computer-Assisted Structure Elucidation of Organic Compounds", for part 6 see lit. [4]
2 Kessler H, Gehrke M, Griesinger C (1988) Angew. Chem., 100:507
3 Robien W (1986) Mikrochim. Acta [Wien] II:271
4 Haas P, Strasser G, Scsibrany H, Kriech M, Robien W (1989) in: Gauglitz G(ed) Software-Entwicklung in der Chemie 3, Springer, Berlin Heidelberg New York
5 Bremser W, Fachinger W (1985) Magn. Res. Chem., 23:1056
6 Zippel M, Mowitz J, Köhler I, Opferkuch HJ (1982) Anal. Chim. Acta 140:123
7 Ayer WA, Browne LM, Lin G (1989) J. Nat. Prod., 52:119
8 Sepulveda-Boza S, Breitmaier E (1987) Chemiker-Zeitung 111:187

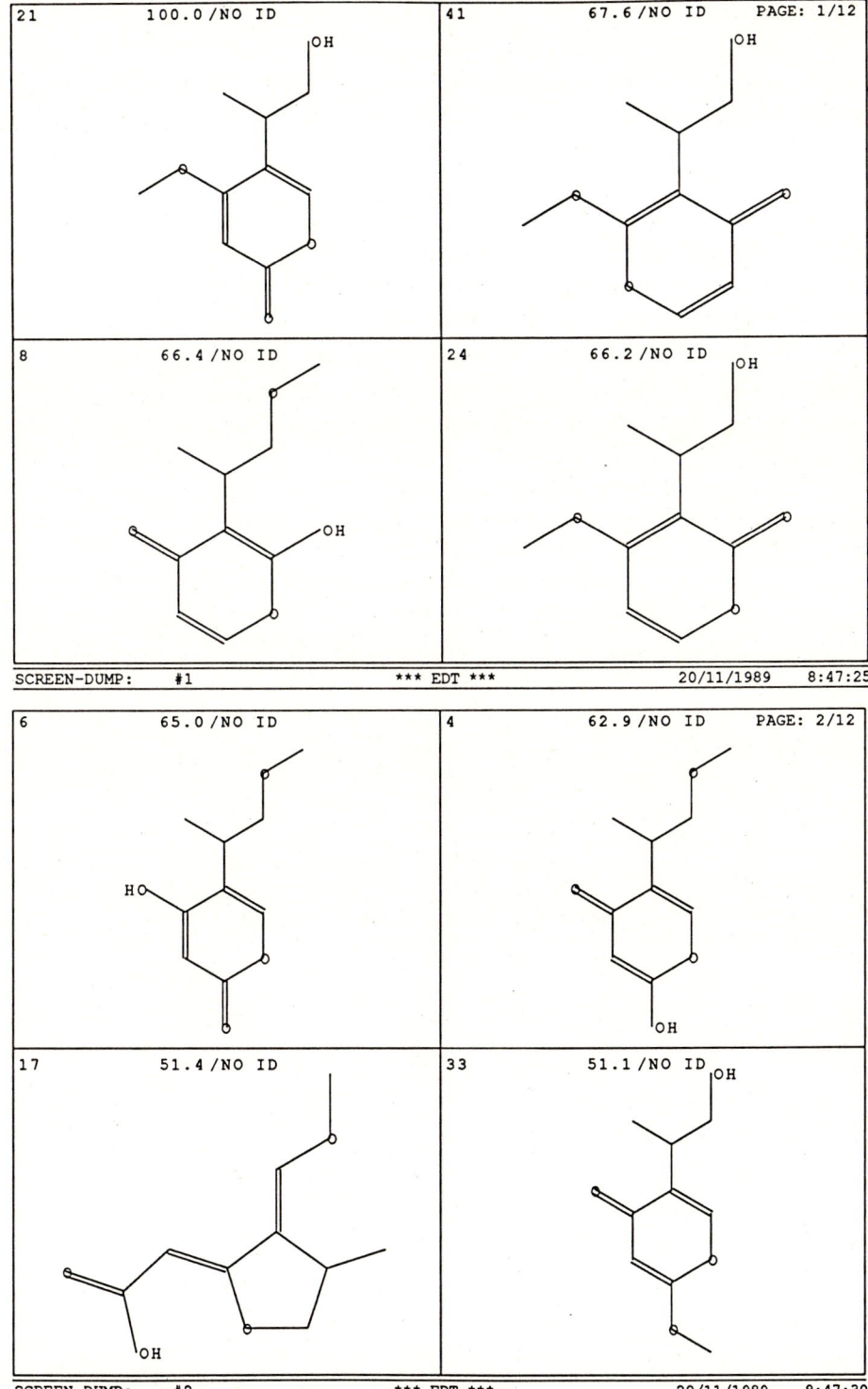

Fig. 1: Interpretation of 1D-^{13}C-NMR data. The 8 best candidate structures are given

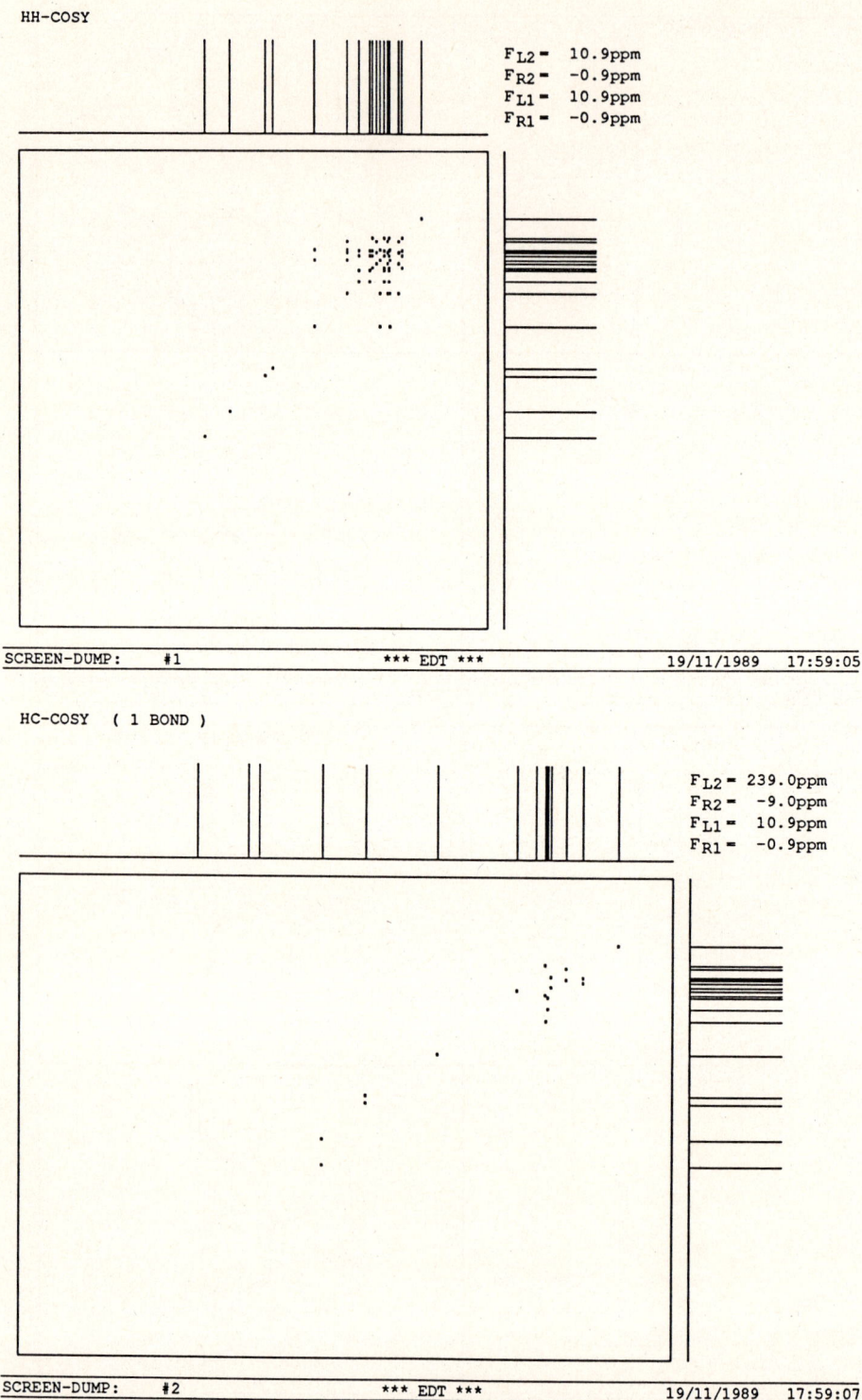

Fig. 2: HH-COSY and HC-COSY of 9ß-hydroxycostus-acid

HC-COSY (LONG RANGE)

F_{L2} = 239.0ppm
F_{R2} = -9.0ppm
F_{L1} = 10.9ppm
F_{R1} = -0.9ppm

SCREEN-DUMP: #3   *** EDT ***   19/11/1989   17:59:08

PROBLEM: 9-HYDROXY-COSTUS-ACID

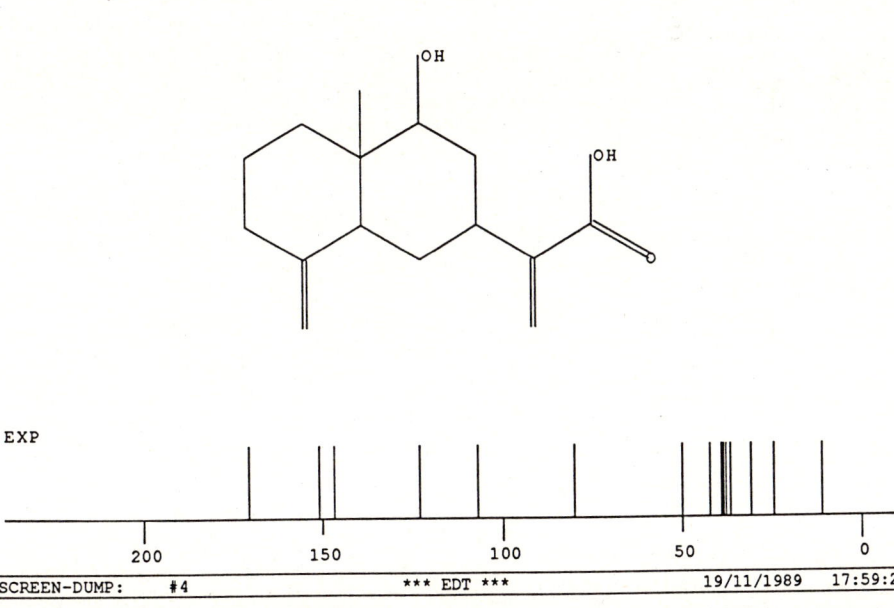

SCREEN-DUMP: #4   *** EDT ***   19/11/1989   17:59:23

Fig. 3: COLOC-spectrum of 9ß-hydroxycostus-acid and correct solution structure

CDDS - A PERSONAL COMPUTER BASED SYSTEM FOR AUTOMATED INTERPRETATION OF GC/MS-ANALYSES

H. Lohninger

Technical University Vienna
Institute for General Chemistry
Lehargasse 4/152
A-1060 Vienna

ABSTRACT

The application of chemometric detectors in gaschromatography/mass spectrometry has been shown to be useful in recent years [1,2]. The most important drawbacks of the systems developed so far are the lack of user friendliness and the complicated development procedure to obtain new chemometric detectors with a new selectivity. The system presented in this paper has a clearly defined and graphically oriented user interface and makes it possible to develop new detectors in a convenient and efficient way.

INTRODUCTION

The principles of chemometric detectors

As the principle of a chemometric detector (CMD) has been described elsewhere [1,2,3] only a short survey on the operation of a CMD is given here.

If the chemometric detector is looked at as a 'black box' (figure 1), the output of this black box exhibits nearly identical features as a conventional selective detector. It yields a signal which detects one or several substances in a selective way. In contrast to this the input signal differs quit a lot. The chemometric detector needs spectral data as input information and therefore can be used only in combination with spectrometers. Although the application of a CMD is not restricted to mass spectral data, the system presented here only deals with such data.

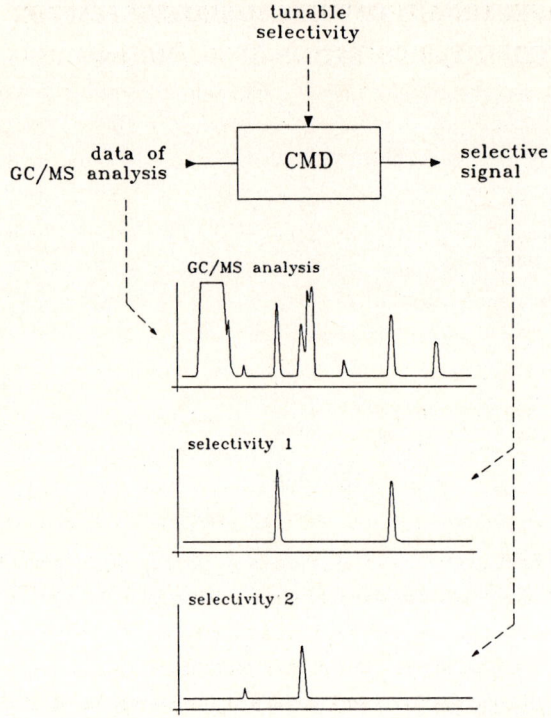

Figure 1: A chemometric detector seen as black box

Spectroscopic measurement methods produce a high amount of raw data (e.g. intensities at mass units) whereas single data points show only little correlation to structural elements of the substance under investigation. Only the combination of these raw data gives a high correlation between spectral data and structures and thus makes it possible to interpret the data.

A spectrum can be seen as a point in a space of N dimensions. The N dimensions are defined by the N single data points (e.g. mass units or segments of wave length). A chemometric detector performs a transformation of the N-dimensional space in a way that the number of dimensions will be reduced to one. The resulting value of this transformation should give a high correlation with the chemical class the substance belongs to. The point of interest during development of a new detector is how to find the transformation which results in an optimum selectivity.

A substantial advantage of a CMD over a conventional detector is its tunable selectivity. The selectivity can be modified by changing the transformation. This task is not quite trivial but can be accomplished by any chemist if a development system provides support. The system CDDS (Chemometric Detector Development System) presented in this paper is a first attempt to implement such a development tool.

Development of a chemometric detector

The development of a new chemometric detector starts with two spectral data sets which hold spectra of substances which belong to two different classes of compounds that should be distinguished by the CMD. The two classes need not to be homogeneous from a structural point of view. They may contain spectra of substances of various classes of compounds or even compounds representing the whole rest of chemistry (e.g alkanes and not alkanes). By using the development system CDDS a transformation can be searched which yields an optimum differentiation between the two groups. If the selectivity of the transformation is sufficiently high it can be used as a chemometric detector and applied to spectral data of unknown substances.

The type of transformation can be selected arbitrarily as far as it is possible to find those parameters of the transformation which result in optimum selectivity. As it is not discernible a priori which sort of transformation yields the best results several kinds of transformation must be selected and compared. Some methods were already proposed and implemented (e.g. multiple linear regression with non linear feature generation [1], heuristic methods which rely on chemical knowledge [4] or logarithmic pattern correlation [5,6]), some methods have to be implemented in future work (e.g. neural networks).

A critical point in the development of new detectors is the quality of the data base which holds both spectral and structural data. In order to get reliable detectors of high selectivity it is very important to have data of approved reliability. It is strongly recommended to check the data which are used in the training phase for correctness.

DESCRIPTION OF CDDS

The system - CDDS - presented here consists of two data bases and five software modules which are integrated by a common user interface. The data bases hold both the spectral and structural data and the parameters of the selective detectors. As the system is currently under construction only a small data base is available (spectra and structural data of 500 steroid spectra). The software modules support both the application of chemometric detectors to GC/MS analyses and the development of new detectors. The internal assembly of the CDDS is shown in figure 2.

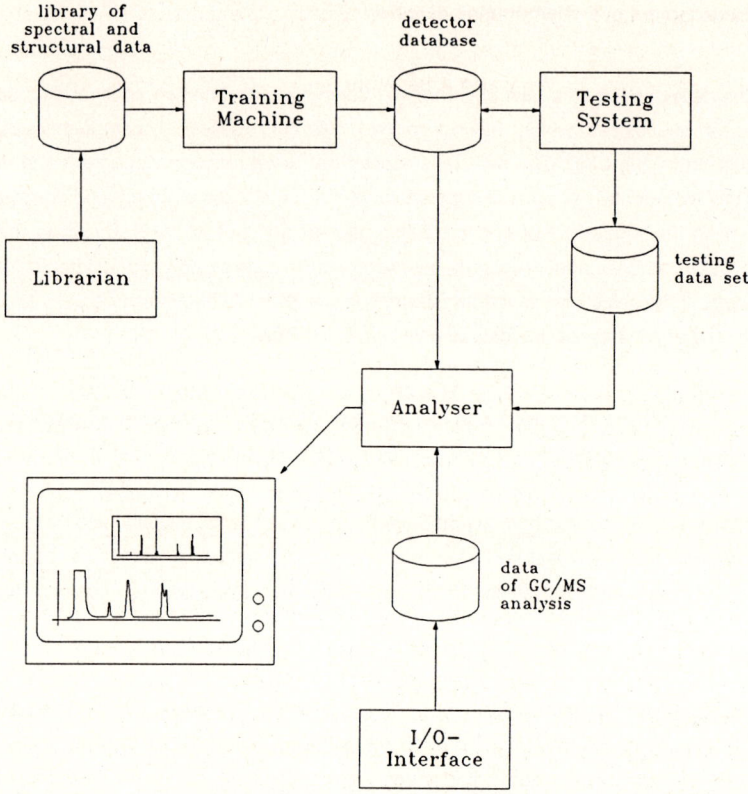

Figure 2: Functional diagram of CDDS

Data bases

The spectral data base holds both spectral data (mass spectra) and structural data. Each entry in the data base contains the following data:

- name of the substance
- mass spectrum
- physical properties (melting point, boiling point, density, refractive index)
- brutto formula
- substructures of the molecule
- connection table
- graphic representation of the molecule

In order to support expansion and maintenance of the data base and retrieval of data a module ('Librarian') has been created which allows to fulfil these tasks in an efficient and flexible way.

The second data base holds the parameters of already existing detectors. This detector data base is stored in a readable format in order to offer the possibility to expand it with detectors which are developed not on CDDS but on external hardware.

I/O-Interface

The I/O-interface allows to exchange data between commercial MS data systems and CDDS. Two possibilities of data exchange are implemented:

1) Exchange of data by utilising the custom data format of the MS data system. In order to transmit data general purpose communications software (e.g. Kermit, Columbia University) is used. After transmission the data are converted from the proprietary format of the MS-data system to the format used by CDDS and then can be processed further by CMDs. Up to now only data which where recorded by the MS data system 'SS300' (Finnigan/MAT) can be exchanged using this method.

2) Exchange of data in a special ASCII format. This method of data exchange requires a small routine on the MS data system which reads the spectra of interest and stores them in this special format on disk or sends them via a serial link to CDDS. In order to support this method CDDS features a built-in terminal emulation. This terminal emulation can be used to connect CDDS to the MS data system and start the necessary conversions. Data which are sent from the MS data system are readily stored on the disk of CDDS and can be converted to the internal format after transmission is complete.

Analyser

The analyser is the most important module as far as the normal user of the CDDS is concerned. It allows the application of chemometric detectors to measured data. After the data have been transmitted to the CDDS the operator can select chemometric detectors of various selectivity to be used in the subsequent analysis. During the analysis the selected detectors are applied to each spectrum of the data file and the results are stored on disk.

After classification has been finished the user can inspect the results by using the graphic interface. This interface is based on multiple windows and offers a wide variety of commands:

Displaying detector results

Up to four traces of the detector results can be displayed in parallel to the chromatogram (integrated intensity). The results can be arbitrarily zoomed and panned.

Displaying spectral data

During examination of the detector results it is often useful to display spectra of the chromatogram on the screen. In order to achieve this the operator selects the spectrum by clicking the mouse at the appropriate area of the chromatogram. A window is then opened which shows the selected spectrum. As an option a background spectrum (e.g. the spectrum at the beginning of the chromatographic peak) can be subtracted.

Set-up of the user interface

The user interface can be adjusted to the needs of the operator. Both colors and default dimensions, and type and number of windows can be adjusted. The set-up can be stored and will be kept till the next change. Each analysis can get a 'personal' set-up in order to be most efficient during interpretation.

Librarian

The module 'Librarian' supports both the creation and maintenance of the data base and the retrieval of spectra. The data of the data base are displayed as card images (figure 3) which hold most of the information available on the specified substance.

Editing the data base

In order to protect the data base from being altered unintentionally a password protection scheme has been established. This scheme allows alteration of the data only if the user knows a password which has been set-up during installation of the system. If the operator does not know this password he can only read the data but not change them.

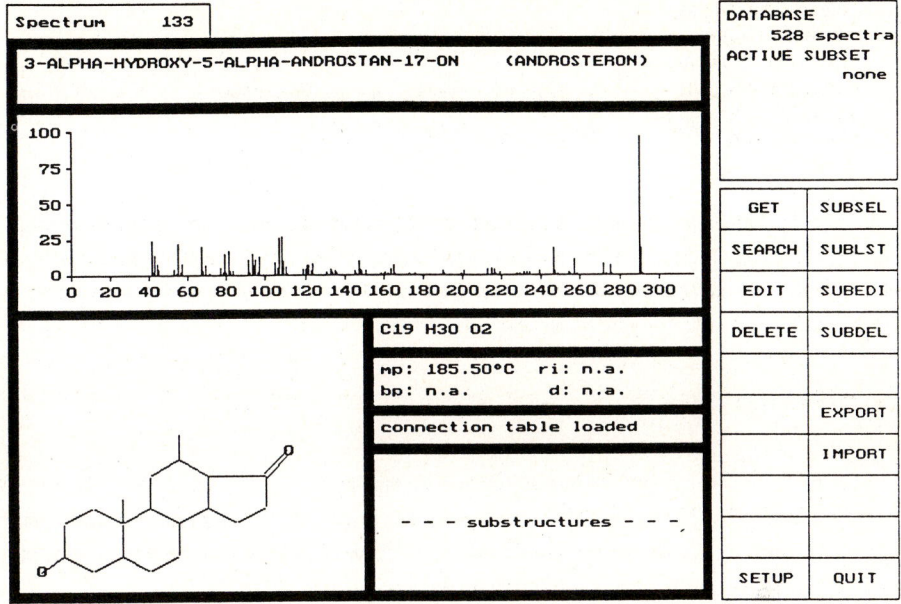

Figure 3: Example of an entry in the data base

All entries in the data base can be changed and expanded in a comfortable manner. In order to support editing of chemical structures a special graphic oriented structure editor has been implemented. This editor allows to draw molecular structures on the screen. After the drawing is finished the editor creates the structural information (connection table) from the drawing. This procedure has two advantages: (1) structural information can be input in a convenient and pleasant manner and (2) a side effect of this procedure is to get the graphic representation of each molecule automatically.

Commands for retrieval of data

In order to combine entries in the data base which belong to a certain class of compounds, subsets can be created. These subsets contain indices into the main data base thus providing an efficient means to collect all information available on a specified class. Using these subsets new chemometric detectors can be developed and tested.

The compilation of spectra can be either managed manually by typing in the numbers of spectra or automatically by using the retrieval system of the librarian. The librarian provides means to search for all data items in the data base. The following search criteria are possible:

name of substance
molecular weight
mass spectral peaks
physical parameters
arbitrary substructures
brutto formula

All search items can be restricted or expanded by looking only for a specified range of values or by using wild-cards. The search for substructures is supported by the structural editor. After drawing the substructure on the screen, a connection table is created and used for the search in the data base.

Training Machine

In order to develop new detectors with new or changed selectivity, CDDS offers a module called 'Training Machine'. This module lets you select from several sorts of detector prototypes and performs a semi-automatic calculation of new detectors with a selectivity based on the chemical classes of two subsets. The following detector models are implemented currently:

- combined mass chromatograms
- linear discriminators with non-linear feature generation
- neural networks

After the selection of the detector prototype and the specification of the two classes of compounds a learning algorithm is executed which leads to a selective detector. The type of the learning algorithm depends on the type of detector selected. If the selectivity of the new detector is good enough this detector can be input to the detector data base and used in further GC/MS analyses by the module 'Analyser'.

Test System

The module 'Test System' supports the operator in testing newly developed chemometric detectors. This module will contain in its final configuration a set of test tools which give information on the performance of a specific detector. At the current time only a GC/MS simulator is implemented, which is used to test the performance of a CMD under chromatographic conditions.

The GC/MS-simulator allows to position up to 100 chromatographic peaks of gaussian shape and of variable width in a chromatogram. Each peak in the chromatogram is assigned to a mass spectrum which can be distorted by adding noise or by multiplying the intensities with a mass dependent factor. The simulated GC/MS is used to determine the performance of a chemometric detector under controllable conditions (distortion, noise, overlap).

EXAMPLE

Figure 4 shows the result of a test run. A test mixture of alkanes and polycyclic aromatic hydrocarbons (PAH) has been analysed by GC/MS and the results were interpreted by two CMDs of different selectivity. The lower part of the figure shows the chromatogram, the upper two traces show the results of two detectors which were selective for alkanes (middle trace) and selective for PAH (upper trace).

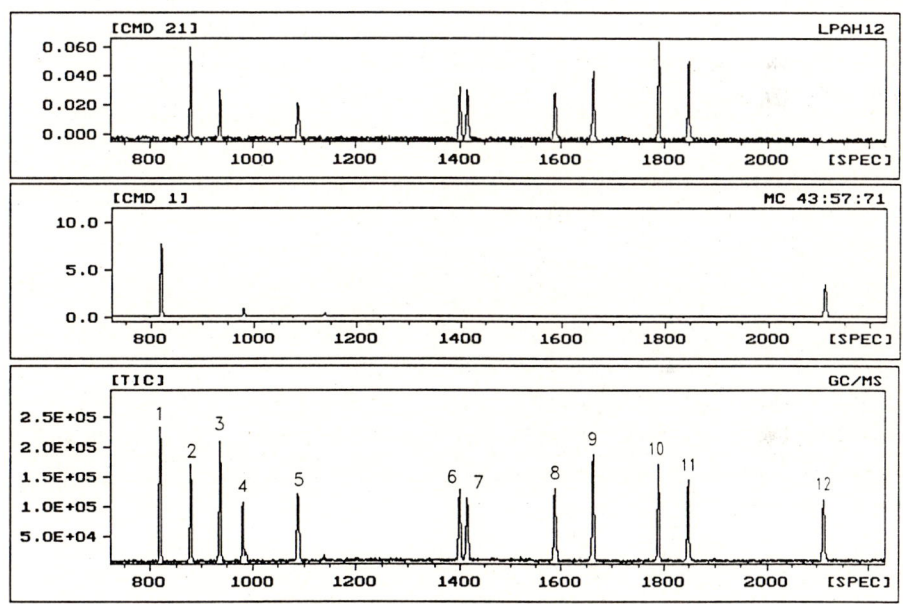

Figure 4: Application of two CMDs with different selectivity on a test mixture. 1=n-C14H30, 2=acenaphtylene, 3=acenaphthene, 4=cyclo-dodecanone, 5=fluorene, 6=phenanthrene, 7=anthracene, 8=2-methyl-anthracene, 9=9-methyl-anthracene, 10=fluoranthene, 11=pyrene, 12=n-C23H48

IMPLEMENTATION

In order to be independent of the wide variety of data systems used by mass spectrometer manufacturers CDDS has been implemented on an IBM compatible personal computer. The program is written in Pascal. In order to use CDDS the following minimum hardware is necessary:

IBM PC/AT or IBM PC/AT-386
640 kByte RAM, 40 MB hard disk
numeric co-processor 80287 or 80387
Hercules, EGA or VGA graphics card
Microsoft compatible mouse
2 serial interfaces

LITERATURE

[1] Lohninger H., Varmuza K.
     Anal. Chem. 59 (1987) p. 236-244
[2] Varmuza K., Lohninger H., Werther W.
     'Softwareentwicklung in der Chemie 2', S. 211, Springer 1988
[3] Lohninger H., Varmuza K.
     Advances in Mass Spectrometry 1985, p. 653, Wiley (1986)
[4] Clerc J.T., Kutter, M., Reinhard M., Schwarzenbach R.
     J. Chromatogr. 123 (1976) 271
[5] Werther W.
     Diplomarbeit, Techn. Univ. Vienna 1986
[6] Varmuza K.
     Fres. Z. Anal. Chem. 322 (1985) 170

Acknowledgement: I would like to thank the 'Fonds zur Förderung der wissenschaftlichen Forschung' for financial support (project P6811C).

EDAS — MS

EXPLORATORY DATA ANALYSIS OF MASS SPECTRA

Werther W. and Varmuza K.

Technical University Vienna, Institute for General Chemistry
Lehargasse 4/152, A-1060 Vienna, Austria

SUMMARY

EDAS-MS is a software-tool, developed for the mass spectroscopic part of the spectroscopic information system SPECINFO. EDAS-MS uses methods from multivariate statistics, especially linear mapping procedures belonging to the field of exploratory data analysis. It supports the chemist in the search for relationships between spectral features und substructures and can be a help to determine the chemical class membership of an unknown compound.

Essential parts of EDAS-MS are: Input routines for mass spectra, procedures for the generation of spectral features, scaling and feature selection methods, principal component analysis and similar mapping methods and statistical substructure analysis. The user interface supports interactive working with the two-dimensional representations of data including access to spectra and chemical structures.

INTRODUCTION

Like other modern analytical instruments the mass spectrometer can produce a large number of measurements for one investigated compound. A set of mass spectra can be handled as a matrix with multivariate data. In mathematics and statistics numerous efficient methods have been developed with the aim to extract relevant information from multivariate data. Some of these methods have been successfully applied to chemical questions, among them the interpretation of spectroscopic data [1,3].

Although the above mentioned conditions seem favourable, multivariate statistical methods are only rarely used in mass spectrometry. Several reasons are responsible for this fact: (1) Structural information supplied by mass spectra is usually not directly encoded in the original data (intensities at individual mass numbers) because the relationships between spectra and structures are more complicated. (2) Living in a three-dimensional world and creating a two-dimensional map of it with his eyes, man is trained to recognize relations in two or three dimensions. A multidimensional view is beyond the ability of his senses. (3) Common software for multivariate statistics requires specialist's knowledge of these methods, which chemists often lack. (4) Convenient interfaces for mass spectroscopic data are often not available. (5) Common software works only with pure numerical data, whereas chemists prefer to work with spectra and chemical structures.

The computer program EDAS-MS, which is desribed in this paper, is an attempt to overcome these difficulties and is intended to support the interpretation of mass spectral data, especially in cases when the unknown is not contained in the library.

THE CONCEPTION OF EDAS

Exploratory Data Analysis of Spectra (EDAS) is based on the application of methods from multivariate statistics, with emphasis on graphical representations of data and interaction of the chemist [3]. This approach for computer assisted spectra interpretation seems useful when no definite rules can be applied, but sets of library spectra are available that are relevant for a given spectroscopic problem. Suitable sets of spectra can be obtained from a library e.g. by substructure search or by spectra comparison.

The computer program EDAS-MS has been developed for the mass spectroscopic part of the data bank SPECINFO [4,5]. A link also exists to the mass spectroscopic library search system MASSLIB [6]. The strategy of EDAS-MS combines the application of spectral features and linear mapping methods (Fig. 1).

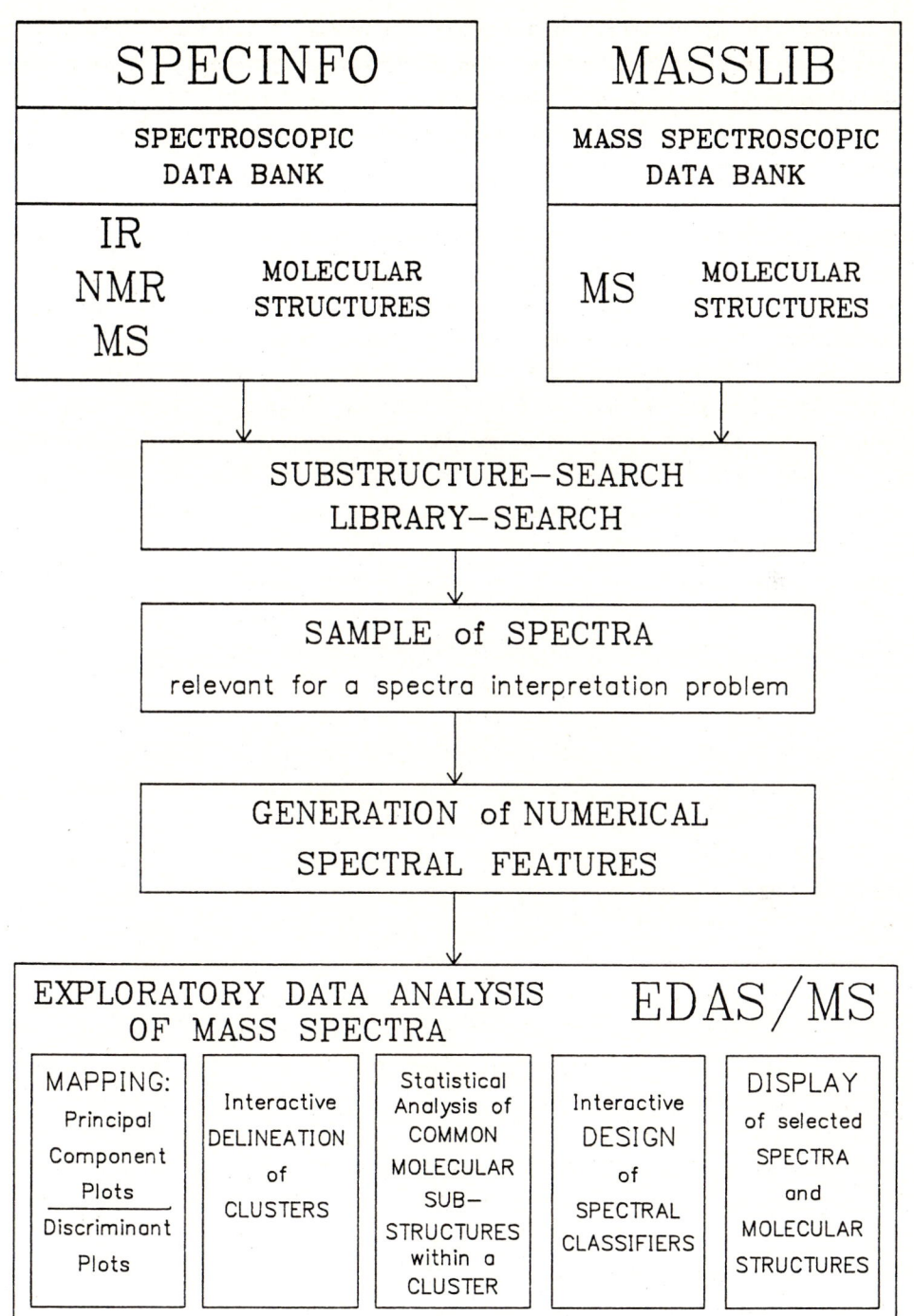

Fig. 1 EDAS-MS is a software tool in connection with SPECINFO.

SPECTRAL FEATURES are figures calculated from the mass spectrum, e.g. modulo-14 spectra, autocorrelation spectra or spectral-type features [7]. Because of the wellknown connections between such spectral features and substructures they seem to be a better data basis for automatic representation of spectra-structure-relationships than the original mass spectrum. Usually a set of 10 to 30 spectral features is calculated for each compound [3,8].

Each compound can then be represented as a point in a multidimensional space with the spectral features being the coordinates of this point. If the used spectral features represent spectra-structure relationships well, compounds with similar chemical structures cluster in this space. By applying linear MAPPING METHODS these multivariate data are projected onto a suitable plane [9]. Now the effective human ability to recognize clusters and to evaluate relative positions of points in the two-dimensional representation of data is used for further interactive processing.

These procedures are realized in EDAS-MS in an user-friendly way: The statistical methods are designed as a 'black box' - the user does not need special knowledge in this field. Input of mass spectra is possible via ASCII-files in EPA-format, spectra from SPECINFO and MASSLIB can be read directly. Chemical structures and the mass spectra can be presented on the screen, so that the chemist can avoid working only with abstract data. A large number of supporting tools is disposable [10].

TECHNICAL DETAILS

EDAS-MS is written in FORTRAN-77 and has been developed at a Microvax-II with the operating system MicroVMS 4.6.

(1) Spectra input:
- direct access to mass spectra, structures and HOSE-substructure-codes in SPECINFO
- direct access to mass spectra in MASSLIB
- input of measured spectra via ASCII-files in EPA-format

(2) Feature generation:
At present following groups of spectral features can be calculated:
- modulo-14 spectra
- spectral-type features
- autocorrelation spectra
- logarithmic intensity quotients

Usual preprocessing (scaling, normalization) and feature-selection methods are available.

(3) Mapping methods:
- principal component mapping
- discriminant component mapping

The result of a mapping is represented as a score-score-plot. Loading-loading-plots are also available.

(4) Tools for interactive data exploration:
- routines for graphic design such as zooming, variable plot-symbols, inserting of unknown objects
- attaching names, mass spectra and chemical structures to the objects
- interactive delineation of clusters
- statistical substructure analysis
- design of linear classifiers
- leave-one-out-test for linear classifiers

INVESTIGATION OF SPECTRA-STRUCTURE-RELATIONSHIPS - AN ITERATIVE PROCESS

Mass spectra contain a wide range of different substructure information, which cannot be detected together in only one mapping. Exploratory data analysis as described above can be imagined as looking at the multivariate data space from a specific point of view. In order to extract different items of information from the data it is reasonable to change this point of view and to inspect the data from different angles. Several strategies realize this aim: (1) Different projection planes may show different substructure information. (2) The application of different spectral features extracts different types of structure similarity. (3) Changing the composition of the investigated spectra set can reveal new substructure clusters in the projection.

Therefore the investigation of spectra-structure relationships within a class of compounds is not only one step, but an iterative process (Fig. 2). The following successive steps should be repeated until no new information about substructures can be found in the spectral data:

The user defines the "point of view" by selecting the spectra set, the spectral features and the projection plane. In the resulting mapping an interactive search for clusters of compounds is performed, supposing, that these clusters contain common substructures. This 'exploratory' step of the investigation is assisted in EDAS-MS by an automatic substructure analysis. The result is a hypothesis about a relationship between the position in the projection and the presence of a certain substructure. The validity of the hypothesis can be visualized by marking all compound containing this substructure. The hypothesis can be tested in another way by a class-conditional mapping, which supplies maximum separation for given classes of compounds in the projection. Subsequently an attempt can be made to explain a revealed spectra-structure relationship in terms of chemical knowledge. Factor loadings of the mapping can often be interpreted and fragmentation rules can be defined [10].

EXAMPLE

This iterative process of observing mass spectral multidimensional data from different points of view is demonstrated by a simple example. The basis data set contains the mass spectra of 60 aliphatic saturated amines (34 primary and 26 secondary amines) with 1 to 10 carbon atoms, selected by substructure search in SPECINFO.

In the FIRST INVESTIGATION all compounds are used. Fourteen spectral features are calculated by a modulo-14 summation starting at mass 39. These features are neither normalized nor scaled. Principal component analysis is applied and the projection plane is spanned by the first and the second principal component (PC1, PC2). Two well separated clusters of objects can be easily recognized (Fig. 3). Automatic substructure analysis shows that all compounds in the left cluster contain the substructure $-CH_2-NH_2$. That means the left cluster consists only of primary amines not branched at

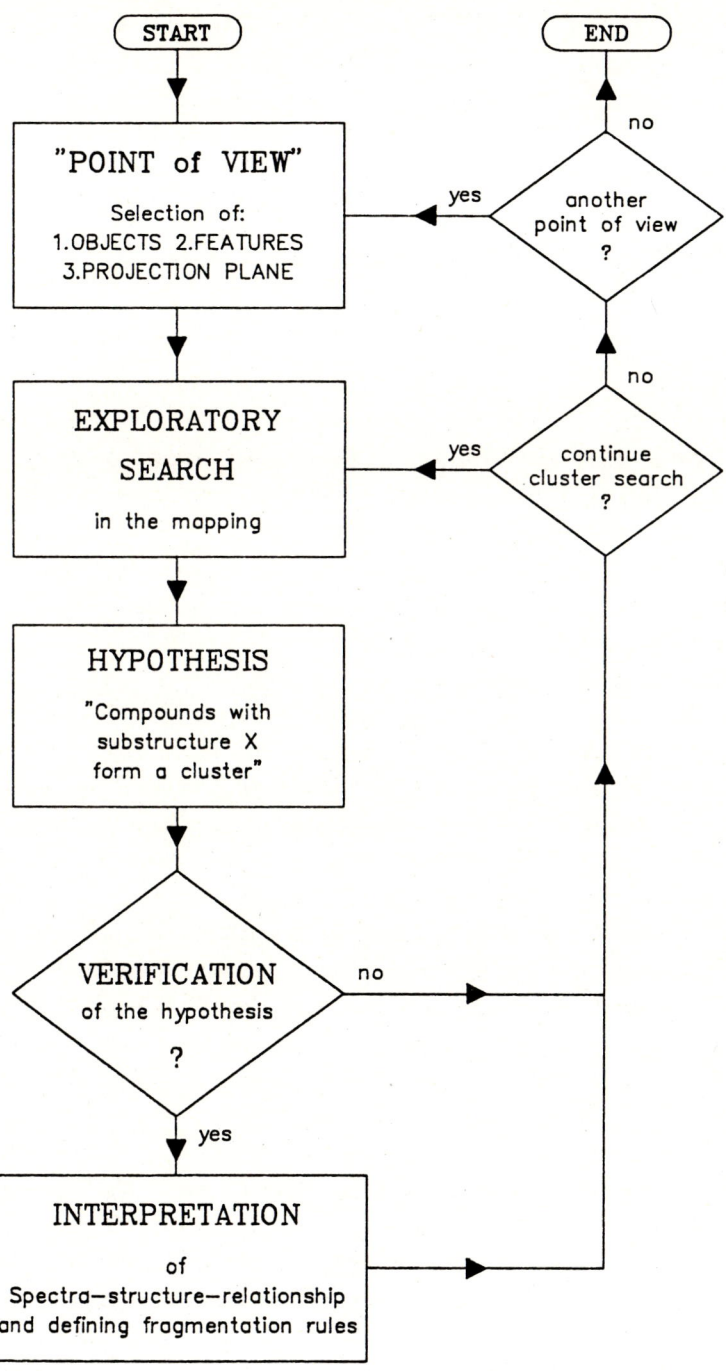

Fig. 2: The investigation of mass spectra with EDAS-MS is an iterative process.

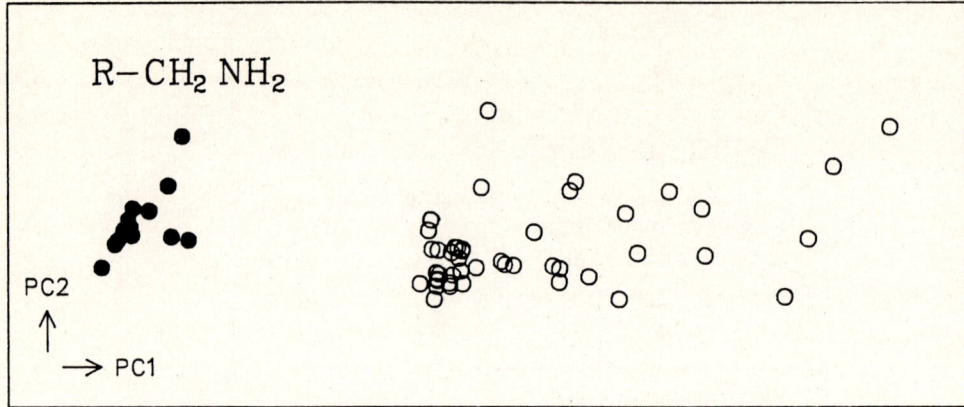

Fig. 3: Principal component mapping (PC1, PC2) of the modulo-14 spectra of 60 aliphatic saturated amines.

the alpha-carbon. Marking this group of compounds in the projection verifies this result.

The separation into these two clusters can be explained by looking at the factor loadings of the first principal component. An outstanding high loading for the feature no. 2 is detected. This feature is the sum of the intensities of the typical ions of aliphatic saturated amines (m/e 30, 44, 58, ...), which are formed by alpha-cleavage or by following olefin loss rearrangement. $CH_2=NH_2+$ (mass 30) is the dominating ion of this series within the group of primary amines not branched at the alpha-carbon. Because modulo-14 summation was started at mass 39 this feature results in a low value, thus separating this group of compounds from the other amines.

To find further structural information from the data set, the point of view is changed in the SECOND INVESTIGATION. The same principal component analysis is used, but now the projection plane is spanned by the second and the third principal component. All compounds with the above mentioned substructure $-CH_2-NH_2$ are removed from the projection. In the new representation (Fig. 4) no distinct clusters are recognizable. Therefore an interactive exploration takes place: One of the attempts to find groups of compounds with common substructures situated closely together has been successful: Statistical substructure analysis resulted in the hypothesis, that this projection may show a separation between primary and secondary amines. This assumption was verified by marking this group of compounds. Furthermore in a discriminant component mapping both classes of compounds could be separated well (a leave-one-out test resulted in only one false classified secondary amine).

High factor loadings are detected for the features no. 1, 3 and 13. The most important ion contributing to the intensity of the feature 3 is the molecular ion. It has been reported [11], that the intensity of M+ for the secondary amines is higher than that for primary amines. No mass spectroscopic explanation could be found for the other two features.

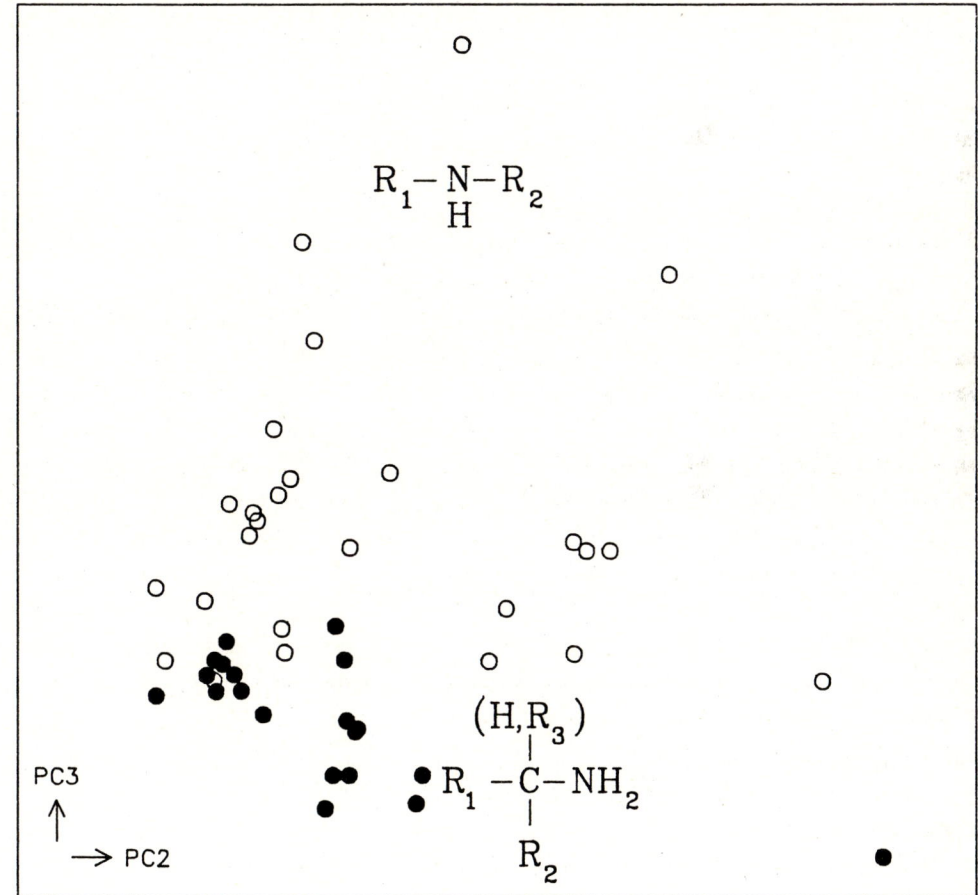

Fig. 4: Principal component mapping (PC2, PC3) of the same features as in Fig. 3. The primary amines not branched at the alpha-carbon atom have been removed from the projection.

In a THIRD INVESTIGATION the data set is reduced to primary amines branched at the alpha-carbon atom. Six other spectral features, the logarithmic intensity quotients I_m/I_{m+1} for m = 39 to 44, are calculated. The projection plane is spanned by the first and second principal component (Fig. 5). The already described iterative process is applied. A separation of alpha-methyl-amines (no further substituent at the alpha-carbon atom) from the other amines of the data set is verified by marking all alph-methyl-amines. A discriminant component mapping with a leave-one-out test resulted in only one compound misclassified. High loadings in the first principal component for the two features calculated from the intensity quotients I_{43}/I_{44} and I_{44}/I_{45} indicate that the ion $CH_3-CH=NH_2^+$ may be the main reason for the separation.

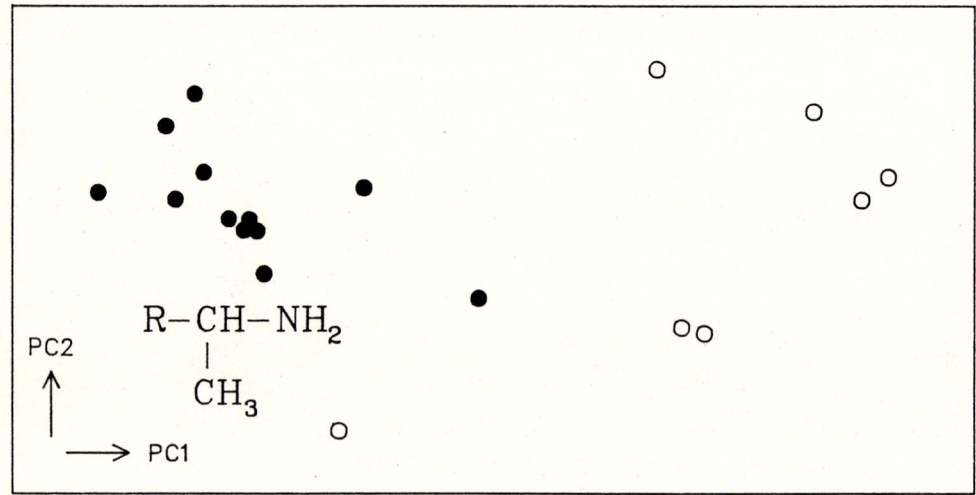

Fig. 5: Principal component mapping (PC1, PC2) of logarithmic intensity quotients of primary amines branched at the alpha-carbon atom.

LITERATURE

1 Massart D.L. et al.: "Chemometrics: a textbook", Elsevier, Amsterdam (1988)
2 Meuzelaar Henk L.C., Isenhour T.L. (eds.): "Computer-Enhanced Analytical Spectroscopy", Plenum Press, New York (1987)
3 Varmuza K., Werther W., Lohninger H.: in "Software-Entwicklung in der Chemie 3" (ed.: Gauglitz G.), p.267, Springer-Verlag, Berlin (1989)
4 Schubert V., Bremser W., Neudert R., Kubinyi H., Gasteiger J., Varmuza K.: Nachr.Chem.Tech.Lab. 37:721 (1988)
5 Bremser W.: Angew.Chem. 100:252 (1988)
6 Henneberg D.: Max-Planck-Institut für Kohlenforschung, Mülheim/Ruhr, FRG
7 Lohninger H., Varmuza K.: Anal.Chem. 59:236 (1987)
8 Varmuza K., Werther W., Lohninger H.: in Advances in Mass Spectrometry Vol.11 (ed.: Longevialle P.), p.1872, Heyden & Son, London (1989)
9 Varmuza K., Lohninger H.: in "PC for Chemists" (ed.: J.Zupan), Elsevier, in press
10 Werther W., Varmuza K.: to be published
11 McLafferty F.W.: "Interpretation of Mass Spectra", University Science Books, Mill Valley, CA, USA (1980)

ACKNOWLEDGEMENTS

This work was supported by the Bundesminister für Forschung und Technologie (Bonn, FRG). The authors thank J. Gasteiger (TU Munich) for software support and W. Bremser, R. Neudert, H. Kubinyi, V. Schubert and D. Henneberg for stimulating discussions.

Towards the Automatic Generation of a Mass Spectrum from the Structure of a Compound

W. Hanebeck, K. Rafeiner, K.-P. Schulz, P. Röse and J. Gasteiger

Organisch-chemisches Institut, Technische Universität München,
Lichtenbergstr. 4, D-8046 Garching, West Germany

Abstract: A system is developed that automatically predicts for a given organic structure the main fragmentation steps and the corresponding peaks in its mass spectrum. This is accomplished by interpretation of fragmentation rules stored in an external file. Physicochemical parameters calculated by empirical methods are used to decide which fragmentations and rearrangements occur preferentially.

Introduction

Mass spectroscopy plays a major role in the elucidations of the structure of organic compounds. In analyzing a mass spectrum possible candidate structures are considered. These structures might have been inferred from other spectroscopic data or from considerations on the course of chemical reactions that led to the compound at hand. For each candidate structure potential fragmentations, rearrangements, and sequences thereof are conceived and evaluated by drawing from our knowledge on the analysis of other mass spectra.

The system that is developed in our group intends to model this process: for a given chemical structure the fragmentation pattern should be derived. Thus, the important features (main peaks and their intensities) of the mass spectrum of this compound should be obtained. This would give a valuable tool to the chemist for evaluating several possible candidate structures.

The breakdown of a molecule in a mass spectrometer constitutes a sequence of chemical reactions. With this in mind we have entered this field. For, we could draw from our long-standing experience in predicting the course of chemical reactions [1]. The design of our most recent version (6.0) of the EROS system [2] proved to be particularly appropriate for reshaping it to the simulation of the reaction events in the mass spectrometer.

THE FOUNDATIONS OF THE APPROACH

Our approach to the prediction of the course of chemical reactions considers each step of a reaction mechanism or fragmentation pattern explicitly. The decision which next reaction step will occur is based on evaluations of electronic and energy effects. The calculation of these physicochemical parameters depend on published algorithms. These parameters include charge distribution[3], inductive effect[4], resonance effect[5], polarisability[6], and bond dissociation energies[7].

Calculations and correlations of physical data have underscored the physical significance of the parameters evaluated by these empirical procedures [8]. It is of particular importance for the problem at hand, the simulation of the fundamental steps in the mass spectrometer — that the above parameters were very useful in correlating quantitative data on fundamental gas phase reactions. These included proton affinities of amines [4,6], of alcohols and ethers, of thiols and thioethers [9], of carbonyl compounds, as well as of data on gas phase acidities of alcohols [10].

A brief analysis should further substantiate the potential of these physicochemical parameters for modelling events in mass spectral fragmentations. Figure 1 shows a data set of ethers. Certain bonds in these molecules were indicated as primary sites of fragmentations (leading to cations and radicals) by marking them with bent arrows (two directions possible).

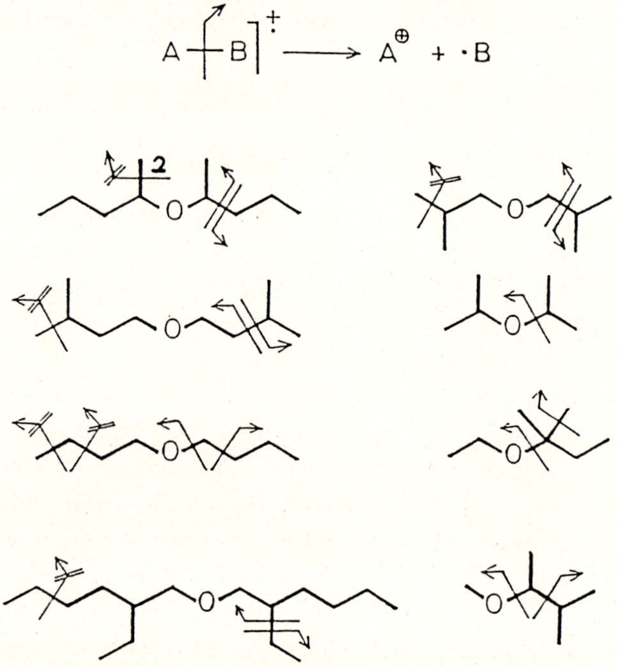

$$A\!-\!\!|\!-\!B \;]^{+\cdot} \longrightarrow A^{\oplus} + \cdot B$$

Figure 1: Sample data set of ethers with bonds classified as breakable (bent arrows) or nonbreakable (arrows crossed out)

Some other bonds were classified as nonbreakable, the likelihood of them being sites for primary fragmentations was considered low. These were all cleavages leading to methyl or ethyl radicals.

The various physicochemical parameters were calculated for all bonds in the data set of ethers. A cluster analysis showed that some of these parameters were good in assembling related bonds in individual clusters. Figure 2 shows a three–dimensional space having bond dissociation energy (BDE), the resonance effect (R), and bond polarity (q_f) as coordinates.

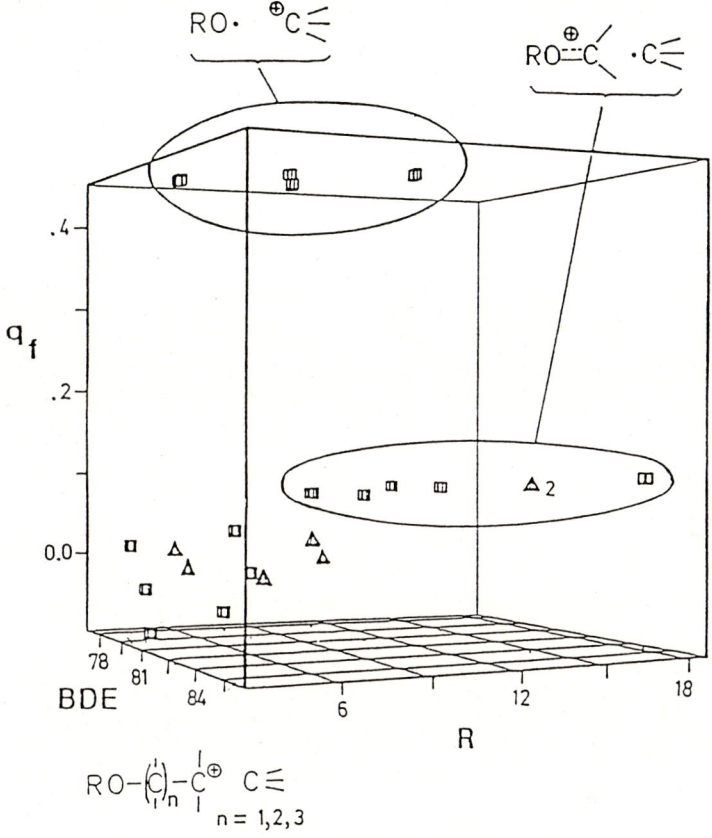

Figure 2: Representation of the fragmentations selected in Figure 1 in a space with bond dissociation energy (BDE), resonance effect (R), and bond polarity (q_f) as coordinates, small cubes indicate easily occurring fragmentation, pyramids nonbreakable bonds

The left uppermost cluster contains bonds to the oxygen atom dissociating in a simple σ–cleavage. The cluster in the middle part of the cube contains all those bond breakings corresponding to an α–cleavage. The pyramid in this cluster (point 2) seems out of place. It was classified as nonbreakable as it corresponds to a bond breakage that leads to a methyl radical (see Figure 1). However, this fragmentation is also an α–cleavage and should therefore be taken as an easily occurring reaction, its classification as nonbreakable was wrong.

The story taken from this study is that:

- the parameters are well suited for representing certain fragmentations of bonds in the mass spectrometer
- similar fragmentation types are characterized by similar values in these parameters

- false classifications of bonds can be recognized by looking at the values of these parameters
- statistical analysis of multidimensional spaces spanned by these parameters should lead to new insights into basic driving forces of fragmentations in the mass spectrometer.

The Design of the System

Our first attempt for the generation of fragmentations in the mass spectrum only used a rather limited set of fundamental reaction processes: ionisation, heterolytic and homolytic bond breaking and the making of bonds between two radicals or between a positively and negatively charged atom [11]. The representation of rearrangements and of more complicated reaction steps became rather artifical and led to difficulties in the evaluation process.

We therefore decided to use all the major electron shifting processes found in fragmentation sequences in the mass spectrum as individual elementary steps. Each such elementary process is represented by an individual entry in an external data file.

The new design of the system for modelling mass spectra (EROS–MS) heavily dwells on the EROS 6.0 version [2] and thus keeps the core system and the knowledge base (data file with elementary processes) clearly apart from each other (Figure 3).

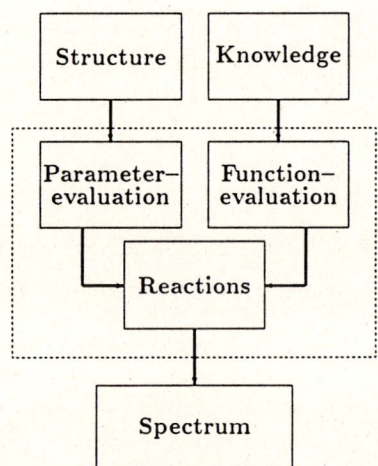

Figure 3: A overview of the system

Figure 4 shows the reaction scheme underlying the α–cleavage as an example of such an elementary process stored in the file. When this prescription for rearranging bonds and electrons is applied onto an ionized amine or ether, the α–cleavage is generated.

The information about the fragmentation patterns used by the program is stored in an external ASCII-File. This file is read as the program starts, and interpreted for internal use. The known reaction patterns are stored in this file.

> **Scheme: α-cleavage**
>
> $$R - \overset{+\bullet}{1} - 2 - 3 - R^1 \longrightarrow$$
> $$R - \overset{+}{1} = 2 \quad + \quad R^1 - 3\bullet$$
>
> **Application of reaction pattern:**
>
> $$R - \overset{+\bullet}{NH} - CH_2 - CH_2 - R^1 \longrightarrow$$
> $$R - \overset{+}{NH} = CH_2 \quad + \quad R^1 - CH_2\bullet$$
>
> $$R - \overset{+\bullet}{\underline{O}} - CH_2 - CH_2 - R^1 \longrightarrow$$
> $$R - \overset{+}{\underline{O}} = CH_2 \quad + \quad R^1 - CH_2\bullet$$

Figure 4: α-cleavage: reaction type and its application in a amine and in an ether

The structure of the knowledge base (called "rule file") is:

- File header, declaration of default values
- Reaction rules (pattern), containing:
  - rule header
  - declaration of reaction pattern (path)
  - declaration of a chain between two reaction centers (for rearrangements)
  - declaration of the conditions the molecule must fulfil for the rule to operate
  - declaration of the conditions the atoms in the path must fulfil for the rule to operate
  - declaration of the conditions the path of bonds must fulfil for the rule to operate
  - declaration of the reaction generator for atoms (changes in charge and electronic structure)
  - declaration of the reaction generator for bonds (changes in bond order)
  - definition of the reactivity function

The basic reaction pattern is stored in the rule file. The program searches the structure given for reaction paths fulfilling all conditions defined in the knowledge base. A reactivity function (see below) is calculated for the reactants. If the function value is in a predefined range, the program applies the reaction generator onto the structure, calculates the physicochemical parameters for the products, and then determines a final value for the reactivity function. If this final value is within an allowed range, the program generates and stores the reaction.

While the program is running each allowed reaction generates a new structure or structures. The positively charged species resulting from this reaction are stored and searched for new possible reactions. Neutral species are eliminated. The program builds a reaction tree (Figure 5) by continuously creating reactions and new structures. The conditions terminate the growth of this tree are:

- the mass of the generated cation is lower then a defined value or

- the intensity of a cation is lower than 1% of the highest intensity calculated

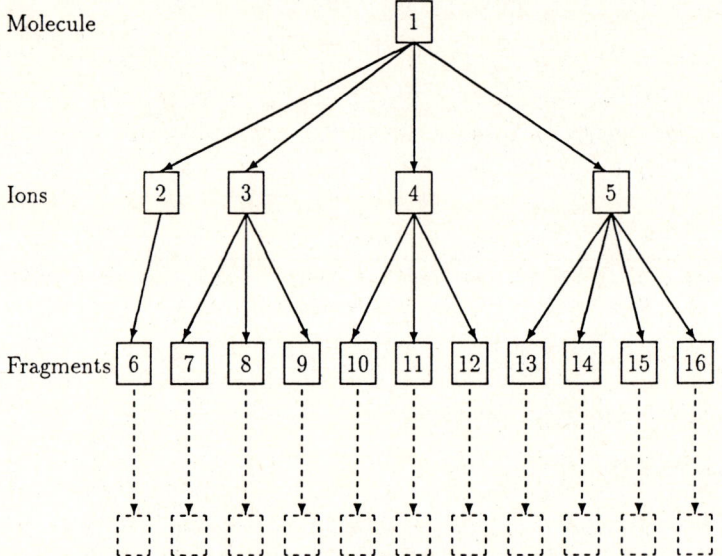

Figure 5: The tree of ions, daughter ions, and neutral losses

The program calculates a reactivity index (RI) using a linear combination of the physico-chemical parameters defined in the rule file. The functions are being obtained through logistic regression analysis with a data set describing one reaction type each. This function value is normalized to one, to give the reactivity index (FP: function value, c_i: coefficients, P_i: parameter value):

$$\mathcal{FP} = c_0 + c_1 P_1 + c_2 P_2 + \ldots$$

$$RI = \frac{1}{1 + e^{-\mathcal{FP}}}$$

The intensity of the signals is calculated using a pseudo-kinetic model including the reactivity index. First the sum of all reaction indices of a node is calculated. The intensity (I_i) of all new nodes is calculated one by one. Then the sum of all new intensities is substracted from the intensity of the node reacting. Therefore the intensity of a fragment depends on the goodness of reactions leading to it and on the ones destroying this node. The formulas are:

$$N = \sum_{i=1}^{n} RI(i)$$

$$I_i = \left[\frac{RI(i)}{N} \times \left(1 - e^{-Nt}\right) \right] \times I_{Pre(old)}$$

$$I_{Pre(new)} = I_{Pre(old)} - \sum I_i$$

Example

On top of Figure the experimental mass spectrum of a 2,6–dimethyloctanoic acid methyl ester taken from a database is given. The base peak m/e 88 represents the McLafferty ion. Using five basic processes in the rule file (ionisation, α–cleavage, McLafferty rearrangement, displacement reaction and CO–elimination) a mass spectrum is generated, that is represented in the middle of Figure .

The intensity of the molecular ion is calculated to have a very low value. Therefore the molecular ion does not appear in the calculated spectrum. This spectrum correctly shows the two most intense peaks of the experimental spectrum. The other two fragmentations, although also present in the experimental spectrum, are given too high an intensity. Clearly many of the other smaller peaks of the experimental spectrum cannot be expected to be modelled with these five elementary processes only.

The picture becomes already better when another (a sixth) process is added to the rule file: the fragmentation of a carbenium ion into another carbenium ion and an olefin. Then the spectrum at the bottom of Figure is obtained. It gives two new peaks also present in the experimental spectrum and reduces one of the unrealistically high peaks to an intensity more corresponding to the observed one (It is clear that the intensity of a peak in a mass spectrum might vary pretty much depending on the conditions of recording). However, the most important features of the mass spectrum are reproduced.

Conclusion

It must be stressed that the purpose of this study was not to completely reproduce the experimental spectrum. Rather, by switching on some of the basic processes in the database the user can determine which processes give peaks also present in the experimental spectrum. Thus an instrument is available that allows the chemist to search for the presence of specific fragmentation processes in a spectrum. This lays the foundation to a better understanding of the fragmentations of a compound in the mass spectrometer and helps in elucidating its structure.

Clearly this is only the beginning of a road that might lead to the simulation of the complete mass spectrum of an organic compound.

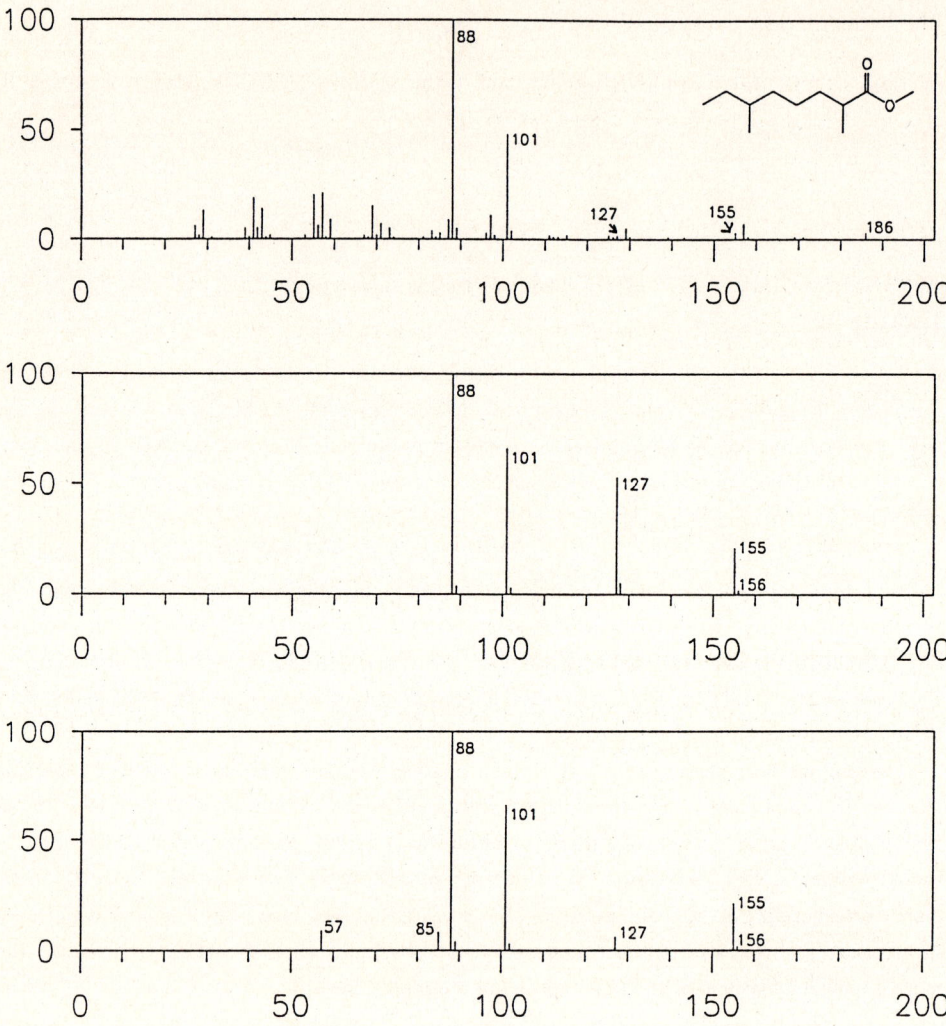

Figure 6: Experimental spectrum (top) and two calculated spectra of a simple ester (with five or six basic processes, resp.; see text)

Acknowledgements

The work reported in this paper was supported by a grant of the Bundesminister für Forschung und Technologie of the FRG (BMFT–project no. 106 321 8–7). Responsibility for the contents of this report is with the author.

We thank the BMFT for the support of our work. We appreciate stimulating discussions with Dr. V. Schubert (GMD-PTF) and Dr. W. Bremser, Dr. R. Neudert, and Prof. H. Kubinyi (all BASF) as well as Prof. K. Varmuza (TU Wien).

References

[1] Gasteiger J, Hutchings MG, Christoph B, Gann L, Hiller C, Löw P, Marsili M, Saller H, and Yuki K (1987) *Topics Curr Chem*, 137:19-73

[2] Röse P and Gasteiger J, this issue

[3] Gasteiger J and Marsili M (1980) *Tetrahedron*, 36:3219-3228

[4] Hutchings MG and Gasteiger J (1983) *Tetrahedron Lett*, 24:2541-2544

[5] Gasteiger J and Saller H (1985) *Angew Chem*, 97:699-701

[6] Gasteiger J and Hutchings MG (1984) *J Chem Soc Perkin 2*, 559-564

[7] Gasteiger J (1978) *Comput Chem*, 2:85-88

[8] Gasteiger J (1988) in *Physical Property Prediction in Organic Chemistry* Springer-Verlag, Heidelberg

[9] Gasteiger J and Hutchings MG (1984) *J Am Chem Soc*, 106:6489-6495

[10] Hutchings MG and Gasteiger J (1986) *J Chem Soc Perkin 2*, 447-454

[11] Hanebeck W, Saller H, and Gasteiger J (1988) In Gasteiger J, (ed), *Software-Entwicklung in der Chemie 2*, Springer-Verlag, Heidelberg

Factor analysis of spectral data from chemical reactions

G. Gauglitz, S. Weiß

Institut für physikalische

und theoretische Chemie

Auf der Morgenstelle 8

7400 Tübingen

Introduction

Factor analysis (FA) is a mathematical method, which was originally developped for examination of psychological tests. It was introduced by Thurstone [1], who also showed, that this method could help the psychologists in evaluation of a large amount of data more objectively.

Nowadays, factor analysis has also been introduced in other scientific areas. It is usually used to evaluate a large number of multivariate data, obtained taking computers for data acquisition. By use of FA the amount of data is drastically reduced without any loss of information. This is an obvious advantage, which is implicit in factor analysis.

In chemistry, factor analysis has already been applied to many subjects: to determine the rank of spectroscopic data from mass-, IR-, UV/Vis-, fluorescence- spectroscopy of multicomponent mixtures. In this application the rank of the data matrix is the number of compounds, which result the mixed spectra. Additional constraints sometimes allow to reconstruct the spectra and thereby to identify the components. In HPLC, overlapping peaks may be resolved by use of diode-arrays for spectral detection. The influence of solvents and substituents on chemical shifts and coupling constants of NMR-features has been studied by FA. In chromatography, solutes and solvents were classified into groups having similar characteristics by factor analysis. A broad overview of these and further applications is given in [2].

Theory

To be factor analyzable, the data matrix must obey the following equation:

$$d_{jk} = \sum_n r_{jn} c_{nk}$$

or

$$\mathbf{D} = \mathbf{R}\,\mathbf{C} \tag{1}$$

Herein, the matrices have the dimensions: $\mathbf{R}$ (rr x rc, rr > rc), $\mathbf{C}$ (cr x cc, cc > cr) and $\mathbf{D}$ (rr x cc). Many experimental data follow this rule of formation or may be supposed to do so.

For demonstrative purpose, a chemical reaction is presented here as an example. The reaction is the well-known photochemical formation of phenanthrene from trans-stilbene:

$$\text{trans-stilbene} \underset{h\nu}{\overset{h\nu}{\rightleftharpoons}} \text{cis-stilbene} \xrightarrow{h\nu} \text{phenanthrene}$$

In this reaction, three compounds take part, despite a short living dihydrophenanthrene. The reaction is observed by measuring complete UV-spectra at various time values during the reaction. Each spectrum therefore is a mixture of three pure spectra, whereby the extinction $d_{\lambda t}$ is the sum over n = 3 absorbing components. These spectra are shown in *fig*. 1. Utilizing the law of Lambert-Beer and suppressing the thickness of the sample, this can be written as

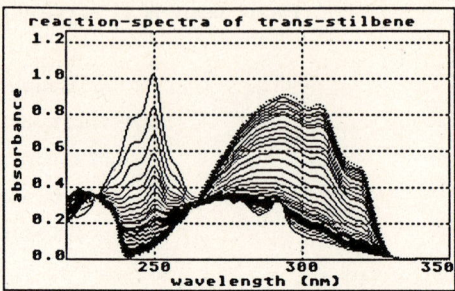

fig. 1: reaction-spectrum of trans-stilbene

$$D = E\ C. \qquad (2)$$

E is the (λ x n) absorptivity matrix, C is the (n x t) concentration matrix of n compounds at t different times. Each column c_k specifies the contribution of the several compounds to the complete mixture. Therefore each row of C can be viewed as a 'modulation' vector for the 'spectra' matrix E. This 'modulation' matrix can arise for example from the individual elution profiles or from the change of concentration during a chemical reaction. In order to be analyzable in principal, there must be sufficient measurements at different wavelengths and times. It is also obvious, that the rows of E and the columns of C may not be all the same or any linear combination. In general, the rank of E and C must equal the number of components n.

After preprocessing the data, the first step of factor analysis would be a principal component analysis (PCA). This task is done by building up the covariance matrix Z

$$Z = D^T D \qquad (3a)$$

or

$$Z = D\ D^T. \qquad (3b)$$

Next, Z is diagonalized such that

$$Z\ Q = Q\ \Lambda \qquad (4)$$

is obtained. One gets the eigenvektor matrix Q and the corresponding eigenvalues λ_i, which are the elements of the digonal matrix Λ. Furtheron, the eigenvectors q_i and the eigenvalues λ_i should be considered to be sorted in decreasing magnitude of the λ_i. Z is symmetric and the eigenvectors are normalized, thus

$$Q^{-1} = Q^T. \qquad (4)$$

Considering the following derivation

$$Q^T Z\ Q = (Q^T\ D^T)(D\ Q)$$
$$= (D\ Q)^T (D\ Q)$$
$$= U\ U^T,$$

where

$$U = D\ Q \qquad (5)$$

or

$$D = U\ Q, \qquad (6)$$

Q^T can be identified with C and U with E.

Because of experimental error, the **U** and **Q** matrices have full rank of **Z**. That is, **Z** can be completely reproduced by **Q** and **U**. By partitioning **Q** into n primary eigenvectors $q_i^{\ddagger}$ and the remaining secondary eigenvectors q_i^{+}, it will be possible, to reproduce **Z** and **D** by $\mathbf{Q}^{\ddagger}$ and $\mathbf{U}^{\ddagger}$ within experimental error. The primary eigenvectors therefore can be seen to describe the data matrix, whereas the secondary eigenvectors describe only the error in these data. The first five eigenvectors of the stilbene reaction are given in fig. 2.

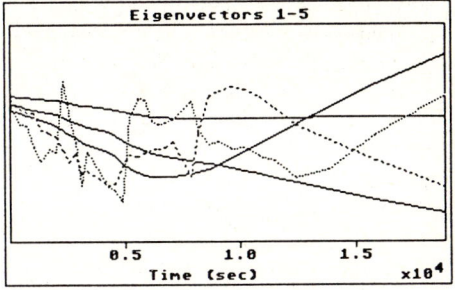

fig. 2: The first five eigenvectors

The central problem within the PCA procedure is to determine the number of primary eigenvectors, which is the real rank of **D** in absence of experimental error. To overcome this problem, various solutions were discussed in the past [2, 3, 4, 5]. Among the methods which do not require any knowledge of the imbedded experimental error, the indicator function introduced by Malinowski [6] is widely applied. This function is defined by

$$IND(p) = \frac{RE(p)}{(c-p)^2} \tag{7}$$

with the real error beeing

$$RE(p) = \left[\frac{\sum_{j=p+1}^{c} \lambda_j}{r(c-p)}\right]^{1/2} \tag{8}$$

Herein, r is the bigger dimension of **D** and c is the smaller one. IND(p) will result in a minimum, if p equals the real rank of **D**. *Fig.* 3 shows, that this function fails in the above mentioned example, since the IND function has a minimum at n = 8. The reason for this misbehavour is mainly because the standard deviation σ_{jk} has not the same value for each datapoint d_{jk}, which is a requirement for this empirically found function.

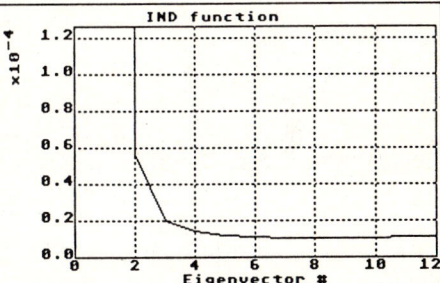

fig. 3: The indicator function.

To describe a new approach, which will not fail in this aplication, the next step of factor analysis must first be discussed. In general, this is called the rotational or transformation step. We will focus again on the rank-determination.

Both the transformation and the rotation of the vectors a_j can be expressed mathematically by the following equation:

$$\mathbf{A}^* = \mathbf{A}\,\mathbf{T} \tag{9}$$

T is the transformation matrix, which leads in the case of rotation to the transformed vectors a_j^*, the length of a_j being unchanged. The rotational methods can be divided into those, which underlie an analytical criterion for the rotational angle, and those, which are forced by empirical a priori knowledge about the system. The former are known as abstract rotations, because the resulting vectors cannot be interpreted by physical or chemical aspects. The chemical reaction provided as an example herein shows some rotated eigenvectors as concentration profiles which have physically suspicious negative values, and therefore may only be considered as abstract concentration vectors.

Among the various rotation methods there are some which lead to orthogonal vectors and others resulting in an oblique vector set. By far the most frequently used criterion for the rotational angle is to maximize the variance **within one** vector. This so-called varimax criterion can be expressed by defining the simplicity s of a vector as the variance of its squared loadings,

$$s_s = \frac{n \sum_j (q_{js}^2)^2 - (\sum_j q_{js}^2)^2}{n^2} \qquad (10)$$

and maximizing the sum over all rotated vectors

$$S = \sum s_s \qquad (11)$$

to reach the maximum simplicity S for the complete vector set [6]. The resulting vectors remain orthogonal under these conditions.

For the example presented here, it doesn't matter whether to rotate the eigenvectors with the varimax criterion or others like quartimax [7] or oblimax [8], where the last one would lead to an oblique set. The usefulness of an abstract rotation will be unfolded later when discussing the iterative target transformation. Common to all rotation procedures is the fact, that rotation is always done with two vectors at a time, giving a new pair of transformed vectors.

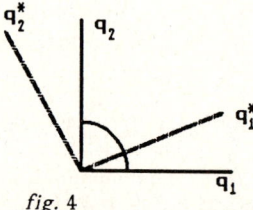

fig. 4

So, the first eigenvector must be rotated successively with the second one, the third one and so forth up to the n-th vector of the primary eigenvector set. Then, the second vector will be rotated together with the third one up to the n-th vector. After having transformed the (n-1)-th with the n-th vector, the whole procedure is repeated until the maximizing criterion is fullfilled.

Consider now one rotational step of a vector pair as in *fig.* 4. It is obvious, that every rotated vector q^* simply is a linear combination of the unrotated vector pair. That is, any 'properties' of the original vectors will be mixed into both rotated vectors.

A special property of any secondary eigenvector is the imbedded noise, coming up from experimental error. This noise can be visualized by regard-

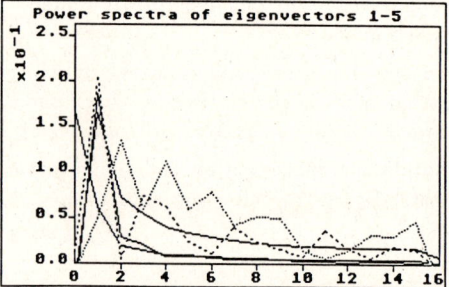

fig. 5: Power-spectra of the first five eigenvectors.

ing the power-spectra of the first eigenvectors (*fig.* 5). The fourth and fifth eigenvector in our example show relatively large fourier coefficients for the higher frequencies on the right-hand part of the diagram. Rossi and Warner [5] therefore considered the fourier spectra of the eigenvectors to estimate the number of significant eigenvectors. They proposed the following slightly modified function

$$A^j_{ulim} = \sum_{i=0}^{ulim} |F^i(q_j)| \qquad (12)$$

which is simply the partial sum over the fourier coefficients of the magnitude spectrum of the j-th eigenvector. By defining T as the total sum over all observable frequencies, the percentage given by

$$\% A^j_{ulim} = \frac{100 \, A^j_{ulim}}{T} \qquad (13)$$

may deal as an estimator for the relative importance of the lower frequency portion in the eigenvector q_j. When plotting %A versus j (*fig.* 6), there should be a drastical decrease at the first eigenvector, which belongs to the secondary set. The reason is, that high frequencies become more important, just describing noise or experimental error like the secondary eigenvectors do. As can be seen in *fig.* 6, the problem is to define a threshold level, which distinguishes the primary from secondary eigenvectors. This threshold level would be about 0.4 for the lower curve in the present example but probably must be altered for other problems.

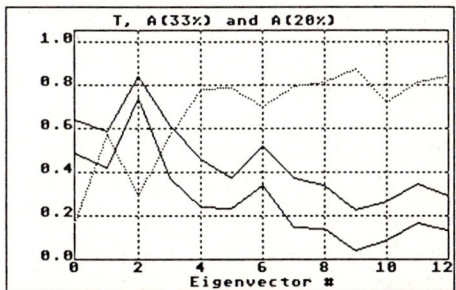

fig. 6: Total sum T (dotted), relative partial sum A(33%) and A(20%) (lower line)

In contrast to the method proposed by Rossi and Warner, the fourier spectra for the varimax rotated vectors are calculated. These spectra differ according to the number of primary eigenvectors, which were used in the complete rotation process. The procedure for the estimation of the real rank then goes as follows: First, the rank n is set to 2. The rotated vectors are calculated together with their appropriate fourier spectra and their relative partial sum over fourier coefficients $\%A_{ulim}$. Next, the rank will be incremented by 1. After rotation of all eigenvectors, the $\%A_{ulim}$ will be computed again. This procedure is repeated until all corresponding $\%A_{ulim}$ values have decreased. This will be the case, if secondary eigenvectors were included into the rotational process which introduce noise into the primary eigenvectors. This results in a decrease of the relative importance of the low frequencies. Therefore, the rank assumed for calculation the last $\%A_{ulim}$ values is 1 greater than the real rank n.

Table 1 shows the A(33%) quotients for the example presented herein. The real rank for the data matrix is known to be 3. When considering column 2 and 3, which were calculated for n = 3 and n = 4 respectively, it can be seen, that all values have decreased for n = 4, indicating the correct number of components beeing 3. This my be visually recognized by regarding *fig.* 8, 9 and 10, which show the power spectra of the first, second and third varimax vectors, both for n = 3 (-) and n = 4(...). If the rank

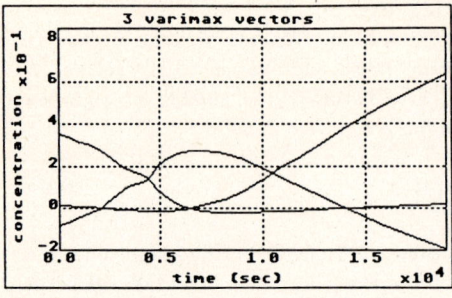
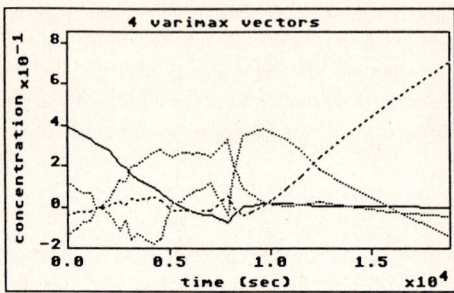

fig. 7: Varimax vectors for n = 3   Varimax vectors for n = 4

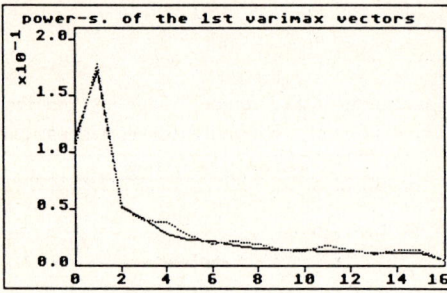
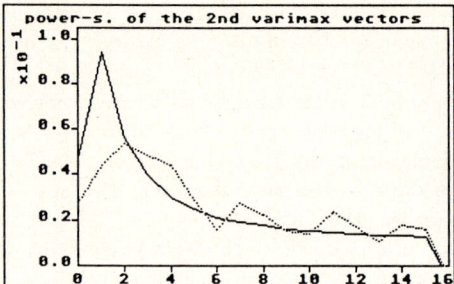

fig. 8: Power-spectra of the 1st varimax-v.   *fig.* 9: Power-spectra of the 2nd varimax-v.

was assumed to be 4, the higher frequencies of the varimax vectors become more important as can be seen by the dotted lines. A discussion concerning the mathematics in detail will be given in another publication, which is in preparation.

To continue with factor analysis, a further transformation of the varimax vectors must be done to get vectors, which are interpretable in a physical or chemical sense.

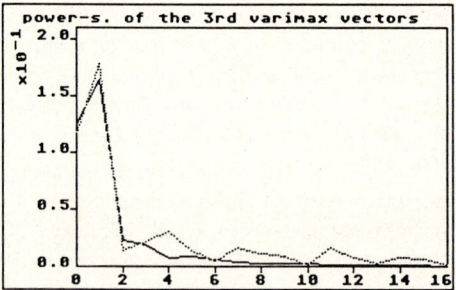

fig. 10: Power-spectra of the 3rd varimax-v.

To overcome this task, additional constraints must be introduced. For our problem, these are twofold: first, the concentration or absorption can not be

n =	2	3	4	5
A^1(33%)	0.59	0.63	0.62	0.58
A^2(33%)	0.57	0.54	0.47	0.48
A^3(33%)		0.84	0.69	0.57
A^4(33%)			0.68	0.62
A^5(33%)				0.54

Table 1: Relative importance of low frequencies of rotated eigenvectors A(33%) dependent on the assumed rank

negative and second, the concentration vector must consist of only one relative maximum. By this means, the special case of oscillating reactions is excluded. It must be emphasized, that namely the second constraint is valuable only for concentration profiles of kinetic and chromatographic data, which is true in the example presented here.

The algorithm that solves the transformation with those additional constraints is known as 'iterative target transformation' and was developped by Vandeginste et al. [9]. The procedure is as follows: a starting vector t^1 is chosen, that confirms with the above mentioned constraints. In our case, this could be a vector with all elements set to zero except at that point, where the varimax vector, which ought to be transformed, has its maximum. For simplicity, the previously rotated eigenvector matrix Q_n^* is furtheron termed as V, which consists of the n column vectors v_j. The projection of the vector t^1 into the space spanned by the v_j involves finding the least square solution for the rotation vector r from the equation $t^1 = V\, r$, giving

$$r = [\, V^T\, V\,]^{-1}\, V^T\, t^1 \tag{14}$$

$$= V^T\, t^1, \tag{14a}$$

and then calculating the projected vector t^{1*}

$$t^{1*} = V\, r \tag{15}$$

The second step in equation (14a) is valid, because the v_j are all orthogonal and normalized. Usually, the projected vector t^{1*} differs from t^1 and therefore does probably not obey the criterions for the special problem. Thus, this new vector has to be modified in such a way, that the constraints are fulfilled again (i.e. set negative values to zero and eliminate maxima besides the main peak). The resulting vector t^2 will be a better approximation for the real concentration profile than t^1 was. Then, the iteration is continued by calculating t^3 starting with t^2, then t^4 and so on. The vectors will converge under these conditions, (see [10] for proof) and therefore the iteration can be terminated, if the correlation coefficient exceeds an max

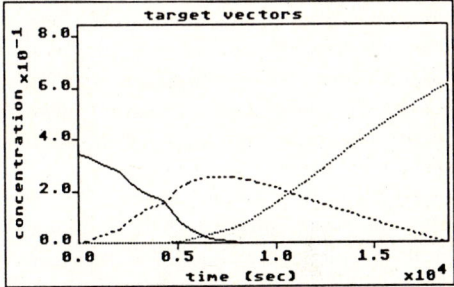

fig. 11: Final target vectors as valuable concentration profiles.

upper limit of say 0.999. The complete procedure must be repeated for all the varivectors of V.

In principal, the iterative target transformation can also be carried out with the eigenvectors itself. In tis case, the problem would arise, how to choose a valuable start vector - for our problem the value of the initial concentration maximum (as a needle peak). By comparing *fig.* 2 with *fig.* 7, it can be seen, that the varimax vectors are more similar to real concentration profiles than the eigenvectors. To understand this, one first must remind the fact, that eigenvectors ordered according to decreasing eigenvalues are principal axes. That means, the first axis (eigenvector) holds the most information of **all** the vectors which span the data space. This is expressed by the largest eigenvalue, which can be understood as a measure for the variance of the projected data points onto the appropriate axis. The motivation for the rotational step is to maximize the variance of other principal axes with smaller eigenvalues and thereby assign more relevance to them. This results in a more likely concentration profile. Because the initial maximum position must not equal the final position after the iterative target transformation, only a rather good guess is required for the start vector. To demonstrate this nice feature, one can set, in a second run, the initial maximum slighly besides the final maximum, known from the previous run. The resulting target vectors will normally coincide.

If the number of factors was not determined correctly, some target vectors would show a rather doubtfull profile, e.g. in a tall peak mostly improbable in chemical reactions. This can serve as a further indication for a wrongly estimated rank.

Factor analysis is accomplished by calculating the so-called 'factor scores'. The factor scores in the presented example are the n spectra of the pure compounds. Up to here, the concentration matrix **C** from eq. (2) is identical with the transposed target vectors **T**:

$$\mathbf{C} = \mathbf{T}^T$$

Thus, the 'spectra matrix' **E** can be computed from eq. (2) by a multilinear regression step:

$$\mathbf{D} = \mathbf{E} \mathbf{T}^T \tag{2a}$$

$$\mathbf{E} = \mathbf{D} \mathbf{T} [\mathbf{T}^T \mathbf{T}]^{-1}. \tag{16}$$

The goal of factor analysis, which was to find the two matrices **E** and **C** just from a given data matrix **D**, is thereby reached. However, when regarding the concentration profiles provided by the target vectors (*fig.* 11), the chemist would reject the interpretation of these target vectors being absolute concentrations of the chemical reaction mentioned here. Whereas the shape of each of the curves might be correct within experimental uncertainty, the law of

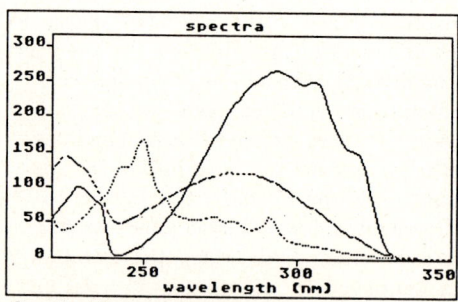

fig. 12: The 'pure' spectra.

mass seems to be violated. In fact, no knowledge about the mass dependencies the reaction mechanism was put into the algorithm. By introducing a scaling matrix **S**, consisting of only diagonal elements s_{ij}, the factor analytical solution can be modified in general by

$$\mathbf{D} = [\mathbf{E} \, \mathbf{S}^{-1}][\mathbf{S} \, \mathbf{T}^T]. \tag{17}$$

Given a reaction model, **S** can be calculated from **T** by use of the law of mass, whereby the relative concentrations will be adjusted for the assumed model. Consequently, the thus scaled spectra matrix $[\mathbf{E} \, \mathbf{S}^{-1}]$ can be considered to be absolute and can show, for example, the isosbestic point for every reaction step.

Conclusion

Factor analysis provides a tool for evaluating a large amount of data, which obey the simple matrix equation (1). It was originally intended for psychologists to have a look 'behind the data', that is, to understand, by which factors the data were generated or influenced. For the chemical reaction that was presented as an example, some a priori knowledge about the system is available. After dedicating the number of compounds, this knowledge was used to split the data matrix into two underlying factor matrices, which are in this case the well interpretable pure spectra and the corresponding concentration profiles. The available knowledge can be introduced by means of the iterative target transformation procedure. In our example, only two constraints suffered for a rather good estimation of the concentration profiles. In consequence, the chemist is able to assign these results to a specific model for a chemical reaction and may prove his assumption by use of further evaluation algorithms like least-square fits or formal integration [11].

The reliability of the results from factor analysis is a serious problem. Many point operations are carried out starting from erroneous data. Indeed, part of the experimental error imbedded in the data matrix is excluded by the PCA, but calculational certainty cannot be guarrantied for the overall process. It is cumbersome, to examine error propagation only by analytical considerations. Simulations must serve as an alternative to get a feeling for the reliability of the resulting matrices. Moreover, the effect of strongly correlated spectra can thus be examined, which should result in a confidence interval for the resulting spectra and concentration profiles. Further research on this topics is a necessary task for the future.

There are still some modifications of the factor analytical approach. First, the PCA may be substituted by a procedure, which is known as singular value decomposition (SVD). This algorithm has a slight advantage concerning the numerical uncertainties of PCA, especially if 64 bit (or even 80 bit) floating point variables are not available on a computer system [12]. Second, the so called 'target testing' procedure instead of the iterative target transformation has to be applied to a completely other kind of problems, where no a priori knowledge is available. This procedure involves validating a hypothesis by means of a target vector, which consists of known physical or chemical properties. In contrast to conventional least-square fitting, any single hypothesis may be tested at a time before accepting or rejecting it. For further reading confer to [2].

Literature:

[1] L.L. Thurstone: "Multiple factor analysis", Chicago: Univ. Chicago Press 1947
[2] E.R. Malinowski, D.G. Howery, "Factor Analysis in Chemistry", Robert E. Krieger Publishing Company, Malabar, Florida (1989)
[3] N. Otha, *Anal. Chem.*, **45** (1973), 553
[4] R.I. Shrager, R.W. Hendler, *Anal. Chem.*, **54** (1982), 1147
[5] T.M. Rossi, I.M. Warner, *Anal. Chem.*, **58** (1986), 810
[6] H.F. Kaiser, *Psychometrica*, **23** (1958), 186
[7] J.B. Carroll, *Psychometrica*, **18** (1953), 23
[8] D.R. Saunders, *Psychometrica*, **26** (1961), 317
[9] B.G.M. Vandeginste, G. Kateman, *Anal. Chim. Acta*, **173** (1985), 253
[10] T.H. Brayden, P.A. Poropatic, J.L. Watanabe, *Anal. Chem.*, **60** (1988), 1154
[11] H. Mauser, "Formale Kinetik", Bertelsmann Universitätsverlag Düsseldorf (1974)
[12] R.I. Shrager, *Chemometrics and Intelligent Laboratory Systems*, **1** (1986), 59

NORMALIZATION OF IN SITU-SPECTRA IN THIN LAYER CHROMATOGRAPHY

S.Ebel, J.S.Kang and W. Windmann

Department of Pharmacy and Foodchemistry
University of Würzburg
Am Hubland, D-8700 Würzburg

Abstract: Normalization is a necessary step in order to compare two spectra. Mostly point normalization at the maximum of absorbance is used. Other methods are area normalization or the use of regression techniques. Reflectance spectra in TLC are not obeying Bouguer-Lambert-Beer's law, therefore normalization has to be done by two-dimensional linear regression or nonlinear regression using a saturation function. Both methods are leading to good results. Nonlinear regression needs more computing time and more sophisticated computer programming.

INTRODUCTION

Normally in routine work in TLC/HPTLC Rf-values compared to a reference are taken for identification. But there is a lack in reproducibility and often this way is impossible. Coupling of spectrophotometric techniques increases the precision of identification of substances after chromatography. Now a days reflectance spectroscopy is done by computer control and it is possible to record spectra directly from the separated spots. But up to this time only a few research work is done in respect to UV/VIS-spectra libraries. Because there are some disadvantages

- normalization of spectra is only possible if there is only a small difference in mass per spot
- problems in linearity
- lack of information in UV/VIS-spectra in general

- lack of information in respect to functional groups of the molecules
- changes in the spectra caused by residues of the eluent

APPARATUS

A schematic diagram of a TLC reflectance spectrophotometer is presented in fig.1. At a blank position of the plate the wavelength depending reflectance of the sorbens layer is taken and stored as reference. The same procedure is done in the centre of the spot to be identified to get the uncorrected reflectance spectrum of the sample.

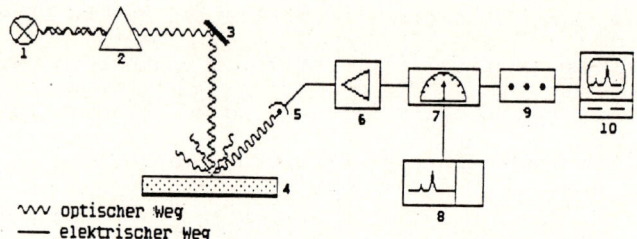

Fig.1:
Schematic diagram of a single beam TLC reflectance spectrophotometer
1 light source   2 monochromator
3 mirror   4 TLC plate
5 Photomultiplier or photodiode   6 amplifier
7,8 analog signal processing (recorder)
9,10 digital signal processing (A/D-converter, Computer)

BACKGROUND-CORRECTION

UV absorbing substance have a wavelength depending absorptivity. If the reflectance is measured, the intensity is lowered by the absorptivity. This will lead to a reflectance spectrum that is similar to an absorption spectrum. The background reflectance is given by eq.(1) [1]. Contrary to absorption, reflectance is not proportional to concentration of substance per area unit. Therefore the decrease in reflectance is not proportional to the amount of substance per spot and will lead to problems in comparison of spectra. Up to this time real theoretical based functions of signal depending on concentration as the wellknown Bouguer-Lambert-Beer's law in absorption of homogeneous solutions are not described in the reflectance mode. One estimate is given by the Kubelka-Munk equation (2) [2]. In practice a transformation similar to Bouguer-Lambert-Beer's law (3) is used (4).

(1) $\quad \Delta R(\lambda) = R_b(\lambda) - R_{b+s}(\lambda)$

$\qquad R_b(\lambda) \quad$ wavelength depending reflectance of the background

$\qquad R_{b+s}(\lambda) \quad$ wavelength depending reflectance of the spot (background diminished by sample absorption)

(2) $\quad \dfrac{\alpha}{\rho}(\lambda) = \dfrac{[1-R_\infty(\lambda)]^2}{2R_\infty}$

$\qquad \dfrac{\alpha}{\rho}(\lambda) \quad$ wavelength depending ratio of absorptivity to reflectance

$\qquad R_\infty \quad$ measured absolute reflectance of an infinite layer

(3) $\quad A(\lambda) = -\log[T(\lambda)] = -\log \dfrac{I_m(\lambda)}{I_0(\lambda)}$

(4) $\quad R(\lambda) = -\log\left(\dfrac{R_{b+s}(\lambda)}{R_b(\lambda)}\right)$

$\qquad A(\lambda)$ wavelength depending absorption
$\qquad T(\lambda)$ wavelength depending transmittance
$\qquad R(\lambda)$ background corrected reflectance

In fig.2 the relative reflectance of the background and the uncorrected spectrum of sudanred are presented. Both spectra are normalized in respect to maximum of background reflectance. The background corrected spectrum differs if eq.(1) ["reflectance spectrum"] or eq.(4) ["absorption spectrum"] is used. This is demonstrated in fig.3.

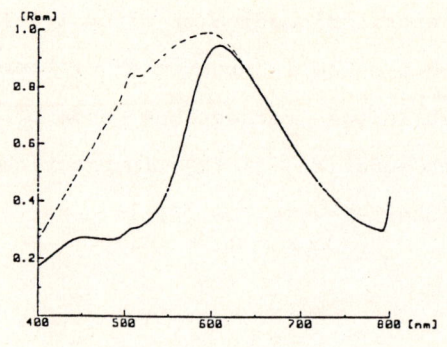

Fig.2:
Raw reflectance spectra of background (---) and within the TLC-spot of sudanred (——)

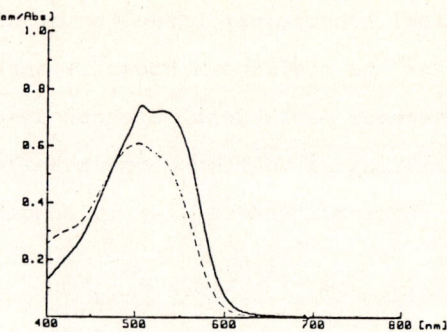

Fig.3:
Background corrected spectra in reflectance (——) (1) and absorption mode (----) (4)

NORMALIZATION OF SPECTRA

Point- and Area-Normalization

The commonly used method in normalization of spectra is the use of the maximum in absorption that is taken as $A_{max}=1$ and all other absorption values are calculates as relative values to this one ("point normalization"). Another way is the use of the area under the spectrum to be normalized which assumed that $\Sigma A_i = 1$ ("area normalization").

In homogeneous solution, in which the Bouguer-Lambert-Beer's law is valid, the ratio of absorption is independent from wavelength and only depending on the ratio of concentration. This is the base of normalization of two spectra (5). Point normalization is possible at each wavelength but from error propagation considerations the best choice is the maximum of absorption.

(5) $$N_p = \frac{c_r}{c_s} = \frac{A_r(\lambda_{max})}{A_s(\lambda_{max})}$$

(6) $$A_N(\lambda_i) = N_p \, A_s(\lambda_i)$$

$A_s(\lambda_i)$ spectrum of sample
$A_r(\lambda_i)$ spectrum of reference
N_p normalization factor
$A_N(\lambda_i)$ normalized spectrum of sample

In point normalization any statistical or systematic error in the maximum of absorbance will influence the whole normalized spectrum as systematic error. In order to avoid this disadvantage all measured values of both spectra may be used (area normalization).

$$(7) \quad N_a = \frac{c_r}{c_s} = \frac{\Sigma A_r(\lambda)}{\Sigma A_s(\lambda)}$$

$$(8) \quad A_N(\lambda) = N_a A_s(\lambda)$$

The normalized sample spectrum is now compared with the reference spectrum.

Normalization by Regression

Usually spectra from spots separated in TLC are not taken in the low range, where reflectance is nearly proportional to concentration. At higher amounts per spot the signal to noise ratio will increase. The similarity of two spectra is visualized if the corresponding absorption values are plotted taking the reference as independent and the sample as dependent variable. In fig.4 the specra to be normalized and the correlation plot is presented. Different spectra (cf. fig.4a) are showing conspicuous deviations from a straight line. If the spectra are more similar the correlation plot changes more and more to a straight line. The correlation plot is a straight line if both spectra are identical or the substances are identical. Contrary to spectroscopy in homogeneous solution this fact is in TLC only valid if the amounts of substance per spot nearly the same (cf. fig.4c). If the amount of substance per spot differs, a typical curved correlation plot is resulting (cf. fig.4b). Such a curved line is caused by nonlinearity of reflectance and concentration, i.e. Bouguer-Lambert-Beer's law is not valid. One model to describe this nonlinearity is to assume a function of saturation as wellknown from enzyme kinetics as Michaelis-Menten kinetic or from adsorption phenomena as Langmuir adsorption isotherm.

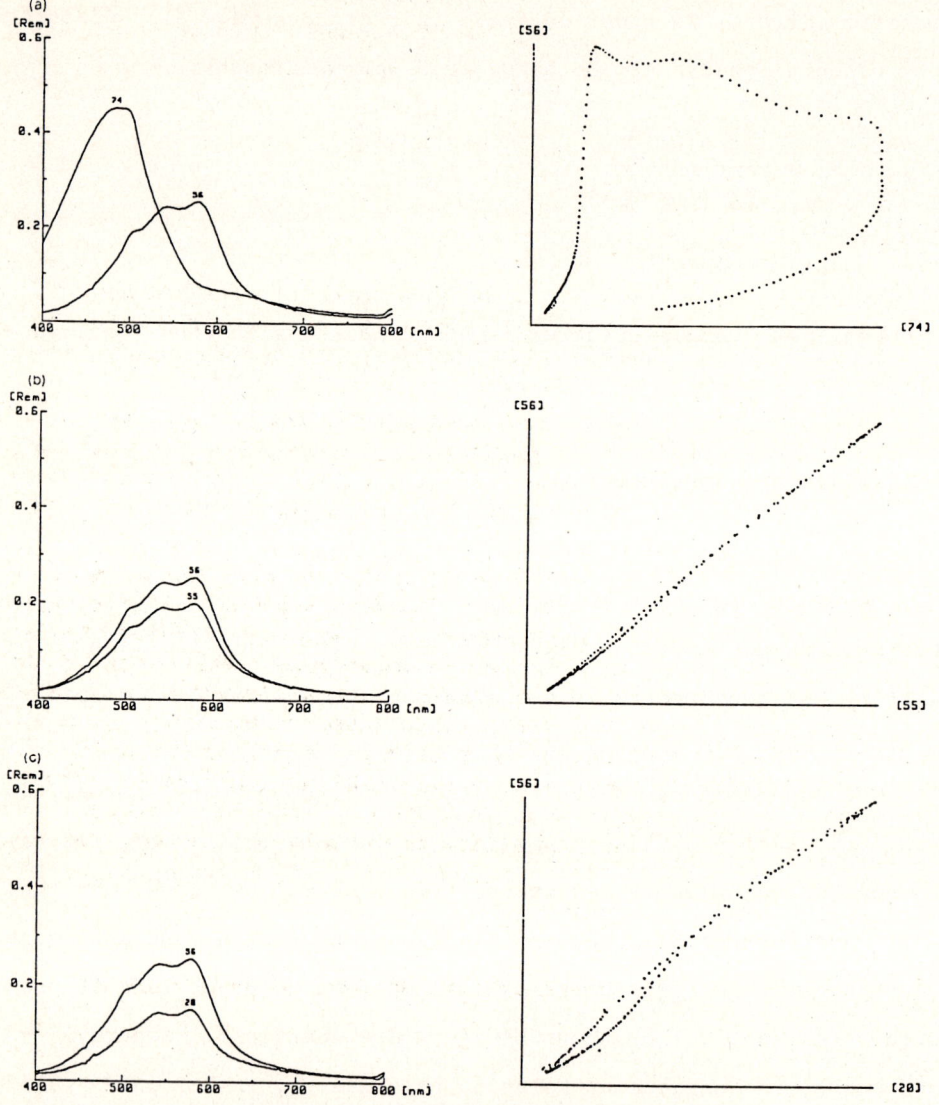

Fig. 4: Various spectra and their correlation plots
    a   Ciba FII / aminoazobenzene
    b   Ciba FII peakmaximum / inflection point
    c   same as b, but spotting amounts 1:5

Instead of point- or area normalization or a normalization by means of linear regression (9) the saturation function (11) or the two-dimensional linear regression (10) may be used as base for normalization.

(9) $$\hat{y}(x) = \bar{y} + a_1(x-\bar{x})$$

(10) $$\hat{y}(x) = \bar{y} + a_1(x-\bar{x}) + a_2(x^2-\overline{x^2})$$

(11) $$\hat{y}(x) = \frac{a_1 x}{a_2 + x}$$

Calculation of the regression coefficients $\bar{y}$, a_1 and a_2 are done by the principle of minimizing the sum of squared errors (12), where in the case of eq.(10) a straightforward [3] and in case of eq.(11) an iterative procedure has to be used [4].

(12) $$S_d = \Sigma[y_i - \hat{y}(x_i)]^2 \stackrel{!}{=} \min$$

$$y_i \equiv A_s(\lambda_i)$$
$$x_i \equiv A_r(\lambda_i)$$

The regression techniques are based on the assumption that the reference is free of errors.

COMPARISON OF SPECTRA

Numerical Comparison

Numerical values for comparison of spectra should take in consideration

- easy to interpret
- summarized content of information
- sensitivity in respect to noise
- sensitivity in respect to statistical distortion of spectra
- sensitivity to slightly shifted spectra

Numerical values are based on the correlation of the nonlinear normalized spectra according to eq.(10) or (11) respectively. To compare both spectra a linear regression is taken with the reference spectrum as independent and the normalized sample spectrum are dependent variable, where N_n is the nonlinear normalization factor based on the used regression function.

(13) $$\hat{y}(x) = \bar{y} + a_1(x-\bar{x})$$

$$y \equiv N_n A_s(\lambda)$$
$$x \equiv A_r(\lambda)$$

The remaining sum of squared errors S_d (12) and the relative standard deviation defined by eq. (14) are calculated. Another numerical value is given by the coefficient of correlation r (15) or from the ratio of the differences of areas R_{da} as defined by eq. (16).

(14) $$\text{relsdv}(y) = \frac{\sqrt{S_d}}{\bar{y}(n-2)}$$

(15) $$r = \frac{\Sigma(x_i - \bar{x})(y_i - \bar{y})}{\sqrt{\Sigma(x_i - \bar{x})^2 \Sigma(y_i - \bar{y})^2}}$$

(16) $$R_{da} = \frac{\Sigma|x_i - y_i|}{\Sigma x_i - \Sigma y_i}$$

Visualization

The calculation of only three numerical values seems dangerous in respect to loss of information. Therefore a graphic visualization of the result of comparison seems to be necessary. The main advantage is the differentiation between systematic and statistic errors.

Direct Comparision of Spectra

The direct graphic comparison of spectra is the mostly used method [5],[6]. In this case the reference and the normalized sample spectrum are overlayed (cf. fig.5). It is to be seen that nonlinear normalization (fig.5c) leads to more comparable spectra than point normalization (fig.5b).

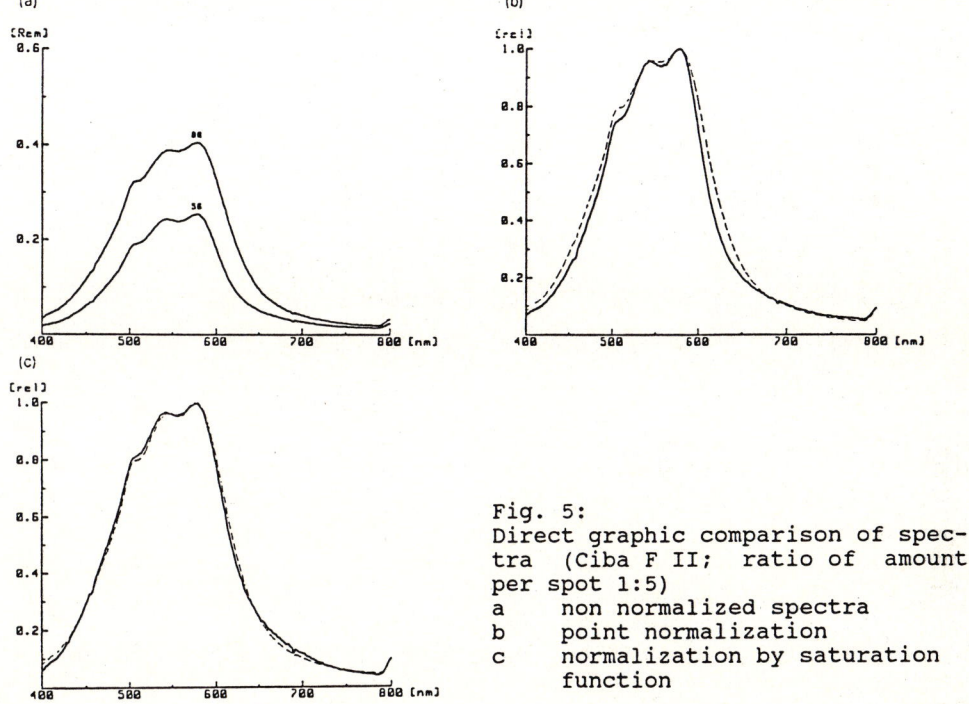

Fig. 5:
Direct graphic comparison of spectra (Ciba F II; ratio of amount per spot 1:5)
a   non normalized spectra
b   point normalization
c   normalization by saturation function

The similarity is also seen from the correlation plot (cf. fig.6). The numerical values are given in table 1.

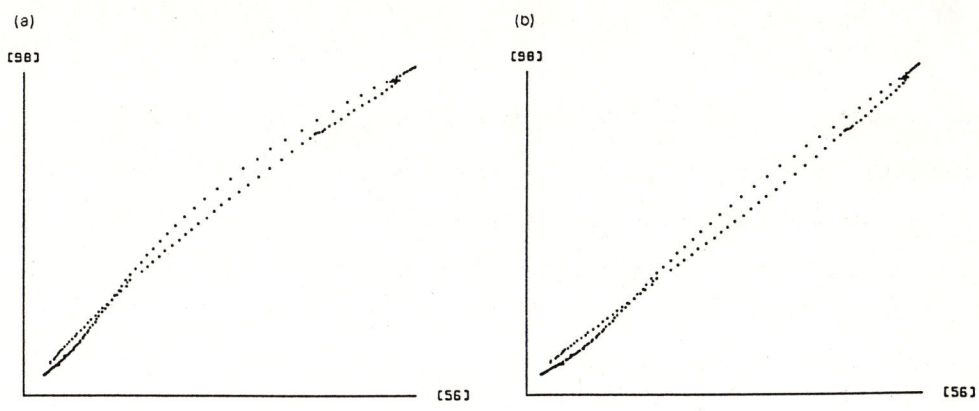

Fig. 6:
Correlation plot (same spectra as in fig.5)
    a   point normalization
    b   normalization by saturation function

Table 1:
Numerical values in spectra comparison
(same example as in fig.5)

normalization procedure	eq.(14) relsdv	eq.(15) r	eq.(16) R_{da}
point	0.0880	0.0410	0.9946
area	0.0881	0.0388	0.9946
linear regression	0.0880	0.0370	0.9946
twodimens.linear.regr.	0.0397	0.0162	0.9989
saturation function	0.0388	0.0159	0.9990

From the given figures and data it is obvious that normalization by two dimensional linear regression and by nonlinear regression using saturation function gives more reliable results and better correlations.

Spectral Differences

Another method to visualize the result of comparison is the graphic representation of the differences between the two normalized spectra as eq.(17).

(17) $\quad \Delta A(\lambda) = A_r(\lambda) - N_p A_s(\lambda)$

The difference spectra using point normalization and nonlinear regression normalization according to the saturation function are demonstrated in fig.7. The differences are smaller in the case of better normalization. Also from this visualization it is evident that not noise or other statistical errors but systematic errors caused by normalization have the larger influence on the remaining differences. Typical curves and not a scatter plot are resulting.

Also this type of comparison may be given as a numerical value, as standard deviation of all differences. In fig.7 both dimensions are given $\Delta A(\lambda)$ in the same scale and $sdv(\Delta A)$ as dotted lines.

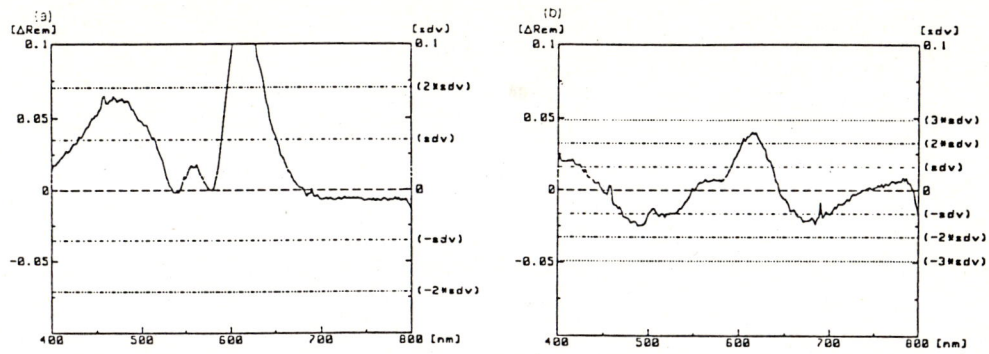

Fig. 7:
Differences between reference spectrum and normalized sample spectrum (same example as fig.5)
    a   point normalization
    b   normalization by saturation function

Table 2:
Numerical values of differnce spectra

normalization	sdv(ΔA)
point	0.0354
area	0.0358
linear regression	0.0331
twodimensional linear regression	0.0166
saturation function	0.0163

Comparison by Derivative Spectra

A comparison of spectra is possible by means of derivative spectroscopy. The wavelengths of maxima and minima and the number of inflection points can be used as criteria for similarity as well as the curves of the derivatives or their differences. This method has one disadvantage: the derivative has a significant lower signal to noise ratio. Smoothing of spectra will decrease the noise, but it is connected with systematic errors. From this point of view information of similarity becomes less reproducible and more biased. In fig.8 the first derivatives of spectra with point normalization and normalizing by saturation function are presented. There is no difference in the result occuring point-, area- or linear regression-normalization because all these methods will have the multiplication with a constant that is only in its numerical value somewhat different. Contrary to these methods nonlinear regression normalization leads to a superimpose of spectrum and normalization function and therefore to a change in the derivative. The advantage of nonlinear regression normalization using the saturation function is evident from fig.8.

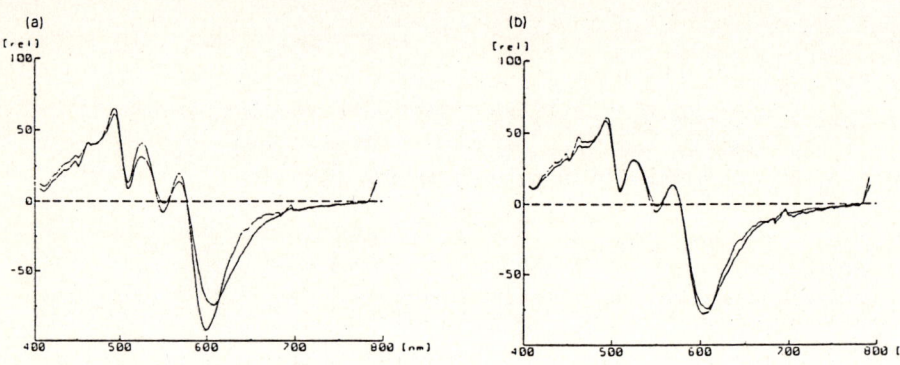

Fig. 8:
Comparison of first derivative spectra
    a   point normalization
    b   normalization by saturation function

Second order or higher order derivatives are of no use in comparison of TLC-in situ-spectra. The noise even in high amounts per spot is too high to give reliable results (fig.9).

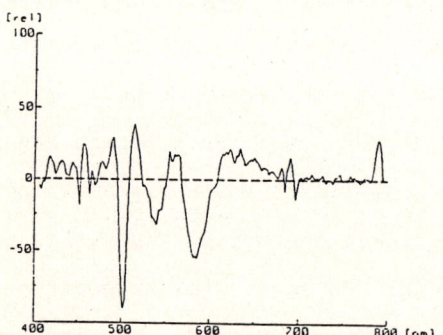

Fig. 9:
Second order derivative spectrum of Ciba FII ($10\mu l$/spot)

CONCLUSION

Computer controlled reflactance spectroscopy in TLC/HPTLC offers the possibility to build up a database of reference spectra. This database can be used for identification purpose or for purity control of separated substances in routine analysis. Normalization of spectra is a necessary step in order to carry out these tests. Normalization using two dimensional linear regression (parabolic regression) is an easy way to eliminate nonlinear effects in absorptivity. Normalization using sturation function and nonlinear regression techniques leads to more reliable results. This in context to nonlinear behaviour of wide range calibration.

Acknowledgement: This work was sponsored by Deutscher Akademischer Austauschdienst and Fonds der Chemischen Industrie.

References

[1] Ebel S, Geitz E, Klarner D (1980) Kontakte (Merck) (1):11-16
[2] Kubelka P, Munk F (1931) Z Techn Phys 12:539
[3] Draper N, Smith H (1981) Applied Regression Analysis. Wiley & Sons, New York
[4] Ebel S, Geitz E, Glaser
[5] Huber L, Drouen A (1988) GIT Fachz Laborat (1):16-19
[6] Miller JC, George SA, Willis BG (1982) Science 218:241-246
[7] Schaefer U (1985) Ph Thesis Würzburg
[8] Kitamura K, Hozumi K (1987) Anal Chim Acta 201:301-304
[9] Tanabe K, Saeki S (1975) Anal Chem 47: 118-122
[10] Weitkamp H, Wortig D (1977) Mikrochim Acta II:315-324
[11] Weitkamp H, Wortig D (1983) Mikrochim Acta II:31-57

ALGORITHMS FOR USE IN PURITY CONTROL OF DRUGS USING UV-SPECTROSCOPY

S. Ebel, C. Urban and S. Windmann

University of Würzburg
Department of Pharmacy and Food Chemistry
Am Hubland, D-8700 Würzburg

Abstract: UV-spectroscopy is a nonspecific analytical method, but in contrast to this it is possible to develop methods for purity control of drugs using this type of spectroscopy. Two algorithms will be described in this paper: a differential method subtracting the spectra of the analysis and a reference followed by nonparametric tests and the rank of a matrix containing both spectra. Base of both methods is a sophisticated comparison of both spectra. With two examples will be confirmed the proposed method. In some cases it is possible to detect impurities below 0.1%.

DIFFERENCES OF SAMPLE AND REFERENCE SPECTRA

This method uses normalization of both spectra and subtraction of both spectra followed by a statistical analysis of the resulting residuals [1]. Some normalization procedures are given in the literature but according to Mück [2],[3] and Kang [4] members of our group only three algorithms may be used in this special case

- point normalization
- normalization by area
- regression techniques (linear or orthogonal)

Orthogonal regression techniques had not been used up to this time. According to Bouguer-Lambert-Beer's law between to spectra of the same substance at two concentrations (c_r concentration of reference, c_s concentration of sample) at each wavelength relation(1)-(3) are valid. N is standing for a normalization-factor.

(1) $\quad A_{i,s} = \epsilon_i c_r$

(2) $\quad A_{i,s} = \epsilon_i c_s$

(3) $\quad A_{i,s} = \dfrac{c_s}{c_r} \epsilon_i = N \epsilon_i$

If both spectra are interpreted as vectors of data $\mathbf{a}_r$ and $\mathbf{a}_s$ eq.(3) changes to (4).

(4) $\quad \mathbf{a}_s = N \mathbf{a}_r$

Point normalization uses only the wavelength at the maximum of absorbance. For use in purity control point normalization gives the most reliable results if the mean of the absorption in the absorbance maximum and of the two further values on each side are taken (5).

(5) $\quad N = \dfrac{\Sigma a_{i,s}}{\Sigma a_{i,r}} = \dfrac{\bar{a}_s}{\bar{a}_r}$

If both spectra are belonging to pure identic substances and the normalization factor N is determined with low error, the residuals d_i of the subtracted normalized spectra (6) should obey a normal distribution with an estimator for the mean as zero (7) and the resulting estimated standard deviation is only given by the errors in measurement.

(6) $\quad \mathbf{d} = \mathbf{a}_r - N \mathbf{a}_s$

(7) $\quad \mathbf{d} \stackrel{\alpha}{=} N[0, \mathrm{sdv}(A)]$

According to a proposal of Weitkamp [5] the sum of the squared errors defines identity or equivalence of the two spectra. In terms of statistics an impurity does not cause a normal distributed error in the spectrum. Changes in the spectrum have to be interpreted as a systematic error. Optimal tests for comparison of two combined series or paired measurements are tests based on the sign of the paired differences. According to Wilcoxon or Dixon and Mood this method is valid for non normal distributed differences too [6].

NONPARAMETRIC TESTS

Wilcoxon Test

This test examines differences of paired data in respect to symmetric distribution to the median. The nullhypothesis is defined by the distribution function F(d) (8) or by the density function f(d) (8a). If H_0 is rejected, the series is not symmetric in respect to the median i.e. med(**d**)≠0.

(8) $H_0 = [F(+d)+F(-d)=1]$
(8a) $H_0 = [f[+d]=f(-d)]$

From all pairs of data the differences are calculated (9),(9a) and in a following step all differences are arranged according their absolute magnitude. The smallest difference is at position j=1, the largest at j=n. Furtheron the sign of the difference is remarked.

(9) $d_i = A_{s,i} - N A_{r,i}$

(9a) $\mathbf{d} = \mathbf{a}_s - N \mathbf{a}_r$

The sum of all positions of differences $d_j>0$ (R_p) and of differences $d_j<0$ (R_n) is calculated. Theoretically eq.(10) has to be fulfilled.

(10) $R_p + R_n = n(n+1)/2$

(10) $P_W = \begin{vmatrix} R_p & : & R_p<R_n \\ R_n & : & R_p>R_n \end{vmatrix}$

The critical value for the Wilcoxon test $T_W(n,\alpha)$ is given by eq.(12) with α as errorprobability and z_α as limits of the standard normal distribution at this α.

Critical values are tabulated [7] in dependence of α and n. In purity control of substances using UV-spectroscopy with more than 25 wavelengths it is possible to use eq.(12) and to compare with critical values of eq.(13) leading directly to the probality of the standard normal distribution.

(12) $\quad T_W(n,\alpha) = n(n+1)/4 - \sqrt{z_\alpha n(n+1)(2n+1)/24}$

(13) $\quad z = \dfrac{|T_W - n(n+1)/4|}{\sqrt{n(n+1)(2n+1)/24}}$

Dixon Mood Test

In this test only the signs of the differences of the two spectra at each wavelength are examined [8]. The probability of the signs is obeying a binomial distribution with $u=v=1/2$. The probability is calculated using eq.(14) where x denotes the number of positive or negative differences and n the number of differences $d_i \neq 0$.

(14) $\quad prob(x) = \dfrac{n!}{x!(n-x)!} u^x v^{n-x}$

Nullhypothesis is rejected if prob(x)<0.05 with x as the smaller number of positive or negative signs. If n>20 it is possible to calculate directly the probality of the standard normal distribution (15).

(15) $\quad z = \dfrac{|2x-n| - 1}{\sqrt{n}}$

The tests according to Wilcoxon or to Dixon-Mood are using median statistics and are therefore robust tests and free of conditions as normaldistribution of **d**.

PARAMETIC TESTS

Analysis of Variances

This two methods proof the systematic difference of the median zero. The differences at all wavelengths are calculated according to eq.(9). The standard deviation for the differences d_i, $sdv(d_i)$ is defined as the maximum deviation. All values higher than the maximum deviation are proofed by tests according to Wilcoxon and to Dixon Mood. From error propagation the variance of this differences should be calculated by using eq.(17).

(9) $\quad d_i = A_{s,i} - N_p A_{r,i}$

(16) $\quad var(d_i) = [\dfrac{\partial d_i}{\partial A_{s,i}}]^2 var(A_{s,i}) + [\dfrac{\partial d_i}{\partial N}]^2 var(N_p) +$

$$+ [\dfrac{\partial d_i}{\partial A_{s,i}}]^2 var(A_{r,i})$$

(17) $\quad var(d_i) = var(A_{s,i}) + A_{r,i}^2 var(N_p) + N_p^2 var(A_{r,i})$

Using point normalization as described above (mean of 5 measured values, $n_p=5$) and simplifying $var(A_{r,i})=var(A_{s,i})=var(A)$ $var(d_i)$ is calculated as (19).

(18) $\quad var(N_p) = n_p\, var(A)\, [\dfrac{1}{\bar{a}_r^2} + \dfrac{\bar{a}_s^{-2}}{\bar{a}_r^{-4}}]$

(19) $\quad var(d_i) = var(A)\, [1 + n_p(\dfrac{1}{\bar{a}_r^{-2}} + \dfrac{\bar{a}_s^{-2}}{\bar{a}_r^{-4}})A_{r,i} + \dfrac{\bar{a}_s^{-2}}{\bar{a}_r^{-2}}]$

The variance of absorptivity var(A) is caused by different sources of error. The pathlength of the used cuvettes is 1.000 ±0.002cm e.g. if

A=1 the error is 0.002 or 0.2%. This error will cause influence to the normalization factor N_p if different cuvettes are used for sample and reference or if reference spectra are taken from a spectral database. Another source of error is the positioning of the cuvettes within the ligthbeams of the spectrometer. A deviation of about 3° leads to an error of 0.1%. Also multiple reflexion leads to further errors. From experiments the error was determined.

(20) $\quad \text{sdv}(A) = \sqrt{\text{var}(A)} \sim 0.002$

In order to decrease errors sample and reference spectra should have similar concentrations. If not and also one solution has a high absorptivity error caused by false light the results are biased.

Rank Analysis

The reference spectrum is determined by the measured Absorptivitis $A_{r,i}$ and the concentration c_r. The vector $\mathbf{a}_r$ may be expressed as (20) or (20a).

(20) $\quad \mathbf{a}_r = |A_1, A_2, A_3 \ldots A_n|_r$

(20a) $\quad \mathbf{a}_r = c_r |\epsilon_1, \epsilon_2, \epsilon_3 \ldots \epsilon_n|$

With the same denotation the sample spectrum is given by (21),(21a).

(21) $\quad \mathbf{a}_s = |A_1, A_2, A_3 \ldots A_n|_s$

(21a) $\quad \mathbf{a}_s = c_s |\epsilon_1, \epsilon_2, \epsilon_3 \ldots \epsilon_n|$

Combining both spectra to a matrix **A**, a matrix with two rows and n columns is obtained.

Two vectors (for instance $\mathbf{a}_r$ and $\mathbf{a}_s$) are linear not independent if any suitable linear combination of both vectors has a socalled nullvector as a result. The rank of a matrix (for instance **A**) is defined as the maximum number of linear independent row- or column vectors [9] (22).

(22) $R_m(A) < \min(r,c)$

R_m rank of the matrix
r number of rows
c number of columns

If $R_m(A)=1$ then both row vectors are linear dependent, i.e. both spectra are identical spectra. This is also valid for nonequal concentrations. In this case the linear relationship is defined by the normalization factor N_p. If the spectra are belonging to different substances with different spectral features the result of rank analysis will be $R_m(A)=2$. If there is an impurity in the sample spectrum $R(A)$ will differ from the value $R_m(A)=1$, i.e. $R_m(A)>1$. Therefore it would be possible to use rank analysis for detection of impurities if these have different spectral features.

The result of the multiplication of the original two row matrix **A** with its transposed is a quadratic and symmetric Matrix **B** (23) with two rows and two columns. In terms of absorptivities this matrix is represented by (24), in terms of elements of the matrix eq.(25) is valid.

(23) $B = (A\, A^T)$

(24) $B = \begin{vmatrix} \Sigma A_{r,i}^2 & \Sigma A_{r,i} A_{s,i} \\ \Sigma A_{r,i} A_{s,i} & \Sigma A_{s,i}^2 \end{vmatrix}$

(25) $B = \begin{vmatrix} b_{1,1} & b_{1,2} \\ b_{1,2} & b_{2,2} \end{vmatrix}$

The rank of the matrix **B** has to be the same as the rank of **A** (26)[9].

(26) $R_m(B) = R_m(A)$

If the determinant of **B** equals zero then the rank of **B** equals one, i.e. the spectra are identically.

(27) $\det(B) = 0 \longrightarrow R_m(B) = 1$

According to the rules of matrix algebra it is evident that the determinant of a matrix equals the product of the eigenvalues. Therefore the eigenvalues have to be calculated. Because all measured data have small errors, all eigenvalues will be larger than zero. Most of the described methods in rank analysis have been developed for use in factor analysis in order to reduce size of very large data sets, to build up similarity models or dimensionality reduction [10],[11]. In case of detection of impurities by means of UV spectra statistic criteria have to be choosen to determine the rank of the matrix A.

According to Shrager and Hendler [12] eq.(28) is used. When comparing two spectra this equation is simplified to (28a) where E_1 and E_2 are the calculated eigenvalues of matrix B and n is the number of wavelengths. The variance var(A) is tanken from experiments (20).

(28) $\sum_{j=r+1}^{m} E_j < n\, m\, \text{var}(A) < \sum_{j=r}^{m} E_j$

(28a) $E_1 < 2n\, \text{var}(A) < (E_1 + E_2)$

If eq.(28a) is fulfilled the rank of the matrix is one, meaning the spectrum of the sample does not differ from that of the reference. No impurity is detected.

EXAMPLES OF PURITY CONTROL

All solutions of standards and samples had been prepared freshly. All measurements had been repeated by other persons to make quite sure the reproducibility. Quite the same concentrations of standards and samples had been taken. Measurements had been done using Hewlett-Packard photodiode spectrometer HP8450. The used program is written in HP-BASIC and using HP computers series 200 or 300.

Metronidazol

Metronidazol contains from the path of synthesis 2-Methyl-5-nitroimidazol (MNI) as impurity. The amount or content is restricted by several pharmacopoeas (DAB 9: 0.3%). The spectra of metronidazol and MNI are similar but not identic.

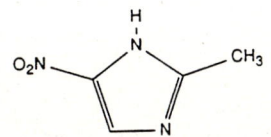

a) Metronidazol   b) 2-Methyl-5-nitroimidazol

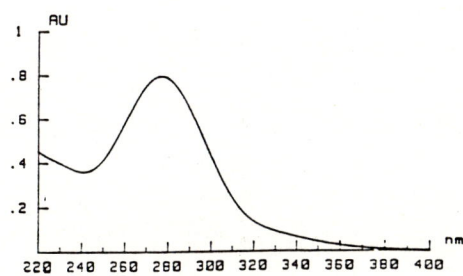 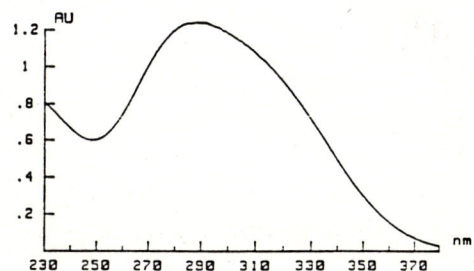

Fig.1:
UV-spectra of metronidazol (a) and the impurity MNI (b).
Solvent: HCl (c = 0.1 mol/L); concentration metronidazol
c = 1.2 E-4 mol/L

The results of both methods are presented in table 1 with the results of the nonparametric tests and rank analysis.

Table 1:
Result of the detection of impurity MNI in metronidazol
Concentration of metronidazol: c = 1.25 E^{-4} mol/L

content of impurity	Detection of impurity	
	Nonparametric tests	rank analysis
0.1%	−	+
0.2%	−	−
0.3%	+	+
0.4%	+	+
0.5%	+	+

Diclofenac

A possible impurity of diclofenac or diclofenac sodium salt is 2-indolinon as a byproduct from synthesis. Routine analysis of indolinon is mostly done by TLC. In this example the spectra differ more than in the case of metronidazol.

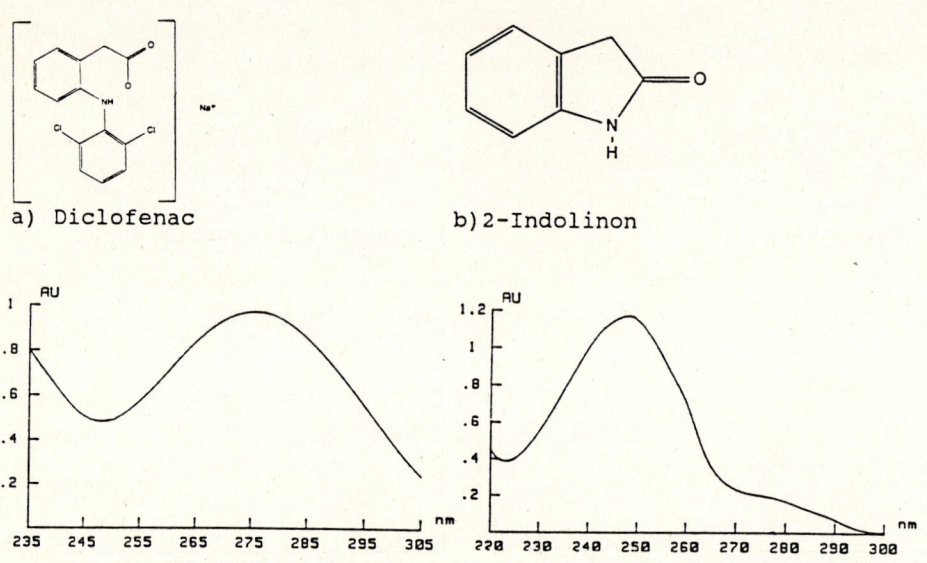

a) Diclofenac   b) 2-Indolinon

Fig.2:
UV spectra of diclofenac (a) and indolinon (b).
Solvent: borate buffer pH = 9.6; concentration of diclofenac:
c = 9.03 E-3 mol/L

In this case it is possible to detect the impurity at a 0.2% level with both methods.

Table 2:
Result of detection of indolinon in diclofenac
concentration of diclofenac: c = 9.03 E-3 mol/L

content of impurity	Detection of impurity	
	Nonparametric tests	rank analysis
0.1%	−	−
0.2%	+	+
0.3%	+	+
0.4%	+	+
0.5%	+	+
0.6%	+	+
0.7%	+	+
0.8%	+	+

CONCLUSION

A method to detect impurities by comparing the UV-spectra of a pure substance as a reference with the sample is described. Although there is no chromatographic separation procedure, the method will give reliable results. The advantages of the described method are that it is less time consuming and that it causes lower costs than corresponding methods in HPLC. The method was tested with two substances with known impurities. Only the results are reported in this paper. The numerical values and further informations shall be published elsewhere. The method does also work with unknown impurities if they have different spectral features compared to the substance to be analyzed.

Acknoledegment: This work was sponsored by Fonds der Chemischen Industrie.

References

[1] Windmann S (1990) Ph D Thesis, Würzburg

[2] Mück W (1987) Datenauswertung in der HPLC-UV-VIS-Kopplung. Ph D Thesis, Würzburg

[3] Ebel S, Mück W (1988) Fresenius Z Anal Chem 331:359

[4] Kang JS (1990) Beiträge zur Analytik von Antibiotika mit Hilfe der Dünnschichtchromatographie. Ph D Thesis, Würzburg

[5] Weitkamp H, Wortig D (1977) Mikrochim Acta II:315

[6] Sachs L (1978) Angewandte Statistik. Springer Verlag, Berlin

[7] McCornack RL (1965) J Amer Assoc 60:864

[8] Dixon WJ, Mood AM (1946) J Amer Statist Assoc 41:557-566

[9] Lipschutz S (1986) Lineare Algebra, McGraw-Hill Book Company, Hamburg - New York

[10] Mellinger M (1987) Chemom Intell Laborat Systems 2:29-36

[11] Wold S, Esbensen K, Geladi P (1987) Chemom Intell Laborat Systems 2:37-52

[12] Shrager RI, Hendler RW (1982) Anal Chem 54:1147-1152

DSYM-PC A NOVEL PROGRAM TO SIMULATE HR-NMR SPECTRA FOR SPINS I=1/2 ON IBM-COMPATIBLE COMPUTERS OF PC/XT/AT TYPE

S. Goudetsidis and G. Hägele

HEINRICH-HEINE-Universität Düsseldorf
Institut für Anorganische Chemie und Strukturchemie
Universitätsstr. 1, D-4000 Düsseldorf 1, GFR

Abstract: DSYM-PC is a novel program to simulate HR-NMR-spectra. Modern menue-driven techniques are used to present screen- and graphical output on HP-Laser Printer in HPGL-standard. DSYM-PC is designed for research and teaching.

PROGRAM DESCRIPTION

Molecular structures in liquid phases are deduced via interpretation of NMR-parameters, accessible by precision analysis of NMR-spectra. A successful spectral analysis leads to a satisfying correspondence of experimental and calculated NMR-spectra in final simulation tests. In addition to an inherent scientific relevance the analysis and simulation of NMR-spectra represents a considerable economic value in industrial laboratories.

DSYMPLOT, a simulation-program, developed by our group for main-frame computers, handles efficiently NMR spectra of practical significance. The centre of DSYMPLOT represents a spectral generator for systems consisting of spins with I=1/2 for isotropic and anisotropic phases. Chemical equivalence is delt with in the most general way. Since DSYMPLOT is not limited to two-fold symmetry operations only (as precursors were) it can be used for more complicated spin systems involving higher symmetry as well. Up to now this program was used on-line on mainframe computers and work-stations. Very recently a version for remote-job-entry handling was introduced to the DFN-user group.

Spectral simulators known up to now - including our DSYMPLOT - made specific demands for hardware facilities, practically limiting these programs to larger computer systems not easily accessible to many potential users. In addition the mainframe version of our DSYMPLOT required some knowledge in group-theory, to set up manual input of symmetric operations. This might be regarded as some kind of disadvantage for speedy NMR studies. The art of programming as used in those previous years neclected ideas concerning comfortable environments dedicated to easy handling of computer-systems and programs. As a consequence quite a number of potential users refrained from more intricate program types.

Within a few years an astoundingly fast development of personal-computers lead into a wide field of applications, previously reserved as domains for mainframe computers only. Comfortable environment and increasing performance guided a growing number of users to search for specific solutions of various scientific problems based on personal computers.

We transformed the precursor mainframe computer program DSYMPLOT into DSYM-PC, a modern and comfortable software-product based on novel PC-supported concepts. Consequently DSYM-PC was developed aiming towards personal-computers of the new generation. Particular attention was paid to graphics leading to convincing presentations, speedy handling and a qualified comfort. DSYM-PC consists of our DSYM-generator mow modified for personal-computers, screen graphics, plotter and printer facilities, accessible in a powerful and comfortable menue technique. Menues provide a wide range of possibilities to solve NMR-simulation problems supporting users with modern window-techniques in a fast and efficient way. Visual control is main-tained throughout all steps of the simulation-process. We wish to point out explicitly, that DSYM-PC does NOT require the input of symmetry operations - as DSYMPLOT did. It is sufficient to supply the name of the appropriate symmetry group to obtain an automatic assignment of symmetry operators. Thus valuable stimulations from former DSYMPLOT-users are verified by now.

SUMMARY OF RELEVANT FEATURES FROM DSYM-PC

The following features mark the actual version of DSYM-PC: Fully menue-directed user-surface and comfortable input-masks. Users may select and change the input parameters supported by input-fields of screen-masks. Numerical output of input-parameters, transition frequencies and intensities are given on screen and printer. Further advantages are: Screen graphics of simulated spectra, zooming with mouse, specific choice of plot-regions, scales, titles, colors. Fast hardcopies from the graphical screen are obtained on IBM- or EPSON-compatible printers. In near future 24-pinwriters will be added. The actual international graphical norm with HPGL-code is implemented, thus comfortable plot-output is accessible via all HP-compatible devices. HPLG-plotfiles may be used for various purposes, e. g. in combination with text-programs like WORD etc..

HARDWARE REQUIREMENTS FOR DSYM-PC

We recommend the following hardware configuration: IBM compatible PC/XT/AT, 640 kByte RAM, graphic-adapter (Hercules, EGA or VGA), mouse (Microsoft or compatible), hard-disc (20 MByte or more), printers or plotters. A HP-Laserprinter will suffice for numerical and graphical output as well.

IGOR2: A PROGRAM SYSTEM FOR GENERATING CHEMICAL REACTIONS AND STRUCTURES.

J. Bauer, Technical University of Munich,
Lichtenbergstraße 4, D-8046 Garching, FRG

Abstract: IGOR2 is a computer program that generates chemical reactions and structures. The development of IGOR2 took place 1986-1989 in several steps. The present status is a PC version with a menu guided user surface and graphic output.

IGOR means   I nteractive
                 G eneration of
                 O rganic
                 R eactions

THE PHILOSOPHIE OF IGOR2 [1]

When molecules are represented by graphs, and reactions are defined as mutual transformations of graphs[2], the following classification of reactions results:

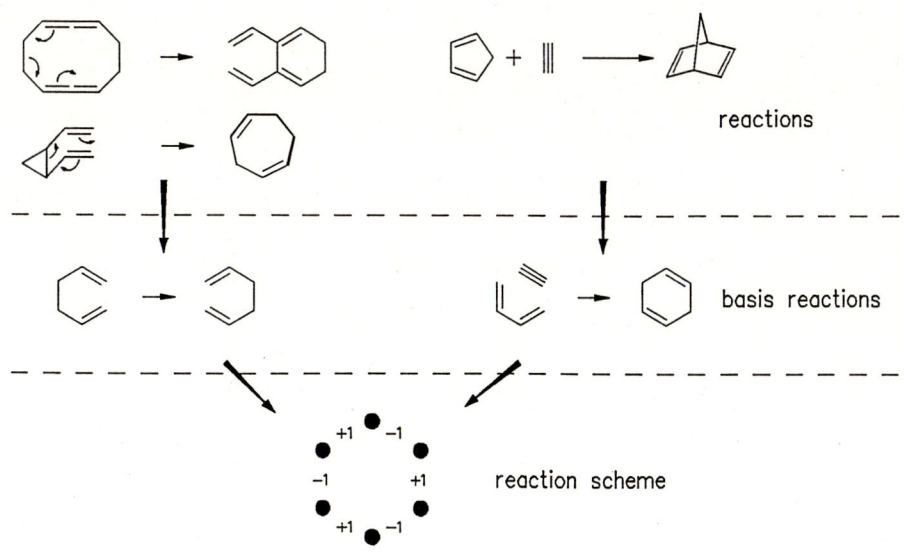

BASIS REACTIONS, REACTION SCHEMES

The first step in the above classification is a mapping of the set of chemical reactions into the set of "basis reactions", i.e. representations of reactive centers whose individual chemical nature is neglected. In a second step the latter set is mapped into the set of reaction schemes. The redistribution of valence electrones during a chemical reaction, represented in terms of changing covalent bond orders between the reactive centers and the numbers of lone electrons of these.

Both of the aforementioned mappings are surjective, i.e. any reaction can be mapped into precisely one basis reaction, and any basis reaction is mapped into precisely one reaction scheme, but the inverse is not necessary true. Accordingly each basis reaction belongs to a corresponding set of reactions, and each reaction scheme belongs to a set of basis reactions. This defines the following jobs:

1. Generate for a given reaction scheme its set of conceivable basis reactions.
2. Generate for a given basis reaction its set of conceivable individual reactions.
3. Generate for a given reaction scheme its conceivable reactions.

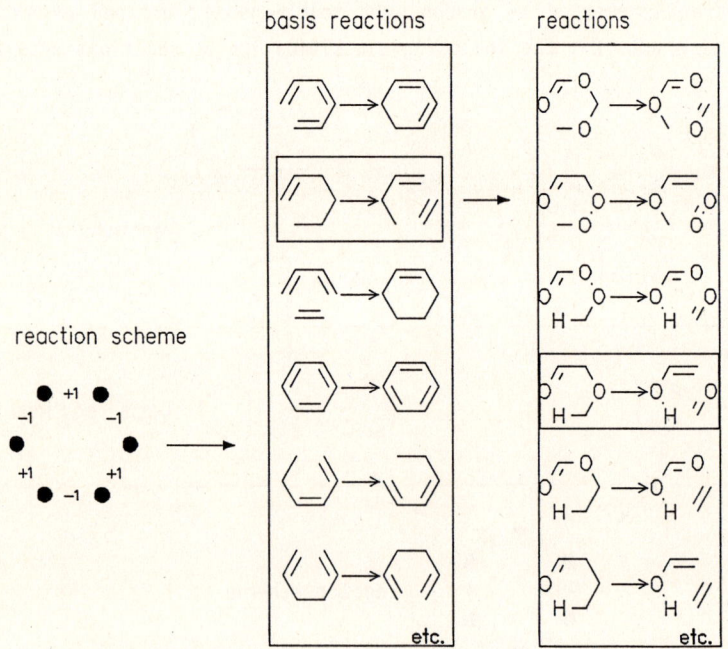

Just these jobs can be executed by IGOR2. The program generates complete sets of the aforementioned type under given boundary conditions.

BONDARY CONDITIONS

The term boundary conditions is now explained in detail.

The only generally imposed restiction is that zero, one, two or three covalent bonds exist between any two atoms.

It is necessary to formulate chemical concept and facts in order to imbed the whole program into chemistry. An adequate representation of the chemical elements and their essential properties is particularly important.
In IGOR2 the chemical elements and their behaviour are described by the so-called "transition tables" that are based on the concept of the valence schemes. A valence scheme indicates one configuration of covalent bonds lone valence electrons and formal charge of a given element.
For example, the user can define the valence chemical properties of the N-atom with the following valence schemes:

$$-\overset{|}{\underset{|}{N}}^{+}- \quad =\overset{/}{\underset{\backslash}{N}}^{+} \quad \equiv N^{+}- \quad -\overset{/}{\underset{\backslash}{N}} \quad =\bar{N}- \quad \equiv NI \quad -\underline{\bar{N}}- \quad =\underline{\bar{N}}^{-}$$

In order to describe a chemical element, a user of IGOR2 defines, first, for a given chemical element the set of all allowable valence schemes in the educts of the chemical reactions, as well as for their products. These valence schemes are now arranged in a matrix where the rows belong to the educts, and the columns are assigned to the products. The entries of the matrix are logical symbols that indicate whether a transition of an educt valence scheme into a product valence scheme is allowable. Such a matrix represents the reaction behaviour of the respective chemical element and is called a transition table.

Transition table of the C-atom.

Carbens are defined as instable !

C	+	=<	==	=−
−∔−	+	+	+	−
=ı	+	+	+	+
+	+	+	+	−
=<	+	+	+	+
>=	+	+	+	+
=−	+	+	+	+

A transition table needs neither be symmetrical, nor need it be square. Thereby we impose a "direction" on reactions.

A user of IGOR2 defines now transition tables for all chemical elements to be used in the reaction generating process. The chemical elements thus described may also be "Macros" (e.g. COOH), or even chemical elements with fictitious properties.

If the user has put in an initial reaction scheme, or a basis reaction, as well as some transition tables that suit the problem, and thus has defined the "chemical horizon" of the upcoming run, then IGOR2 is equipped to generate all of the chemical reactions that are conceivable under the input conditions. Generally, one finds that the set of potential results will be excessively large, unless further restrictions are imposed. Accordingly, IGOR2 offers a variety of further optional limitations, in order to reduce the set of results in the most meaningful and problem-oriented way.

It is not necessary to define all of the boundary conditions to be imposed at the beginning of a run. It has been found that an interactive mode of operation is particularly advantaguous, wher the user is alternating between checking results and setting new boundary conditions.

The following types of boundary conditions have been shown to be particularly beneficial:

1. Molecular Topology
   a) Restriction of the number of cyclic structures of some specified ring-size.
   b) Restriction of the number of disjoint molecules in the educts/products.
   c) Predefinition of certain skeletons in the educts/products

2. Distribution of formal electrical charges
   a) Restriction of total electrical charge
   b) Restriction of the number of charged atoms

3. Occurrence of Substructures
   a) Statement of negative/positive lists of substructures

OUTPUT OF THE RESULTS

IGOR2 contains a module that establishes for the generated molecules 2D coordinates, as a basis for graphic output of the results. The results can also be stored, such that search in data bases is possible.

IGOR2 AS A STRUCTURE GENERATOR

IGOR2 can also be used for generating molecular structures. Here, the same restrictions and options can be used as in reaction generating. One of the application is generating all structures that belong to a given empirical formula.

CONCLUSION

IGOR2 does not elucidate reaction pathways, nor does it elaborate synthesis of a given target. Thus IGOR2 differs from the other chemical computer programs like synthesis design programs. The objective of IGOR2 is to find new reactions and new molecular structures. So IGOR2 is a tool that provides an experienced chemist with some scientific imagination for exploring new fields in chemistry in some kind of computer guided brainstorming process.

ACKNOWLEDGEMENT

We gratefully acknowledge the financial support of this work by the "Fonds der Chemischen Industrie".

REFERENCES

1. Bauer J.; Herges R.; Fontain E.; Ugi I. "IGOR and Computer assisted Innovation in Chemistry". Chimia 1985, 39, 43/53
2. Dugundji J.; Ugi,I. " An Algebraic Model of Constitutional Chemistry as a Basis for Chemical Computer Programs". Top. Curr. Chem. 1973, 39, 19-64.

REACTION DATABASES IN A UNIVERSITY CHEMISTRY DEPARTMENT - ONLINE OR IN-HOUSE ?

Engelbert Zass

Laboratorium für organische Chemie
ETH Zürich
Universitätstr. 16
CH-8092 Zürich

Chemical reactions [1] are described by chemists in a variety of ways, with graphical descriptions using structural formulae dominating. Figure 1 shows the information elements commonly used in such descriptions (graphics enclosed in boxes): the main elements are **chemical structures with their roles**. Reaction information sources should permit queries for any of the elements given in fig. 1, singly or in combination, because questions addressed to them show a great variety which cannot really be accomodated by anything less.

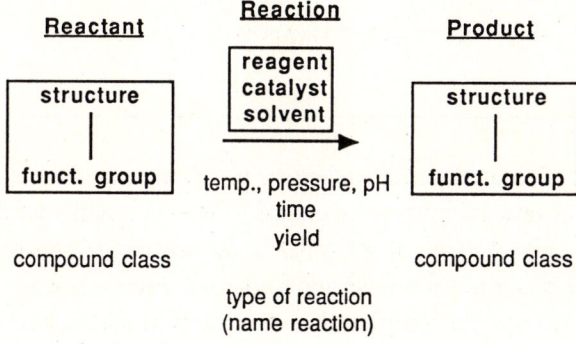

Fig. 1

Chemical Abstracts (CA) in its traditional printed form or as a database is probably the most important single source of chemical information. Therefore, an analysis of its capability to solve queries for chemical reactions [2] must preceed a discussion of specialized reaction databases (the many printed sources, still very important for chemical reaction information, will not be discussed here). Although concepts are indexed, too, *Chemical Abstracts* has a compound-centered indexing policy. This by itself does not disqualify it as a reaction database proper, because reactions can be described fundamentally by compounds and roles. But *CA* lacks precise, reproducible (in the search process) role indicators, and attempts to circumvent this by use of index terms often show deficiencies like the ones in the example in table 1, based on a search in the *STN CA File*.

Table 1
Reaction Indexing in CA
Homologation of Benzyl Alcohol

benzyl alcohol → CO (catal.) → phenethyl alcohol
[100-51-6]   [60-12-8]

References retrieved	total	relevant
registry no.[reactant] (L) trivial name [product]	2	1
registry no.[product] (L) trivial name [reactant]	5	3
indexed with registry no.[reagent]	0	
trivial name [reagent]	1	1
Subtotal	5	3
reaction type (L) registry no.[reactant]	5	4
registry no.[product]	4	4
Total	9	5

This illustrates that even products and reactants do not always get indexed in relation to each other, i.e., *CAS* registry number (database)/*CAS* systematic name (printed index) of the one with the trivial name of the other in the text modification phrase, and that reagents, solvents etc. are not routinely indexed at all unless they are new or find new applications. **Preparation of a compound**, a special and important case of a reaction, can be searched online in *CA* via *CAS* registry number with a P suffix in most host implementations. Analysis of examples showed large discrepancies in the results and therefore in the different (unpublished) algorithms used by the hosts (table 2). The number of irrelevant citations can be large in some cases, as in the last one in *STN*: of the 276 citations for "preparation" of nitrate, more than 100 each were either about (non-preparative) formation or about leaching! This basically simple route to preparative literature thus leaves something to be desired.

Nevertheless, we remain dependent on some reaction and particular preparation searching in *CA* because of the limits in coverage, both in time and by content, of the more suitable sources shown in table 3 and discussed in the following.

Table 2
Preparation of Compounds
(P Algorithm)

Data-Star	DIALOG	ORBIT	Questel	STN

References found with **[registry number]P** (relevant):

Meldrum's acid [2033-24-1]

23 (13)	14 (13)	not available	12 (11)	13 (12)

Adenine [73-24-5]

231	171	not available	342	167

Nitrate ion [14797-55-8]

123	124	not available	431	276

Table 3
Reaction Databases and Systems

name version company, available since	no. of reactions
online	
CRDS (Chemical Reaction Documentation Service) Derwent, 1975	78'000
CASREACT Chemical Abstracts Service, 1988	740'000 (58'500 documents)
in-house	
REACCS 7.1 (Reaction Access System) Molecular Design Ltd., 1981	200'000
SYNLIB 2.2 (Synthesis Library) Still, Chodosh et al.(Distributed Chemical Graphics Inc.),1983	57'500
ORAC 7.5 (Organic Reaction Access by Computer) Johnson et al. (ORAC Ltd.), 1984	100'000

Being aware of the shortcomings of reaction indexing in the *CA File* and realizing the need for reliable reaction information sources, *CAS* embarked on an ambitious, labour-intensive reaction database project which resulted in the availability of the *CASREACT* database in spring 1988 [3]. Figure 2 shows an online display of one of the 26 relevant reactions (out of a total of 33 reactions in 13 documents) found for **alkali-metal** reduction of (aryl) sulfonamides in liquid ammonia [4]. First step in this search was a substructure query in the *CAS Registry File* for arylsulfonamides; the retrieved 3951 compounds were limited to those 642 actually occuring in *CASREACT*, automatically transferred to this database, searched there with the **role indicator** RCT (ReaCtanT) and linked with the registry numbers for ammonia as solvent and either Li, Na, or K as reagent to give the aforementioned result.

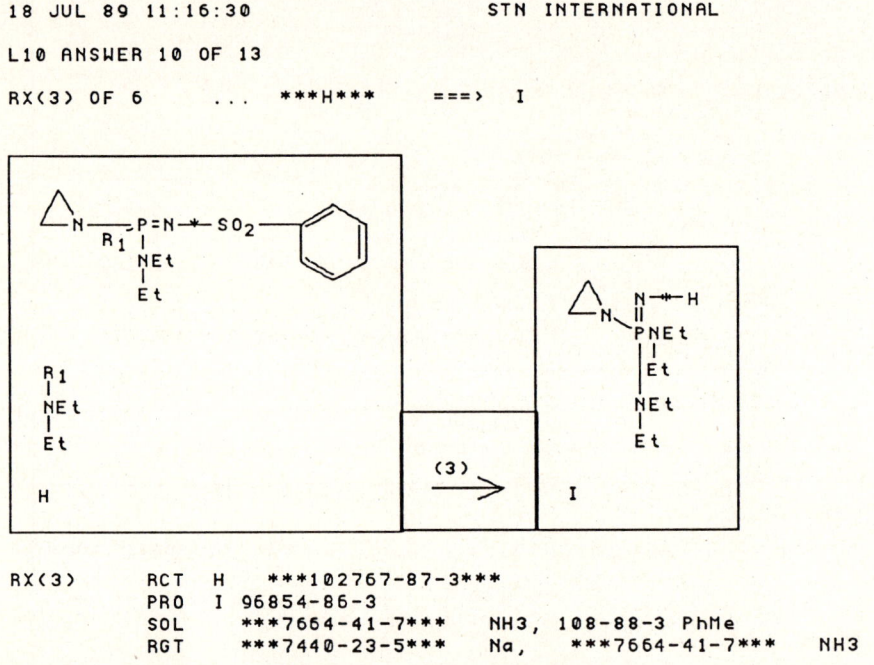

Fig. 2 (© American Chemical Society)

This example shows some important aspects of *CASREACT*: it is not a database with reaction records (cf. table 4), but a literature database with reaction display and enhanced (relative to *CA*) indexing to facilitate search for all reaction partners via the corresponding registry numbers and their roles in the reaction. This implies retrieval of such registry numbers outside *CASREACT*, i.e. in reactions with substructures via one or several expensive substructure searches in the *Registry File*. Because of the limits in this file, very general searches like functional group transformations are often not possible. 'False hits' are not uncommon (and costly to weed out via online display) because reaction centers are displayed and marked with an asterisk, but not yet searchable for enhanced precision in this large and fast growing file. The graphic display (cf. fig. 2) is somewhat inferior to that of the in-house systems shown below, and often continues over several screens. The large variety of display formats, some danger of using one like 'HIT' or 'RX' which may flood you with information without proper warning, and the cost incurred lead often to the display of just the references which then must be checked in a library - not exactly what we expect from a true reaction database, and in our experience to complicated for end-users. But *CASREACT* is the only reaction database with graphic in/output that is publicly available online at present, because the other, older (cf. table 3) public reaction datababase *CRDS* [5] demands

```
        YOU ARE NOW CONNECTED TO THE CRDS DATABASE.
        BIBLIO/KEYWORD DATA AVAILABLE FOR E75751 THRU E76000 (8916);
        MP FROM 75751E THRU 76000E (8916).

        SS 1:   SULFONAMIDE/SM   (223)
        SS 2:   SODIUM)/RT AND 1   (2)
        SS 3:   POTASSIUM)/RT AND 1    (0)
        SS 4:   LITHIUM)/RT AND 1    (0)

        -1-
        AN   - 75356Y HR
        TI   - /INTRAMOLECULAR AMINATION OF ALPHA,BETA-ETHYLENEKETONES VIA
               REDUCTIVE N-DESULFONATION . OCTAHYDROINDOL-4-ONES/ NC-E-S / /
        CI   - Tetrahedron Letters 24, No.6, 551-4 (1983).
        KW   - /ETHYLENE-C/ /SULFONAMIDE/ /CYCLOALKENE/ GIVES *AMINE,CYCLIC* EG
               *PYRROLE* *TETRAHYDRO* *COND-C* *UNFUNCT* *(KETONE,CYCLIC)*
               *CYCLOALKANE* *(CYCLOHEXANE)* *INDOLE* RING-CLOSURE,NC
               METAL-REDN,ALKALI (SODIUM) SOLV=13 (AMMONIA,LIQ) *OXYCARBONYL-3*
               INTRAMOLECULAR 1,4-ADDN (MICHAEL)

        -2-
        AN   - 75014X RY
        TI   - /ETHYLENE DERIVS . FROM BETA,GAMMA-ETHYLENESULFONIC ACID AMIDES
               WITHOUT ALLYL REARRANGEMENT/ HC-X-S / /
        CI   - ANGEW.CHEM.INTERN.ED. 20, NO.12, 1057-58 (1981).
        KW   - /SULFONAMIDE/ /ALLYL-S/ GIVES *CH-ALLYLIC* *(ETHYLENE)* *VINYL*
               SOLV=13 SOLV=4 TEMP=2 (SODIUM) (AMMONIA,LIQ) BASE=5
               METAL-REDN,ALKALI (DIBENZO-18-CROWN-7-POLYETHER) ETHER,CROWN
               PHASE-TRANSFER-CAT


        SS 5:   201/MP AND 207/MP AND 210/MP AND 216/MP AND 64-/MP AND 712/MP
        (4)

        AN   - 77254S R
        TI   - /N-DETOSYLATION . SULFOXIMINES/ HN-X-S / /
        CI   - SYNTHESIS 1977, NO.9, 650-51.
        KW   - /S-IMINE+/ /N-SULFONYL/ GIVES *AMINO,SEC* *S-IMINE*
               *(SULFONE-ANALOG)* *(UNSYM.SUBST)* METAL-REDN,ALKALI DETHIATION
               SOLV=13 (AMMONIA,LIQ) TEMP=2 (SODIUM)

        AN   - 10038I D
        TI   - /WILL FOLLOW SOON/ -- / /
        CI   - WILL FOLLOW SOON.
        KW   - /WILL FOLLOW SOON/

        ELAPSED TIME ON CRDS: 0.19 HRS.
        $19.95 ESTIMATED COST CONNECT TIME.
        $0.00 ESTIMATED COST OFFLINE PRINTS: 0
        $4.25 ESTIMATED COST ONLINE PRINTS: 17
        $24.20 ESTIMATED TOTAL COST THIS CRDS SESSION.
```

Fig. 3 (© Derwent Publications Ltd.)

coded input (controlled descriptors and three-digit codes) and gives only references but no graphic reaction display as shown for the same reaction problem in fig. 3. Again, this is not a system suitable for end-users. The information it contains - the *Theilheimer* handbook and the *Journal of Synthetic Methods* - is also available in-house via *REACSS* (the former now via *ORAC*, too), but only a few universities do at present have access to such systems, and even for those the *JSM* database license fee is probably too high. *CRDS* may therefore be the only source of **intellectually selected, representative** reactions from the primary literature including patents for many chemists, because *CASREACT* is comprehensive within a defined area (reactions from about the 100 most important journals in the *CA* organic chemistry sections) instead of representative.

Representative examples of reactions via intellectual selection, easy-to-use yet powerful user interfaces and in-house availability are the hallmark of the reaction database systems *REACSS*, *ORAC*, and *SYNLIB* (cf. table 3) [6]. The last one has a different 'philosophy' as a 'browsing' system and will not be discussed here.

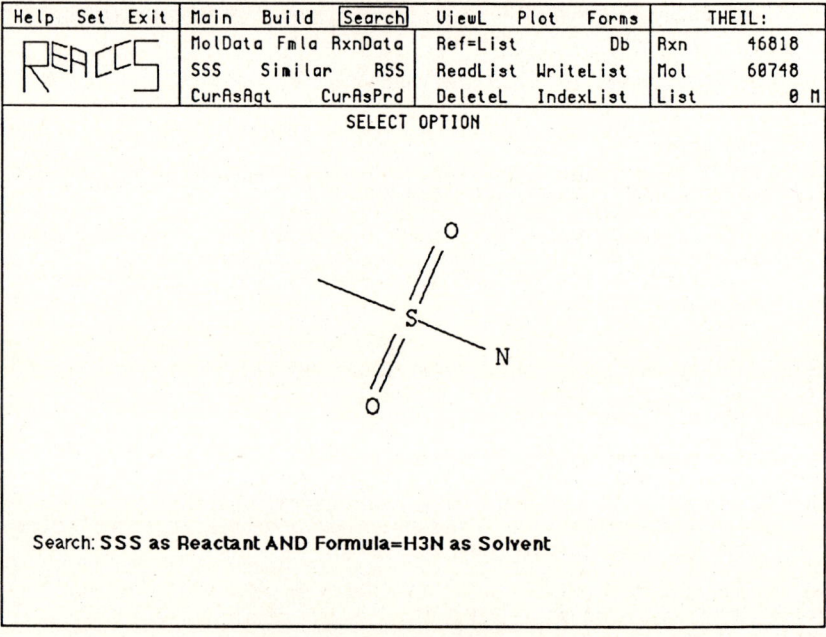

Fig. 4 (© Molecular Design Ltd.)

Figure 4 shows one of several possible queries to answer the question for reduction of sulfonamides in *REACCS*. The combination of menu-drawn structure and command-driven search (keyboard input in the lower part of fig. 4) will most probably be preferred by an experienced user, while a novice could search stepwise and entirely menu-driven to get the same answers; one of four from the *Theilheimer* database is displayed

Fig. 5 (© Karger/Molecular Design Ltd.)

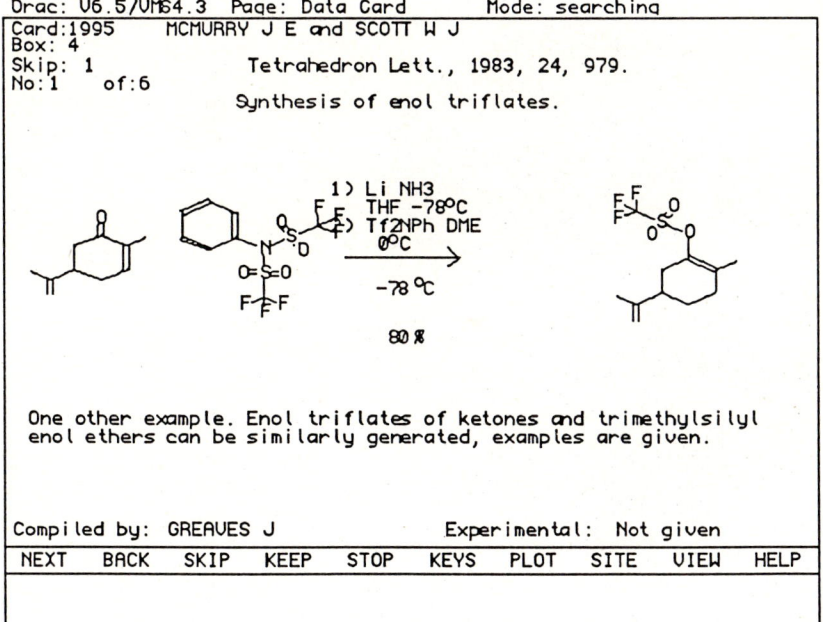

Fig. 6 (© ORAC Ltd.)

in fig. 5. One of the reactions retrieved for the same question in the system *ORAC* is reproduced in fig. 6.

In a **university chemistry department**, reaction searching like all chemical information retrieval must serve two purposes: first, one needs information for ongoing research projects much as in industry; second, chemists must be acquainted with modern methods of information retrieval. In our experience, for both purposes the in-house systems are first choice, because chemists/chemistry students can use them directly by virtue of their user-friendliness (relative to current public online retrieval systems) and their usage-independent costs. That does not imply that one may practically ignore the public databases like *CASREACT* at a university. These are needed, for those problems that do not give satisfactory answers in the in-house systems which still have only a relatively small number of selected reactions, for comprehensive searches, and for demonstrations in chemical information education. At the ETH Zürich chemistry department, more than 500 accesses to *REACCS* alone were registered in less than three months, while in a similar period, only about half a dozen reaction searches were done in the public databases. Figure 7 illustrates this with a breakdown by type of question for the 929 online searches (excluding in-house systems) executed in the chemistry library from January until August 1989.

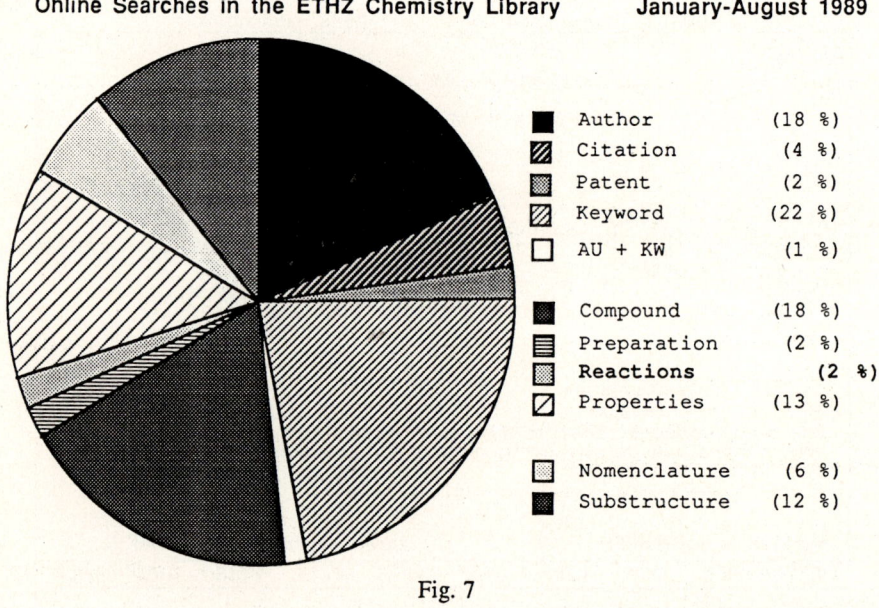

Fig. 7

In-house reaction databases are useful for a third purpose, namely, **education in chemical reactivity**. Students can search examples for simple transformations like the one depicted in *REACCS* in fig. 8 and learn chemistry and database searching at the same time.

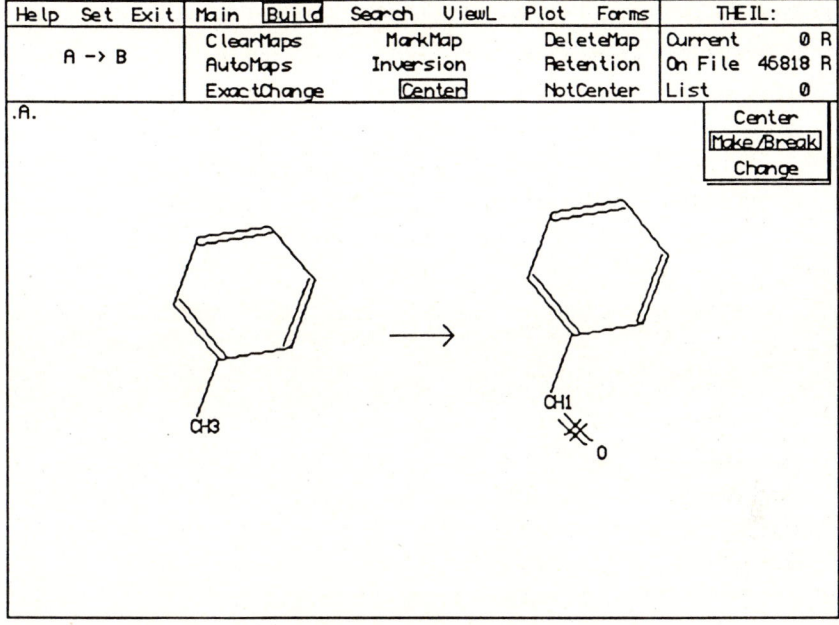

Fig. 8 (© Molecular Design Ltd.)

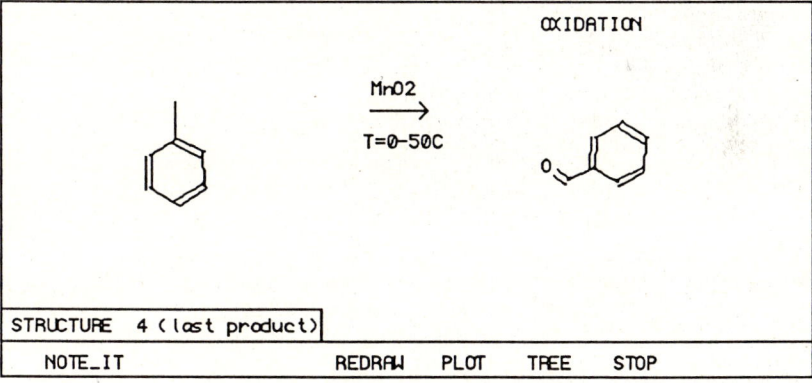

Fig. 9 (© Purdue Research Foundation)

Such an 'educational' search might be supplemented or preceeded by **reaction simulation** in *CAMEO* (Computer-Assisted Mechanistic Evaluation of Organic reactions) developed by *Jorgensen et al.* [7]. Utilization of this menu-driven program is very similar to that of the in-house reaction databases: one draws the starting material with the help of a menu, selects "oxidation/reduction" as one of several basic reaction types from a menu, and continues then by selecting reagents and reaction conditions like temperature, stoichiometry. The result for an oxidation of toluene with MnO_2 is reproduced in fig. 9. Figure 10 shows the complete 'synthesis tree' for running this example when CrO_3 and $KMnO_4$ had been tried before MnO_2; both gave the acid (2=3) and not the aldehyde (4), however.

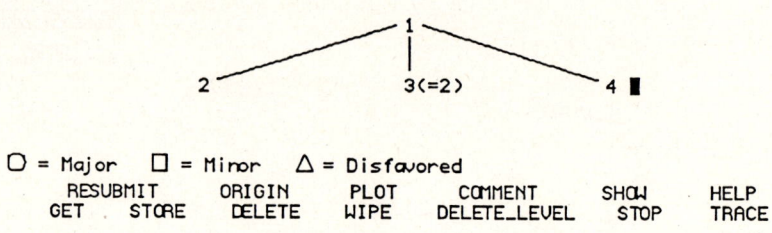

Fig. 10 (© Purdue Research Foundation)

Table 4 Reaction Databases and Systems Summary				
	CRDS	CASREACT	REACCS	ORAC
no. of reactions (x1000)	78	**740**	200	100
coverage since	1942	**1985**	1942 (1869)	(1881)
record type	reaction	**document**	reaction	reaction
search control	command	command	**menu**/command	**menu**
main input	**code**	(graphics)/text	graphics	graphics
main output	**reference**	reaction	reaction	reaction
		---------(plus reference)---------		
search facilities for	[- none, (-) very limited, (+) restricted, + good, ++ excellent]			
compounds	-	++	+	+
substructures	+	(+)	++	+
stereochemistry	-	-	+	+
reaction centers	(+)	-	+	++
reaction conditions	(+)	(-)	++	+
reaction types	++	(-)	+	++

References

1. Willett P (ed) (1986) Modern Approaches to Chemical Reaction Searching. Gower, Aldershot UK
2. Beach AJ, Dabek jr. HF, Hosansky NL (1979) J Chem Inf Comput Sci 19:149
3. Blower PE jr, Chapman SW, Dana RC, Erisman HJ, Hartzler DE (1988) In: Warr WA (ed) Chemical Structures. Springer, Berlin, p 399
4. For a comparison of manual and online searches about this problem, see chs. 1 & 2 in Loewenthal E (in press) A Guide for the Perplexed Organic Experimentalist, 2nd ed, Wiley, London
5. Finch AF (1986) J Chem Inf Comput Sci 26:17
6. Zass E, Müller S (1986) Chimia 40:38
   Borkent JH, Oukes F, Noordik JH (1988) J Chem Inf Comput Sci 28:148
7. Salatin TD, Jorgensen WL (1980) J Org Chem 45:2043
   Paderes GD, Jorgensen WL (1989) J Org Chem 54:2058

Acknowledgments. We thank for the support provided by the respective 'academic programs' for *REACCS* (Molecular Design Ltd. San Leandro U.S.A.), *ORAC* (Dr. A.P. Johnson, ORAC Ltd., Leeds, UK), *CAMEO* (Prof. W.L. Jorgensen, Purdue Univ., West Lafayette, U.S.A.), and *Chemical Abstracts Service* databases.

Towards Synthesis Planning Aids

Through Databank Analysis

Edward Blurock and Thomas Strelow
Research Institute for Symbolic Computation, A-4232 Hagenberg, Austria

Abstract

Since the late sixties Synthesis Planning Programs have been a major field of research in computational chemistry. Usually one finds two opposite approaches: A combinatorial or an empirical approach. The Chemical Synthetic Planning System currently being developed at RISC-LINZ, with a design philosphy based on the powerful methods offered by higher mathematics and artificial intelligence, is of semiempirical nature, and therefore requires the analysis of reaction databanks. The system in its present prototype form is made up of more than 30,000 CommonLISP lines of code. The system is built around the philosophy of flexibility in which, through, for example, its preprocessing stage, the essential chemical information is abstracted from specific information in existing databanks. The ORGSYN© database has been analysed and a pool of transforms automatically generated. The performance of the algorithm is described and examples given. The set of transforms is then used in a Synthesis Planning Aid called "RETROSYN", which is described.

Introduction

Much attention has been focused on the application of computers in chemistry[1]. One of its main uses is the storage and retrieval of vast amounts of chemical information[2]. But its use for more elaborate tasks, is not so widely accepted. Usually the users of a databank have to analyse the information provided. This can be tedious and demanding. It is not the lack of information which makes the use of databases a demanding task, but the lack of organization. Nowadays the chemist often finds himself confronted with the task of finding the proverbial needle in the haystack. This situation will become increasingly worse in the future and the chemist will have to employ advanced techniques to keep the chaos at bay. In this situation the use of artificial intelligence techniques will prove to be of invaluable help[5]. In the course of the development of a synthesis planning aid[9], the chemistry group at RISC-Linz [1] chose to acquire a knowledge base of synthetic Organic Chemistry through the *automatic* analysis of reaction data banks. This knowledge base is one of the vital ingredients of the synthesis planning system, which is of semiempirical nature, combining elements from combinatorial and empirical approaches in this field.

---

[1]The chemistry group consists of people coming from different fields: Quantum chemistry, organic chemistry, mathematics and computer science. Therefore real interdisciplinary work is the Institute's strong point

The Concept

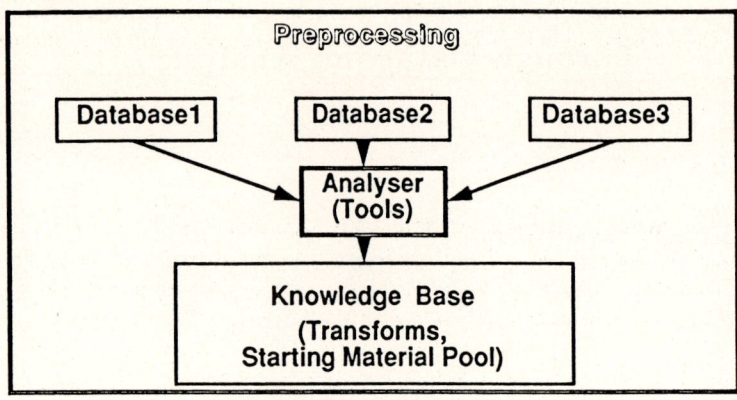

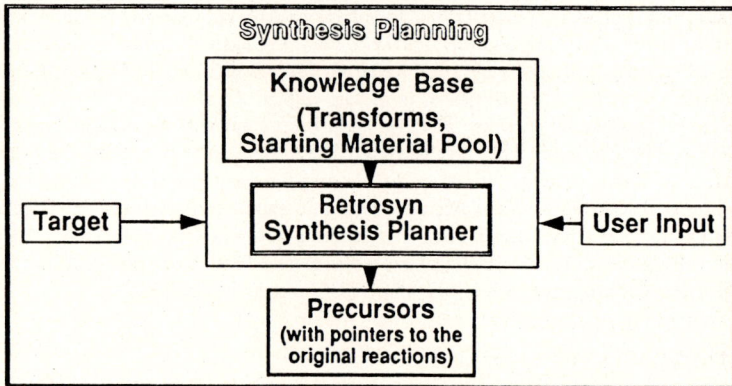

From a set of reaction databases a knowledge base is generated by an analysing module. This is done in a preprocessing step before the synthesis planner is actually used. Therefore the analysis may incorporate time- consuming calculations without affecting the actual planner performance. This together with a suitable interface allows the use of proprietary programs like AMPAC[6], MM2[7] or others.

It has always been a problem for the development of computer programs for chemists, that the terminology used by organic chemists is not suitable for use in computers. Computers do not thrive on a diet of traditional nomenclature and reactions named after a famous chemist . On the other hand organic chemists cannot deal with only numbers. One has to find a representation, which is equally transparent for both sides. This is of vital importance especially in the analysis of huge databanks, which is a tedious task by visual inspection. When expert systems for computer assisted organic synthesis for the retrosynthetic approach demanded the extraction of the transforms from the domain of organic chemistry, it proved to be and still is one of the bottlenecks in the development of these systems. One way out of this dilemma would be the automatic generation of the transforms from reaction databanks , which would dramatically ease the pain of feeding CAOS systems. Unfortunately this seems to be very difficult, because these transforms resemble the way organic chemists think and not the way computers act.

Working on the development of a CAOS system, we decided to follow a semiempirical approach, which involved the analysis of reaction databanks. Due to the ever changing nature of chemistry we were adamant to do it *automatic*. As a result we had to develope computational tools. The importance of the classification of reactions has been of some concern recently[4][5]. The classification system being used at RISC and its *automatic* execution will be published elsewhere.

Automatic Analysis of databanks

The Database

The first database to be analysed was the "ORGSYN" database of the REACCS© system[2]. It was choosen, because it is a representation of basic organic chemistry, without being too large. The analysis started off with a graph-difference calculation of the starting materials and and product graphs[3] . The question was: "What has actually changed during the reaction, regardless of mechanism, catalysts, solvent, temperature and number of steps ?" In this way the information on the reactions could be separated from the random structural features of individual molecules. The resulting difference between the two coloured graphs, we deliberately dubbed "reaction pattern". It may partly resemble some traditional features of reaction representation, but nevertheless one has to keep in mind , that it is an artifact.

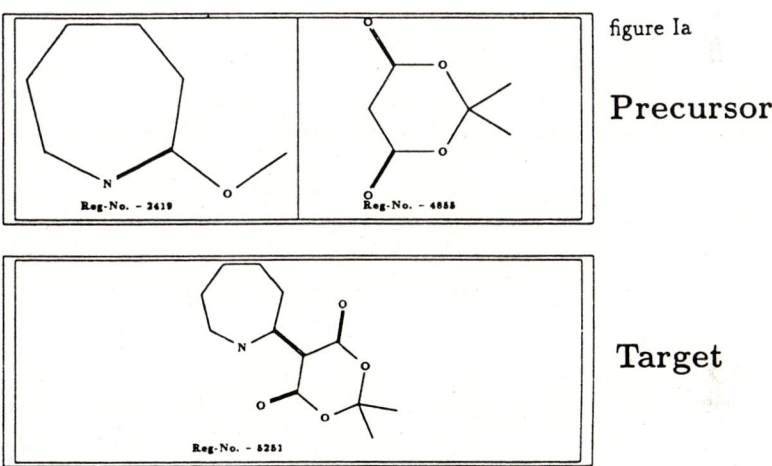

figure Ia

Precursor

Target

The Reaction Pattern

The methods of substructure search, maximum common substructure search and graph differential calculation yield a set of graphs, which we call the reaction pattern. This reaction pattern is the equivalent to a transform, without necessarily representing a single chemical reaction but a more general reaction class. Reactive intermediates are not considered and the reaction patterns tell nothing about the mechanisms of the reactions it stems from. Even multiple step sequences can be treated as one reaction and linked to a reaction pattern. For a start it seemed reasonable

---

[2]MDL Inc. San Leandro CA
[3]A graph is a collection of nodes representing the atoms and edges representing the bonds. A coloured graph has, in our case, atomic information such as atomic number , valence and charge

to document the general feasibility of a reaction in a pattern without considering protection - deprotection or activation-deactivation steps.

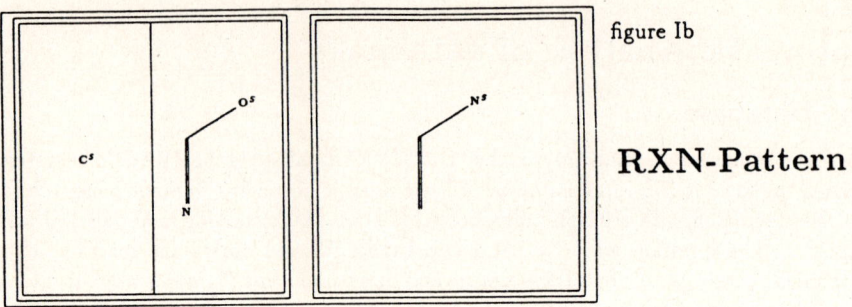

figure Ib

RXN-Pattern

Refinements

It soon became apparent, that the initial classification required some fine structure to represent the reaction class completely i.e. sometimes activating groups do not change during the reaction, but have a strong influence on the applicability of the corresponding pattern. One can describe the atoms in a pattern further by their physical properties or by graph theory. In the case of graph theory one has to expand the pattern in a certain manner. One tries to include the least information necessary to describe the system. Usually these are nodes, which exhibit strong influences on the physical properties of the atoms in the reaction pattern i.e. heteroatoms, double bonds etc.

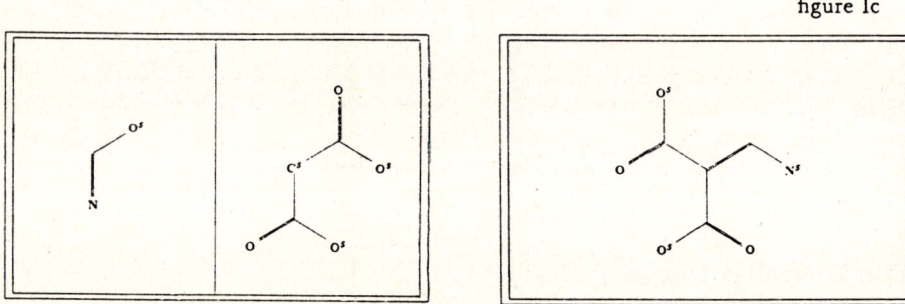

figure Ic

Expanded RXN-pattern

Help from Mathematics

As implied by its name, the Research Institute for Symbolic Computation is devoted to symbolic computation. In the traditional sense computers deal with numbers, which is the area where it is mostly used for. But computers can also be coaxed into working with symbols. This appears to be of increasing importance and many technical applications like computer algebra, expert systems, robotics, CAD/CAM etc. have been found. As a chemist usually thinks and articulates himself in symbols, it seems only natural to use symbolic computation for solving problems in chemistry. It has been shown[3] that the application of graph theory to chemical formulas can lead to perception of structural features. In order to characterize a chemical reaction, we had to devise an effective algorithm, which unequivocally applies graph theory to starting materials and products. The details of the algorithm are published elsewhere[8], but the main ideas are as follows:

- Perform a maximum common subgraph search of reactants and products. This gives a list of correspondences, which is converted into a list of bond changes. The simpliest pattern is choosen as the transform - this is actually the minimum chemical distance principle[11]

- Expand the transform so that important features of the neighbors not undergoing changes in the reaction are included

- Apply heuristics to sort the transforms of the set of given reactants in the order of synthetic significance

Example

An example is given in figure I:

The reaction shown in figure I shows a reaction of meldrum's acid with the iminoether on the left in a knoevenagel-like condensation to give the target compound. Graph difference calculation yields the simple pattern shown below. Clearly the singly bonded carbon atom on the meldrum acid side is a too general description of the situation. If this transform is applied to a different target it would generate many precursors in which the carbon is not activated by electron-withdrawing groups. Hence we should include the groups activating the carbon. This can be done by the calculation of the physical properties involved and thereby colouring the graph or one can use graphical expansion of the pattern in order to include the relevant atoms. The expanded pattern is shown below. The same arguments apply to the example shown in figure II: The methylation of the ring carbon in the precursor will certainly not commence if the substrate carbon is not activated; for instance if it were a member of an aliphatic chain. Expansion of the reaction pattern includes the aromatic system and the nitrogen atom on the other side.

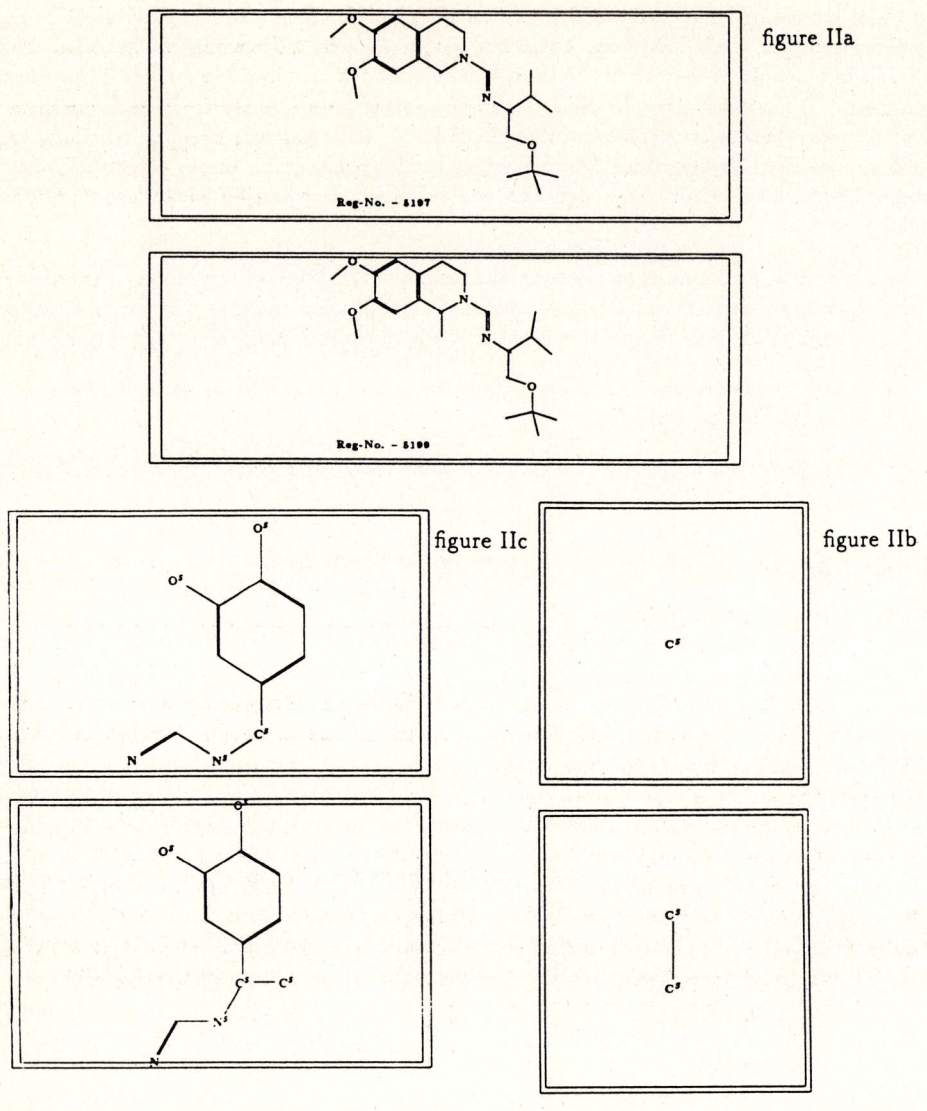

figure IIa

figure IIc

figure IIb

The Knowledge Base

The whole "ORGSYN©" database (5018 reactions) was analysed by the methods mentioned above. The analyses were made on a Vax station 2000; the consumption of CPU-time for the calculation of the simple pattern was as follows:

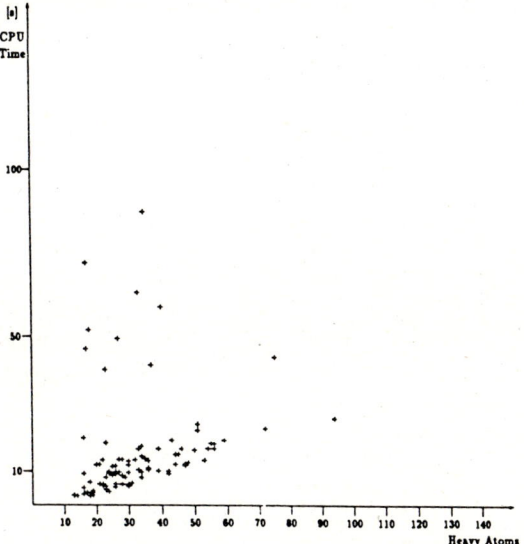

Creation of pattern lies typically between 5 and 20 CPU seconds. The dependence of the CPU-time of the number of heavy atoms is not very pronounced. Roughly one can say that it is a linear (about 0.3s/atom) rather than an exponential dependence. The outliers requiring considerably more time are typically coordination compounds or highly symmetrical compounds. In these cases the algorithm has several possibilities to match atoms from product and precursor i.e. in cooordination compounds the coordination bonds are not designated as such. Therefore the ligands are all alike and not distinguishable by graph theory. Overall time for the top stage analyses of the whole database was about a week on a comparatively slow Vax 2000, which appears to be a reasonable time for a preprocessing step. It cannot be overemphasized, that the incorporation of a preprocessing step makes the system very flexible. Any reaction data bank and updates thereof can be analysed regardless of manufacturer or kind of internal representation used. This enables the user to create knowledge bases according to his needs. This is very important, because it is unreasonable to assume that a system or knowledge base should be able to handle the entire spectrum of chemistry.

Inheritance

In the knowledge bases created the transforms still carry pointers from which individual reactions it emerged. If knowledge bases and databanks can be kept in the same computer, the reactions can be immediately retrieved from the databank. If for storage space reasons the knowledge base has to be decoupled from the databank, the original reactions can nevertheless easily be retrieved by their registry numbers. It is regarded as a central feature that the database is an integral part of the system. In other Synthesis Planning Programs these links are not existent and it's very difficult to establish this later.

The "RETROSYN" program

A synthesis planning aid called "RETROSYN" exists as a prototype program [10] and represents the short term goal of the research project. It consists of about 30,000 lines of Lisp code, which incorporates a user-friendly mouse driven user interface. As the knowledge base currently the transforms extracted from ORGSYN© are used as a trial set. The improvement of this knowledge base is now our primary concern. The operation of the program is roughly as follows:

1. Draw the target molecule on the screen with the help of the mouse

2. Specify the bond to be broken in the target in a retrosynthetic analysis. Thus the user has to provide the strategies involved

3. The program will search its transform base for suitable transforms. If any exist, it will generate the principal pathways and the precursors. These can be viewed by a list manager in the program

4. The items of interest can be printed on a laserprinter (a LaTeX file is generated).

5. In the near future a similarity search procedure will be implemented, which will be able to extract starting materials from a starting material database by comparison with the hypothetical precursors generated.

Conclusion

The software tools created at RISC Linz provide a way of automatically analysing reaction databanks. This lead to a synthesis planning aid, which represents a higher level databank search facility. The possibility to ask higher level questions in databank searches is obviously one way to help the chemist to find his way through databanks. The ultimate goal is a complete synthesis planner, and is the goal of the RISC chemistry group.

References

[1] R.Barone and M.Chanon in G.Vernin and M.Chanon "Computer Aids to Chemistry"; p.19; Ellis Horwood Ltd., Chichester 1986

[2] A.J.Kos and G.Grethe Nachr.Chem.Tech.Lab. **35** 586 (1987)

[3] A.T.Balaban: Chemical Applications of Graph Theory Academic Press 1976

[4] a.)
S.Fujita, J.Chem.Soc.Perkin Trans. II **1988** 597 b.)J.Chem.Inf.Comput.Sci.,1986,**26**,(a)205; (b) 212;(c)224;(d)231;(e)238; 1987,**27**,(f)99;(g)104;(h)111;(i)115;(j)120. c.) Pure Appl. Chem. **61** 605 (1989) d.) Bull.Chem.Soc.Jpn. **61**, 4189 (1988)

[5] C.S.Wilcox and R.A.Levinson Am.Chem.Soc.Symp.Ser.306 (1985) p.209

[6] M.J.S.Dewar, E.G.Zoebisch, E.F.Healy, and J.J.P.Stewart, J.Am.Chem.Soc.,1985,**107**, 3209; QCPE Program No.506, Bloomington, Indiana, USA.

[7] R.B.Nachbar and K.Mislow; QCPE program No. 514 (1986)

[8] E.Blurock und P.Paule RISC Technical Report 1988

[9] a.) E.Blurock, Hierarchial Planning in Automatic Synthetic Chemistry , Software-Entwicklung in der Chemie 2, Springer 1988 b.)E.Blurock and T.Strelow, Automatic Chemical Synthesis Planning, Dechema Monograph vol.116 1989 S.531 c.) E.Blurock und T.Strelow, Ö. Chem.Z. 13 (1989)

[10] E.Blurock und T.Strelow "RETROSYN" Programmbeschreibung 1989

[11] M.Wochner, J.Brandt, A.v.Scholley and I.Ugi, Chimia **42** 217 (1988)

IMPLEMENTATION OF SYNTHESIS STRATEGIES IN PROLOG

M. Wagener and J. Gasteiger

Organisch-chemisches Institut, Technical University München,
D-8046 Garching, West Germany

Abstract: The program STRATOS is presented, that generates synthesis plans for aliphatic compounds. They are made without the help of a reaction database, but solely by examination of the relationships between functional groups in the molecules. Implementation of the program in PROLOG and FORTRAN77 is discussed.

INTRODUCTION

In the design of organic syntheses strategies have to be developed to find pathways from the target structure to available starting materials. These strategies have to involve considerations on the availability and price of compounds, convergence of a synthesis plan, simplification of a structure, feasibility and yield of reactions, etc.

In this paper we will only deal with strategies that are based on evaluations of the feasibility of a reaction. These evaluations have to include considerations of the reaction mechanism from the synthesis precursors to the reaction product. Furthermore, polyfunctional aliphatic compounds will be put in the center of the discussion.

Chemists have learned to recognize certain subunits in molecular structures, which are easily accessible and which can be connected by known reactions. In synthesis planing it is helpful to break down the molecule considered into such fragments in a formal way [1]. For that purpose bonds between perceived subunits are disconnected in a homo- or heterolytic manner. The fragments formed are of purely hypothetic nature. Their only function is to help to analyze the synthesis problem. Such fragments are called synthons [2]. The synthons must then be replaced by reagents for practical use. Fig 1 illustrates this process for finding a synthesis of t-butanol from a grignard reaction with acetone.

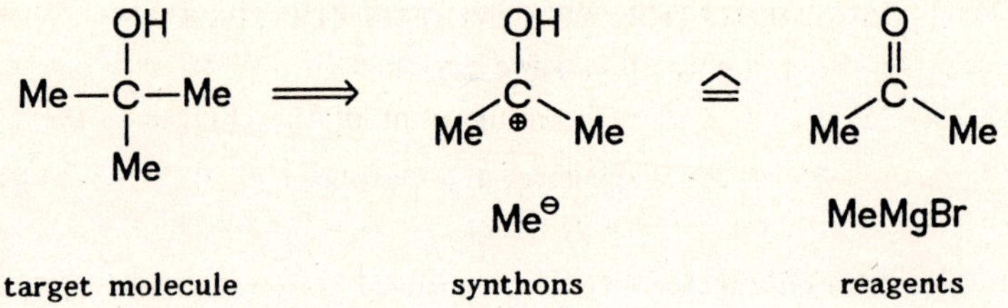

$$\text{Me}-\underset{\underset{\text{Me}}{|}}{\overset{\overset{\text{OH}}{|}}{\text{C}}}-\text{Me} \implies \underset{\text{Me}^{\ominus}}{\underset{\text{Me}^{\oplus}}{\overset{\overset{\text{OH}}{|}}{\text{C}}}\diagdown\text{Me}} \; \hat{=} \; \underset{\text{MeMgBr}}{\overset{\overset{\text{O}}{||}}{\text{C}}\diagdown\text{Me}}$$

target molecule　　　　synthons　　　　reagents

Figure 1: Disconnection of a target molecule into synthons that lead to reagents.

PATTERNS OF REACTIVITY IN ALIPHATIC COMPOUNDS

An important strategy for the synthesis of polyfunctional compounds is to disconnect the molecule between two functional groups. In this way the influence of these functional groups onto the reactivity can be used during synthesis. The largest number of synthetically useful reactions are of polar nature. To find those reactions, the target molecule has to be disconnected heterolytically between the functional groups. This introduces charges in the resulting synthons.

The influence of heteroatoms onto the reactivity in a carbon skeleton can be accounted for by assigning charges to the carbon atoms in an alternating manner. Or, equivalently, the carbon atoms of a chain can be considered as having centers of donor and acceptor properties in alternating sequence (Fig. 2).

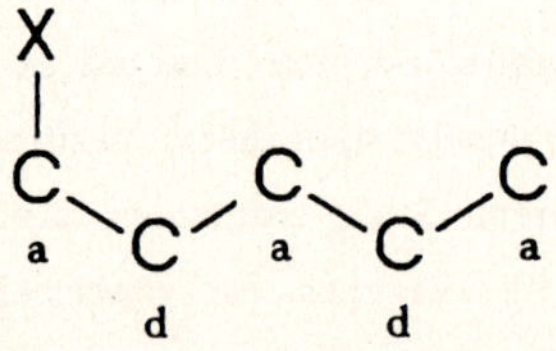

Figure 2: A heteroatom (X) induces an alternating sequence of donor (d) and acceptor (a) centers.

In polyfunctional compounds these sequences emanating from each heteroatom can agree (consonant) or disagree (dissonant) with each other. When disconneting a carbon chain between two functional groups the assignment of the charges in the synthons has to be made in the proper manner. If the positive charge generated in the two synthons is located on an acceptor center and the negative charge is located on a donor center the two functional groups have a consonant relationship. If on the other hand a charge can only be placed on a center of opposite reactivity, the relationship between the two functional groups is called dissonant.

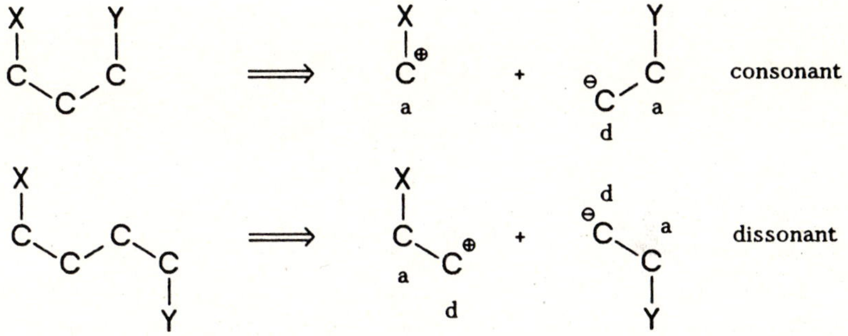

Figure 3: Consonant and dissonant relationships.

IMPLEMENTATION IN THE PROGRAM STRATOS

The principle of consonant and dissonant relationships is used in the program STRATOS to generate synthesis proposals. The program proceeds in four main steps:

a) Search for paths between functional groups in the target molecule; heteroatoms and double bonds are considered as functional groups.

b) Rate the paths found.

c) Generate the synthons based on these ratings.

d) Form the reagents corresponding to these synthons.

Paths with three or five carbon atoms in the chain between two functional groups are classified as consonant. The strategies dealing with them determine a bond in the path for heterolytic disconnection. To decide, which bond in the path should be disconnected, resonance stabilization of charges calculated according to Saller and Gasteiger [3] is considered. That bond will be disconnected, that leads, when it is broken, to the best stabilization of the resulting charges. This defines the direction of the disconnection, too: from the two posibilities that one is chosen, that leads to the better stabilization.

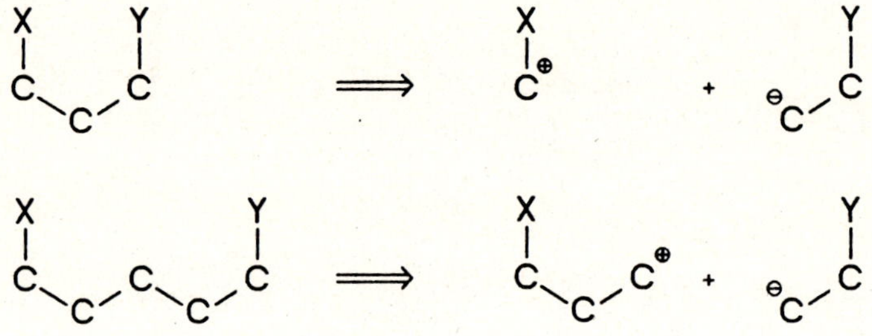

Figure 4: Examples for the disconnection of three- and five-membered paths.

Reagents corresponding to the synthons must be found in a further step. Two approaches for the generation of reagents from positively charged synthons are implemented in STRATOS. The first method is based on elimination of a proton from the center adjacent to the positive charge. The second method involves the addition of an ethoxide anion to the carbon atom that bears the positive charge. Analysis of test cases demonstrated that the partial charge at the electrophilic center in the target structure, calculated according to Marsili and Gasteiger [4], can be used to decide between the two methods. If the partial charge in the molecule not yet disconnected is above 0.1 e, ethoxide is added to the synthon. In the other case, i.e. if the charge is below 0.1 e, the reagent is obtained by elimination of a proton. For negatively charged synthons a proton is added to the center with the negative charge.

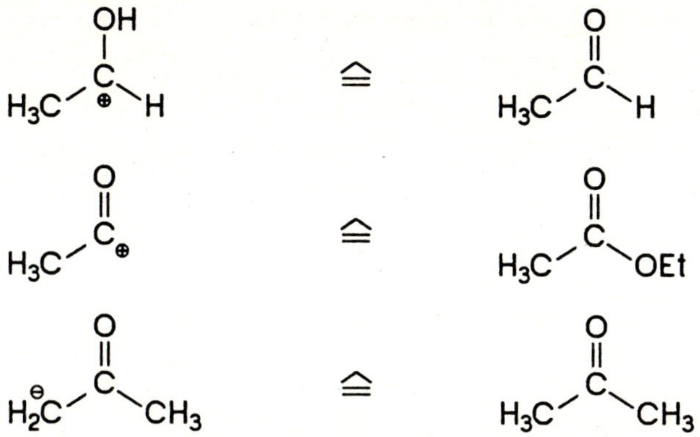

Figure 5: Deriving reagents from the synthons.

Only one tactics is implemented for double bonds: Water is added in the retrosynthetic direction (functional group interchange). This corresponds to an elimination of water during synthesis. The hydroxy group introduced in this way can function as the starting point for other important paths. The addition has to take place in accordance with the natural reactivity, i.e. the hydroxy group has to be added to a center with acceptor properties.

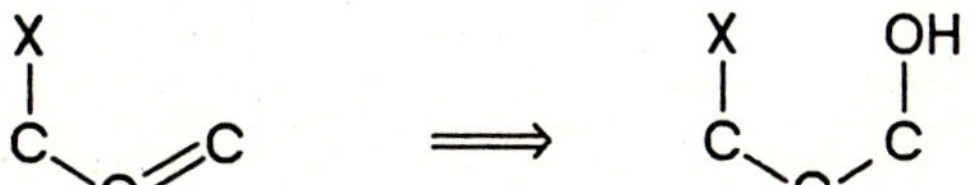

Figure 6: Addition of water to a double bond at the acceptor center.

Strategies that handle dissonant relationships have only been touched in STRATOS. Dissonant relationships ask for 'umpolung' at a donor or acceptor center. This can be accomplished by various groups corresponding to different reagents. Some of these methods have been incorporated into STRATOS. Further work on systematization is however necessary.

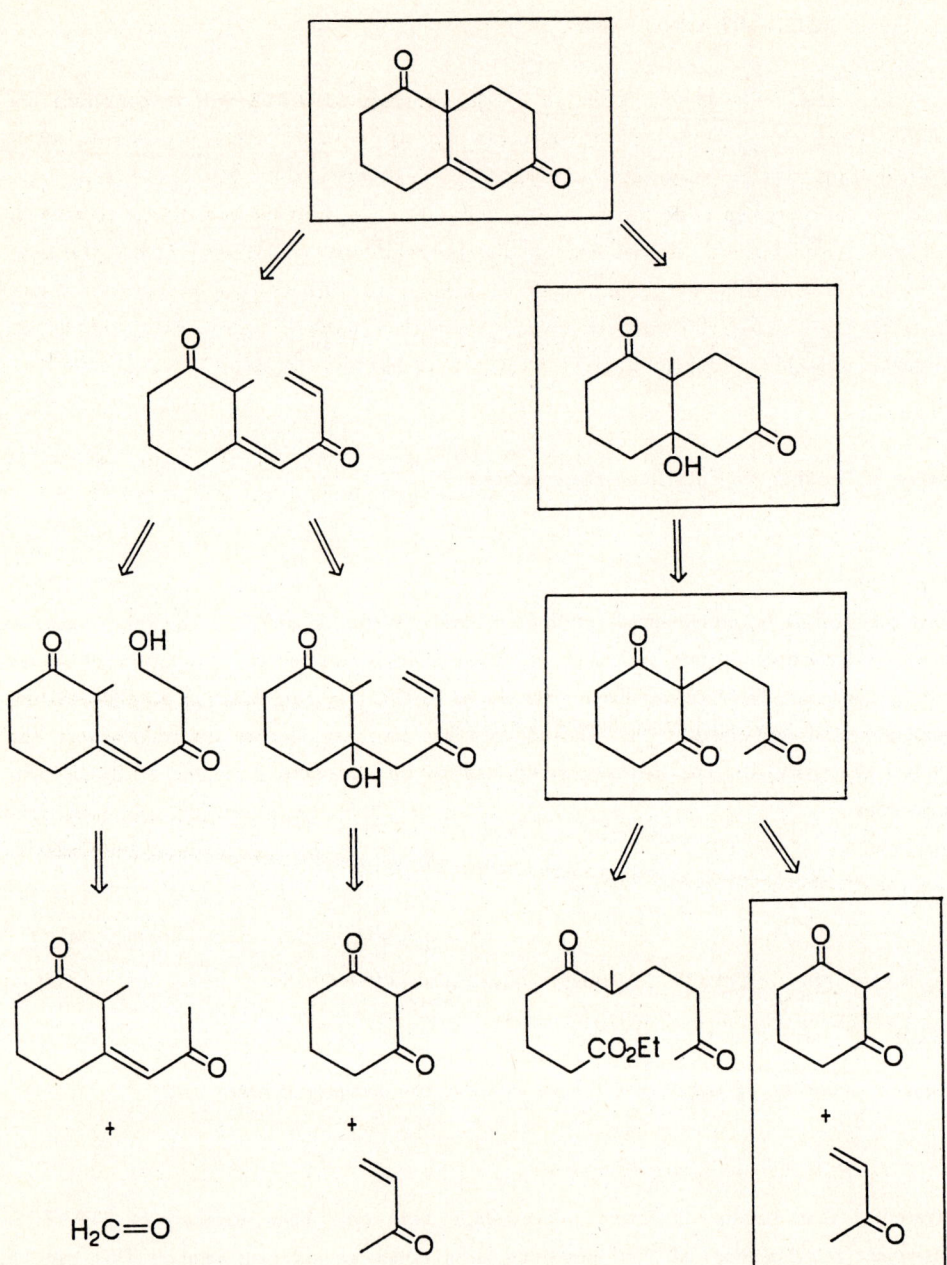

Figure 7: Synthesis plan generated by STRATOS.

EXAMPLE: ROBINSON ANNELATION

The following synthesis plan produced by the program STRATOS will demonstrate the results achieved. Fig. 7 shows suggestions for the synthesis of a bicyclic system, which is an intermediate in the synthesis of steroids. The search, that was confined to three levels, found four synthesis paths. Among them was the Robinson annelation (molecules in boxes), i.e. a Michael addition with a subsequent aldol condensation. This is the usual way of synthesizing such compounds. Another route begins with the same starting materials with an inverted sequence of the steps involved. A subsequent search in the synthetic direction, i.e. a reaction prediction, could decide between the two possibilities.

It has to be stressed, that all these plans and reaction sequences were generated without a database of reactions. They were obtained solely by examination of the relationships between functional groups in the molecules.

THE USE OF PROLOG

The program STRATOS has been coded in FORTRAN77. The results obtained with this system were quite satisfactory. However the program was difficult to maintain and to understand because the strategies were directly translated into subroutines. To improve that situation, it was necessary to separate the strategies and the program that uses them. We have therefore chosen to switch to a language, that already has built-in rules and methods to apply them: PROLOG [5].

The main idea of PROLOG is, that rules can have a procedural meaning and deduction can be viewed as a form of computation [6,7]. The declarative statement:

' A if B, C and D '

can also be interpreted procedurally as:

' To execute A, execute B, C and D '

This allows an easy formulation of synthesis strategies in PROLOG. For example the treatment of consonant paths can be formulated in PROLOG in the following manner:

```
disconnect(Bond,Path) :-
     which-path(Bond, Path),
     consonant(Path),
     best-resonance-stabilization(Bond, Path).
```

Procedurally this reads:
To decide whether the bond can be disconnected,
calculate the path which it is part of,
test whether the path is consonant and then,
test that the disconnection of the bond produces the synthon with the best resonance stabilization.

The declarative reading states a rule for synthesis planning:
A bond can be disconnected,
if it is part of a path which is consonant and
the disconnection leads to the synthons
which have the best resonance stabilization.

This declarative interpretation helps to develop thoughts and concepts. Seen in this way, PROLOG is not just another program language, but a tool for thinking.

ACKNOWLEDGEMENT

M Wagener wishes to thank the Technical University München for the financial support by the 'Stipendium zur Förderung des wissenschaftlichen und künstlerischen Nachwuchses'.

REFERENCES

1. Warren S, (1982) Organic Synthesis: The Disconnection Approach, Wiley & Sons, New York
2. Corey EJ, (1967) Pure Appl Chem 14:19
3. Saller H, (1985) thesis, TU München
4. Gasteiger J, Marsili M, (1980) Tetrahedron 36:3219
5. IF/Prolog Version 3.4, InterFace Computer GmbH, D-8000 München, Garmischer Straße 4
6. Clocksin WF, Mellish CS, (1984) Programming in Prolog, Springer Verlag, New York
7. Sterling L, Shapiro E, (1986) The Art of Prolog, MIT Press

EROS 6.0, A KNOWLEDGE BASED SYSTEM FOR REACTION PREDICTION – APPLICATION TO THE REGIOSELECTIVITY OF THE DIELS-ALDER REACTION

Peter Röse and Johann Gasteiger

Organisch-chemisches Institut, Technische Universität München,
Lichtenbergstr. 4, D-8046 Garching, FRG

Abstract: The EROS 6.0 system is used to analyse the effects determining reactivity in organic reactions and this knowledge can be applied to the prediction of still unknown reactions. Reactions are predicted by modelling the mechanism, the simulation of elementary reaction steps. Each reaction type is described by a reaction rule. Deduction or induction can be used for the derivation of reaction rules. In the inductive way, methods of machine learning and statistics are used to derive rules from databases of example reactions. Application of the system is demonstrated by modelling the regioselectivity in the Diels-Alder reaction.

THE REACTION PREDICTION SYSTEM

Since more than a decade work has been going on to develop EROS (Elaboration of Reactions for Organic Synthesis), a system for the prediction of organic reactions and the design of syntheses. At times the various stages of development have been consolidated into distinct versions of the system. Each version represents a characteristic approach and the state of the art to the tasks of reaction prediction and synthesis design and can be used as a tool for solving these problems.

EROS 5.2 predicts polar reaction mechanisms as a sequence of only two elementary steps: polar bond breaking and bond making [1]. The most recent version under development, EROS 6.0, is different in many aspects from previous versions. Many types of elementary reactions, i. e., heterolysis, homolysis, polar bond making, pericyclic reactions, oxidation, reduction and ionization are possible with EROS 6.0. Each elementary reaction is modelled as a concerted rearrangement of electrons. Knowledge about chemical reactions is now separated from the core program. The influence of reaction conditions like temperature, pH-value and solvent properties can be considered in the calculation of reactivity.

Figure 1 shows the basic design of the EROS 6.0 system. It is made up of four components (fig. 2) and an additional component for the machine learning of rules (fig. 3). The user interface allows the graphical input of the reactants, specification of reaction conditions and the graphical output of the predicted reactions. The knowledge necessary for the prediction of reactions, for example the concerted shift of six π-electrons in a Diels-Alder reaction, is contained in a reaction rule. All reaction rules together constitute the knowledge base of the system.

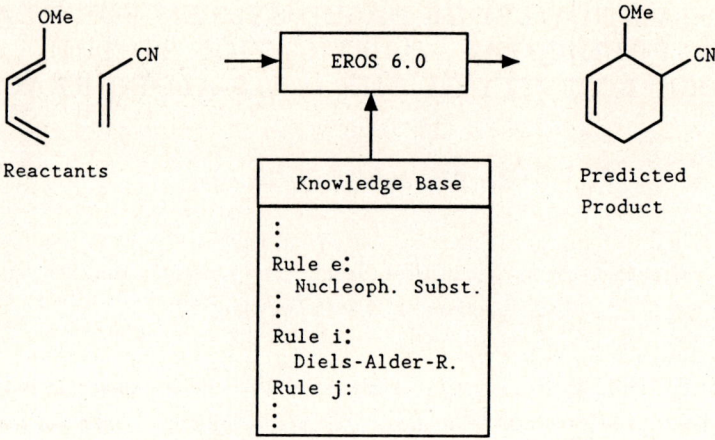

Figure 1: Reaction prediction with a knowledge base

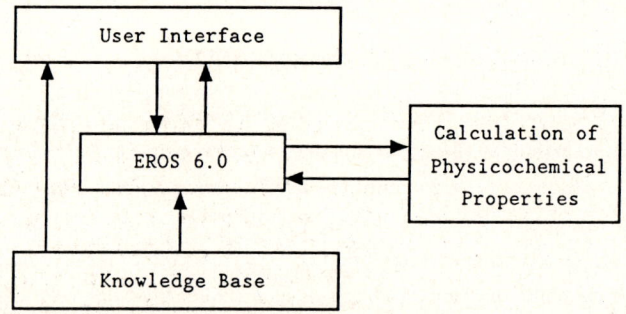

Figure 2: The EROS 6.0 system

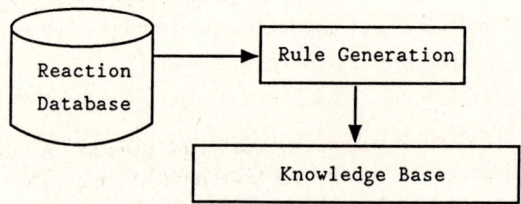

Figure 3: The learning component

The structure of a reaction rule is as follows

- A description of the reaction substructure

- Valence-, topological and physicochemical constraints under which the reaction proceeds

- A reaction generator which describes the shift of electrons during the reaction

- A reactivity function which gives the relationship between reactivity (e. g., rate constant) and calculated physicochemical properties of reactants and products

The EROS 6.0 program applies the reaction rules to the reactants. The major tasks of this program are:

- Search for reaction substructures

- Check of the various constraints given in the reaction rules

- Estimation of reactivity (e. g., relative or absolute rate constant)

- Generation of products and construction of a reaction network

- Calculation of product concentrations

The calculation of reactivity uses basic knowledge about chemical effects, which is provided through an interface to the EROS 6.0 program. This basic knowledge is available in the form of procedures for the calculation of physicochemical properties such as charge distribution, inductive, resonance, and polarizability effect, heat of formation and bond dissociation energy, already contained in previous EROS versions. Empirical methods are used for the calculation of these properties [1,2]. Other methods like quantum or molecular mechanical methods can be used as well, if sufficient computing power is available.

BUILDING THE KNOWLEGE BASE

Reaction rules can generated by the application of two different principles. In the deductive method an initial general rule is used. This rule is further specialized until it can describe certain reaction types. Specialization is done by including contraints to the rule or by enforcing existing constraints. Rules generated by deduction are useful for the generation of isomers (e. g., the isomers of chlorinated dibenzo-dioxine), for the generation of all formally possible reactions and reaction products of given reactants and for searching for new reactions.

In the inductive method rules are derived from example reactions. This is the usual way humans learn about organic reactions. One tries to find the common features of a reaction type and the major factors which influence reactivity. Reaction databases are the starting point for the rule generation in the EROS 6.0 system .

These databases contain:

- Reference to the literature for the reaction

- Constitution and configuration of reactants and products

- Reaction conditions: solvent, temperature, pH-value, etc.

- Reaction data (if available): rate constant, selectivity, activation energy and entropy, thermochemical data, etc.

The derivation of knowledge from this database is shown in figure 4 and is described in [3]. The rule generator is the learning component of the system. The rule generation involves:

- Recognition of the reaction substructure

- Derivation of constraints, under which the reaction proceeds

- Assignment of the reaction generator

- Preparation for the structure-reactivity analysis

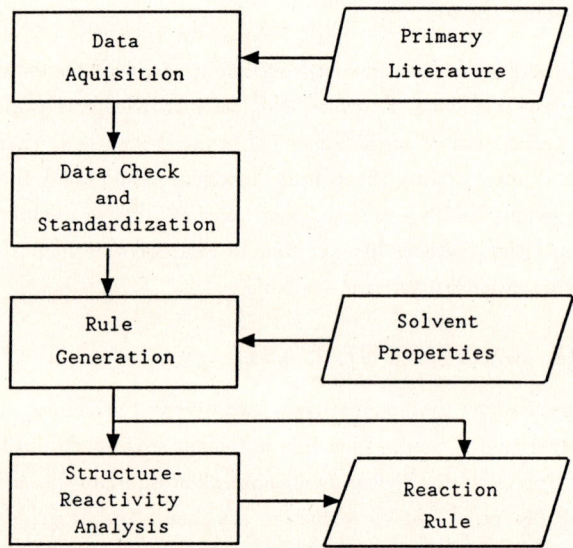

Figure 4: Derivation of reaction rules by induction

The structure-reactivity analysis is accomplished by statistical methods. Experimental rate constants, activation energies or selectivities are correlated with calculated physicochemical properties. As a result linear or nonlinear functions are derived, which can estimate reactivity or selectivity. Such a function is added to the reaction rule. Note that these functions are part of the knowledge base and are kept apart from the EROS 6.0 core program in separate files.

Once a rule has been set up, it is tested. In the reproduction test the reactants are taken from the database and the predicted products are compared with the experimentally observed products given in the database. Each predicted reaction network is further analysed and classified concerning the formation of regioisomers, backward reactions, the number of side reactions and the number of consecutive reactions. In a second test, the test for predictive power, the reaction rule is applied to reactants, which were not used for the derivation of the rule. This test shows how good a rule is for extrapolating knowledge to new reactions.

IMPLEMENTATION

The EROS 6.0 system is highly portable, because it is entirely written in Standard FORTRAN 77. The graphics parts of the system are based on GKS (Graphical Kernel System) [4]. The system has been implemented under MS-DOS and VAX/VMS. Reaction databases have been generated with ChemBase [5] on a PC.

APPLICATION: REGIOSELECTIVITY OF THE DIELS-ALDER REACTION

Derivation of a reaction rule

In the Diels-Alder reaction of an unsymmetrically substituted diene with an unsymmetrically substituted dienophile two regioisomers are possible (fig. 5). In most cases one isomer is preferred. For the derivation of a rule which predicts the major product, a database of

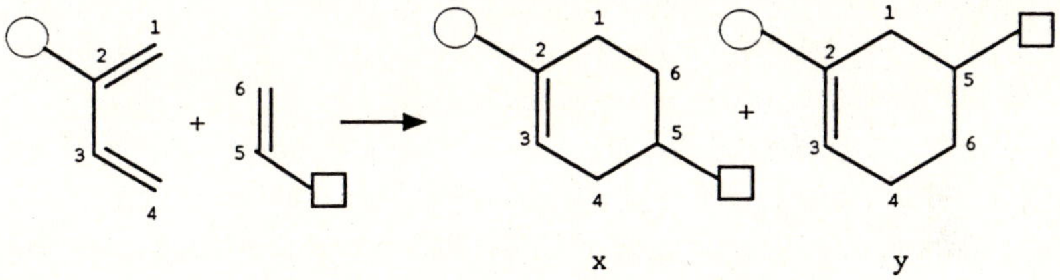

Figure 5: Formation of regio isomers

approximately 150 Diels-Alder reactions was built, including reactions with hetero atoms in the diene (N, O) and the dienophile (N, O, S). For each reaction the reactants, the major product under kinetic control, the regioselectivity, the reaction temperature, the solvent and the yield for the formation of both regioisomers was stored. The regioselectivity in % was calculated from the product ratios.

$$Sel = \frac{[x] - [y]}{[x] + [y]} \cdot 100 \qquad (1)$$

In the case where only one regioisomer was observed, a regioselectivity of 99.9 % was assigned arbitrarily. This is frequently the case, if hetero atoms are included in the diene.

These data were fed into the rule generator, which derived the reaction substructure, the boundary constraints under which the reaction proceeds and the reaction generator. In the next step the data for structure-selectivity analysis were prepared. From the selectivity the difference in the free activation enthalpy was calculated.

$$\Delta\Delta G^{\ddagger} = -RT \ln \frac{100 + Sel}{100 - Sel} \qquad (2)$$

There are almost no data available about the temperature dependency of the product ratios. Therefore $\Delta\Delta G^{\ddagger}$ had to be assumed to be temperature independent. In general, the major product does not change by variation of the reaction temperature, as long as kinetic control is guaranteed.

Another problem is, that the yields are not very high in a number of cases. These product ratios have to be dealt with caution [6]. Therefore the data were weighted, by taking the yield as a weight in the structure-selectivity analysis.

Structure-selectivity analysis

The regioselectivity depends on the asymmetry of the electron distribution in the diene and in the dienophile and on steric hindrance in the transition state. In the preliminary model the influence of inductive and resonance effects was investigated.

Procedures for the quantification of these effects were developed a few years ago [1,2]. The concept of orbital electronegativity [7,8] led to the PEOE (partial equalization of orbital electronegativity) method for the rapid calculation of sigma partial charges and residual sigma electronegativities [9]. The residual sigma electronegativity χ is a good measure of the inductive effect [10]. +I effects lower χ and −I effect raise χ.

Resonance effects are used to explain the stabilization of formal charges in ionic resonance structures. A quantitative measure of the resonance effect is obtained by generating all resonance structures for the delocalization of formal charges and then weighting these structures by π-electronegativity [11]. The stabilization of positive charges by hyperconjugation is also considered.

To develop a mathematical function which gives an estimate for $\Delta\Delta G^{\ddagger}$, the asymmetry in the diene and the dienophile is expressed by differences of electronic effects between the positions 1-4 and 2-3 in the diene (fig. 6) and positions 5-6 in the dienophile (fig. 7). For substituents with a resonance effect only the +M effect in the diene and the −M effect in the dienophile is considered in this study. This corresponds to the typical substitution pattern for Diels-Alder reactions with normal electron demand. Also steric effects and different regioselectivities in exo and endo adducts, i. e., secondary interactions are not handled within the current model.

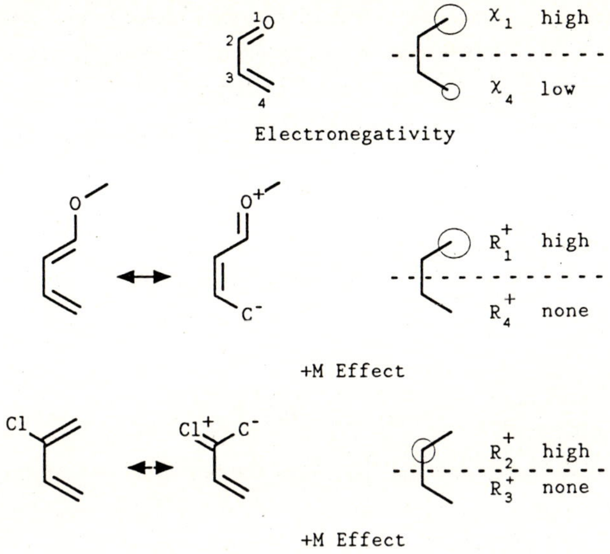

Figure 6: Electronic effects in the diene

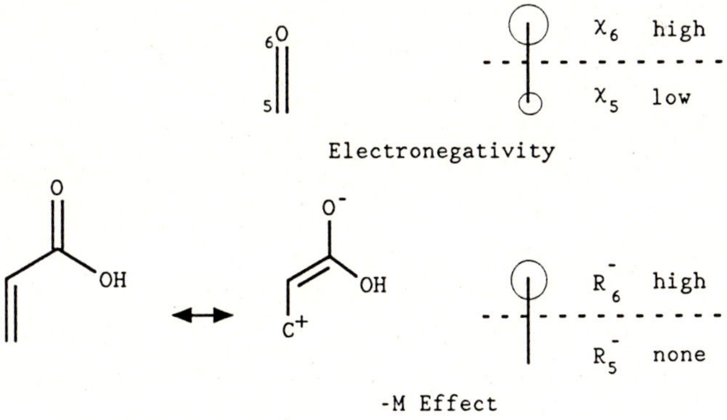

Figure 7: Electronic effects in the dienophile

$\Delta\Delta G^{\ddagger}$ can be separated into two parts: the sign and the absolut value. The sign is responsible for the binary decision which is the major and which is the minor product. If x is the major product (Fig. 5), thus $[x] > [y]$, then $\Delta\Delta G^{\ddagger}$ is negative

$$\Delta\Delta G^{\ddagger}_x = -RT \ln \frac{[x]}{[y]} \qquad (3)$$

and $\Delta\Delta G^{\ddagger}$ is positive for the minor product y

$$\Delta\Delta G^{\ddagger}_y = -RT \ln \frac{[y]}{[x]} = RT \ln \frac{[x]}{[y]} \qquad (4)$$

and therefore

$$\Delta\Delta G^{\ddagger}_x = -\Delta\Delta G^{\ddagger}_y \qquad (5)$$

As the sign of $\Delta\Delta G^{\ddagger}$ changes if either the diene or the dienophile is switched around, the same is true for differences of electronic properties. For example $\Delta\chi_{5,6} = \chi_5 - \chi_6 = -(\chi_6 - \chi_5)$.

The absolute value of $\Delta\Delta G^{\ddagger}$ determines the product ratio and is a measure of the directing power of the diene and the dienophile. The directing power of the diene and the dienophile is expressed by differences in electronic properties. The difference in an electronic property of the dienophile $D_{2\pi}$ is now taken as the x-coordinate and the difference in an electronic property of the diene $D_{4\pi}$ is taken as the y-coordinate of a three dimensional space (fig. 8). According

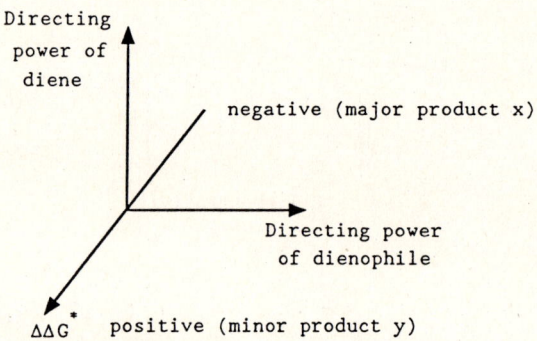

Figure 8: Representation of the directing power in a selectivity space

to our definition of a reactivity space [12], we call this space a selectivity space. If we take $\Delta\Delta G^{\ddagger}$ as the z-coordinate, the x-y plane splits the space into two regions. Points with negative z-values represent the major products and points with positive z-values represent the minor products.

Now we build the vector product

$$\vec{V} = \vec{D_{2\pi}} \times \vec{D_{4\pi}} \tag{6}$$

which only has a z-component. The vector product possesses two important properties. First, if either the sign of $D_{2\pi}$ or $D_{4\pi}$ changes, the sign of the z-component of $\vec{V}$ changes and secondly, the vector product is zero if either the diene or the dienophile or both are symmetric. Then, only one isomer is formed. The vector product is assumed to be proportional to $\Delta\Delta G^\ddagger$. As the regioselectivity is influenced by several effects at the same time, a sum of vector products is used

$$\Delta\Delta G^\ddagger = \sum_{i=1}^{n} c_i V_i \tag{7}$$

Taking the experimental $\Delta\Delta G^\ddagger$ and the calculated vector products, the coefficients c_i were fitted by multiple regression. In the analysis it turned out that three vector products were sufficient to describe the regiochemistry (fig. 9).

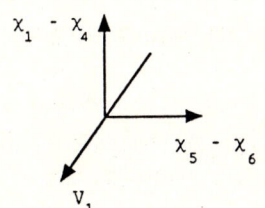

Inductive effect in diene and dienophile

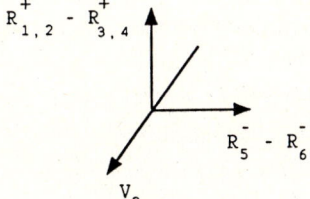

Resonance effect in diene and dienophile

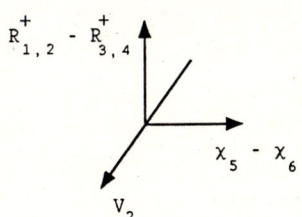

Resonance effect in diene and inductive effect in dienophile

Figure 9: Outline of vector products used

Setting the terms into equation 7 the following equation was obtained

$$\Delta\Delta G^{\ddagger} = c_1(\chi_5 - \chi_6)(\chi_1 - \chi_4) + \\ c_2(\chi_5 - \chi_6)(R^+_{1,2} - R^+_{3,4}) + \\ c_3(R^-_5 - R^-_6)(R^+_{1,2} - R^+_{3,4}) \quad (8)$$

Although the correlation was very rough (R=0.78, S=8 kJ/mol), for 139 of 148 reactions the major product was predicted correctly. This function was then added to the reaction rule. Figure 10 compares some reactions predicted by EROS 6.0 with literature data [13,14,15,16,17] from the database. The application of EROS 6.0 to new reactions is shown in figure 11. The second reaction is a keystep in the synthesis of the antitumor agent streptonigrin [18], where the observed major product was correctly predicted.

Wrong major products resulted for reactions of 1-tert-butyl substituted dienes (omission of steric effects), $O=C(CO_2Et)_2$ and some reactions of $O=S=N$-Tos as dienophile. Some reasons for the rather moderate correlation are: (1) $\Delta\Delta G^{\ddagger}$ was assumed to be temperature independent, (2) a great part of the the product ratios is at best semiquantitative; some were determined by series of degradation reactions of the regioisomers, (3) for many experiments there is no proof that the reaction took place under kinetic control, (4) $\Delta\Delta G^{\ddagger}$ is unknown for regiospecific reactions (99.9 % regioselectivity was assumed), (5) the influence of the solvent was not considered, (6) secondary interactions, steric and other effects were not considered and (7) the electronic parameters used might have some deficiencies.

In discussing the prediction of regioselectivity it was expressed that the orientation phenomena cannot be explained by electronic effects [19]. Only by the use of perturbational molecular orbital methods, such as FMO theory, the preferred regioisomers can be predicted in large number of cases.

Nevertheless this preliminary study suggests that the regioselectivity can be explained, at least to an important part, on the basis of inductive and resonance effects. As we have procedures for the estimation of quantitative values for these effects, we consider several effects at the same time. For the chemist who can only give a qualitative or at best semiquantitative estimate for these effects, it is difficult to predict the synergism or antagonism between the competing effects.

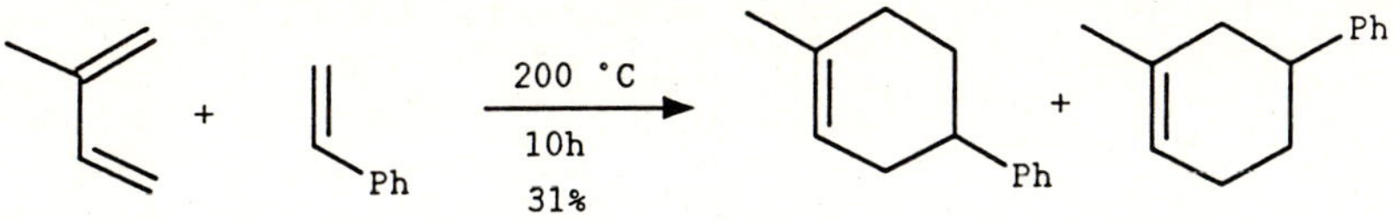

exp.	3.5	:	1
EROS6	2.2	:	1

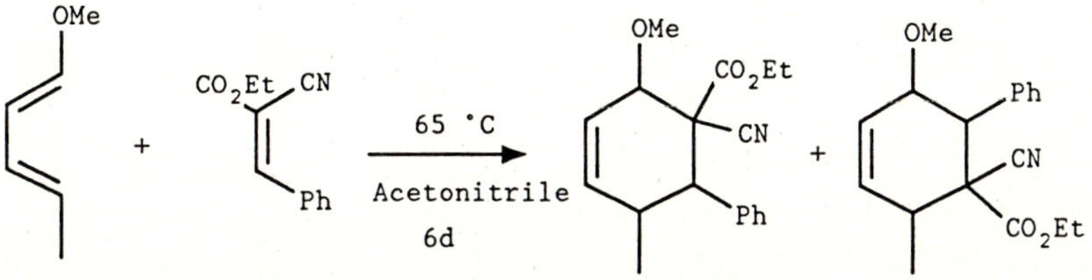

exp.	90%		---
EROS6	0.99	:	0.01

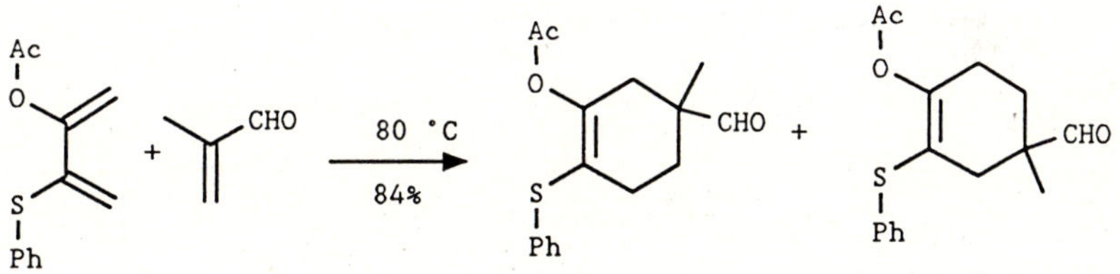

exp.	10	:	1
EROS6	1.07	:	1

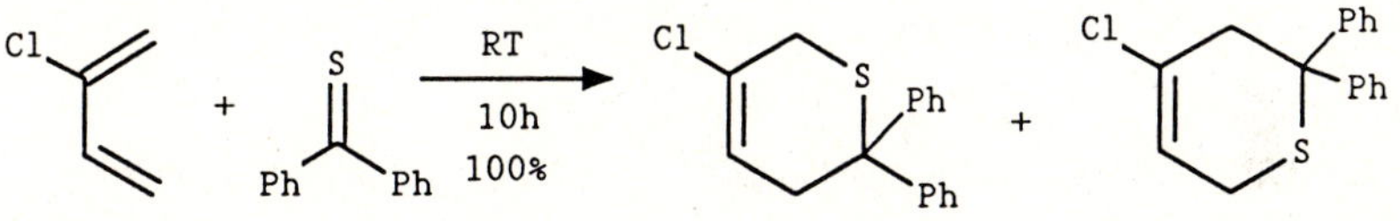

exp.	1.8	:	1
EROS6	2.7	:	1

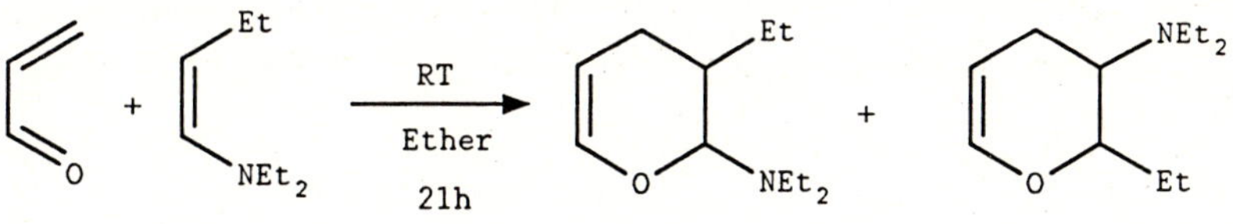

exp.	53%		---
EROS6	1	:	0

Figure 10: Reproduction test

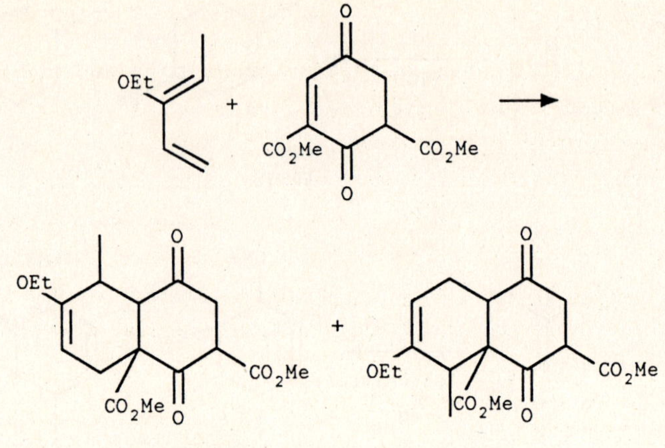

	major product		
exp.			---
EROS6	0.84	:	0.16

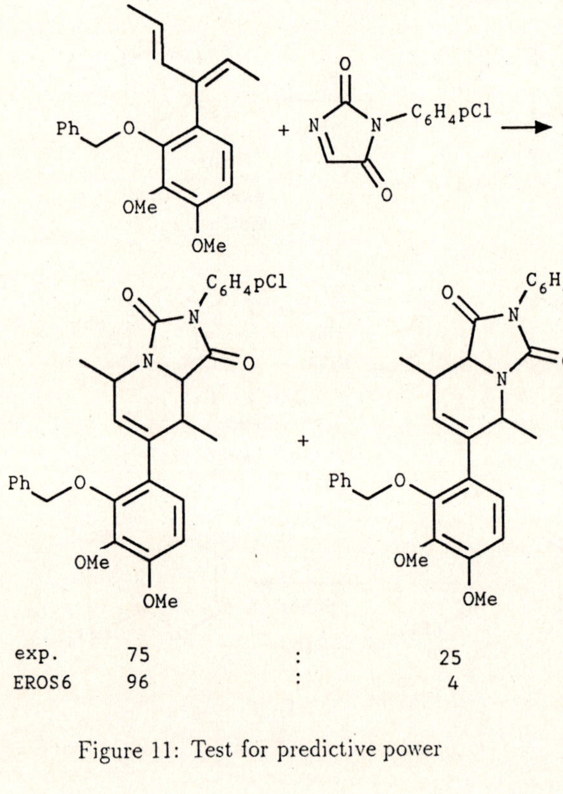

exp.	75	:	25
EROS6	96	:	4

Figure 11: Test for predictive power

CONCLUSION

Reactivity and selectivity is influenced by several effects at the same time. As enough experimental data are available about a reaction type, the EROS 6.0 system can be used to analyse which effects are responsible for reactivity. The knowledge obtained can be applied to the prediction of unknown reactions. Instead of using a database of reactions directly, which can only be used for retrieval, the knowledge contained in a database is extracted by methods of machine learning and statistics and accumulated in a reaction rule. The EROS 6.0 system is a tool, not a complete system for the prediction of reactions. As more and more reaction rules are generated, the system can predict entire reaction sequences and the competition between alternative reaction paths. Due to the separation of the knowledge base and the problem solving component, the system is very flexible. Different knowledge bases can be set up for special problems and modification or extension of a knowledge base does not require any programming. For example the same concept is used for the prediction of fragmentation reactions in a mass spectrometer [20]. Further, several people can work out new rules independent of each other.

ACKNOWLEDGEMENTS

We gratefully acknowledge support of this work by a Kekulé fellowship to P.R. of the Stiftung Volkswagenwerk and the Stiftung Stipendien-Fonds des Verbandes der Chemischen Industrie.

References

[1] Gasteiger J, Hutchings M G, Christoph B, Gann L, Hiller C, Löw P, Marsili M, Saller H, Yuki K (1987) Topics Curr Chem **137**: 19

[2] Gasteiger J (1988) In: Jochum C, Hicks M G, Sunkel J (eds), *Physical Property Prediction in Organic Chemistry*. Springer, Berlin

[3] Röse P, Gasteiger J, Anal chim Acta, in press

[4] Enderle G, Kansy K, Pfaff G (1987) *Computer Graphics Programming, GKS – The Graphics Standard*. Springer, Berlin

[5] ChemBase, Molecular Design Limited, 2132 Farallon Drive, San Leandro, CA 94577

[6] Sauer J (1967) Angew Chem **79**: 76

[7] Hinze J, Whitehead M A, Jaffé H H (1963) J Amer chem Soc **85**: 148

[8] Sen K D, Jørgensen C K (eds) (1987) *Electronegativity*. Structure and Bonding 66

[9] Gasteiger J, Marsili M (1980) Tetrahedron **36**: 3219

[10] Hutchings M G, Gasteiger J (1983) Tetrahedron Lett **24**: 2541

[11] Gasteiger J, Saller H (1985) Angew Chem **97**: 699; (1985) Angew Chem Intern Ed Engl **24**: 687

[12] Gasteiger J, Röse P, Saller H (1988) J Mol Graphics **6**: 87-92, 97

[13] Titov Yu A (1962) Russ Chem Rev **31**: 267

[14] Trost B M, Ippen J, Vladuchick W C (1977) J Amer chem Soc **99**: 8116

[15] Broekhuis A A, Scheeren J W, Nivard R J F (1980) Recl Trav Chim **99**: 6

[16] Ohno A, Ohnishi Y, Tsuchihashi G (1969) Tetrahedron **25**: 871

[17] Opitz G, Holtmann H (1965) Liebigs Ann Chem **684**: 79

[18] Weinreb S M, Basha F Z, Hibino S, Khatri N A, Kim D, Pye W E, Wu T (1982) J Amer chem Soc **104**: 536

[19] Eisenstein O, Lefour J M, Ahn N T, Hudson R F (1977) Tetrahedron **33**: 523

[20] Hanebeck W, Rafeiner K, Schulz K P, Röse P, Gasteiger J, this issue

C A R S A

(Computer-Assisted Research in Synthesis and Application)

R. Moll

VEB Chemiekombinat Bitterfeld
Research and Development, Dpt. FO/S
GDR - 4400 Bitterfeld

Abstract: CKB's computer chemistry project CARSA (Computer-Assisted Research in Synthesis and Application) is a modular structured system, which at present consists of the following parts:
- WIFODATA (substance - and finding - documentation; database)
- QSAR (programs for quantitative structure-activity-relationships)
- RDSS (Reaction Design by Synthon Substitution; program package for synthesis planning)

The system is devised as a tool for synthesis chemistry and of special importance in industrial research.

INTRODUCTION

At the beginning of the seventies it was started at CKB to use computer techniques in pesticide research. In a rather close relation to the chemistry information system SPRESI (1) the database WIFODATA was set up and QSAR - methods were introduced. In 1985 the first version of RDSS (Reaction Design by Synthon Substitution) was installed and all parts were combined to the project CARSA (2)(3)(4). In figure 1 the present components and their relations are outlined. CARSA is an expandable project, wich is enlarged now by methods of molecular modelling and reaction databases. In the following the main parts are described more in detail.

WIFODATA

W. is a system for storing and processing of all types of data, occuring in different fields of pesticide and drug research. It is an open system, which can be adapted to the specific demands of the user.
Content, memory structure, and extent are variable. The quantity of data stored in the database is only limited by the hardware. Chemical and biological facts can be stored together or separately; later corrections are always possible (s. figure 2).
The database installed at CKB comprises
- bibliographical facts

- structures
- physical parameters
- biological results

Accordingly the following information can be searched:
- single structures (stored topologically)
- substructures and fragments
- biological and other data
- data of internal importance (e.g. name of chemist, date of synthesis and test).

The registration of data is made by coworkers with special reference cards. Before the computer processing the information are encoded according to a specific scheme, in which only letter- and figure-terms are used. For this reason an independence of program languages is attained, being one of the great advantages of the system. Another one is the fact, that the biological test results are described in original values. The searches are made interactive (menu s. figure 3). It has to be mentioned, that the system can also be installed on PC's.

QSAR

By means of QSAR-methods the relations between chemical constitution and biological activity are investigated. The methods are mathematical-statistical techniques for the description of biological activity of tested compounds and, on the basis of it, the prediction of potential active structures. In the frame of pesticide research at CKB such methods are applied rather successfully for some time (examples s. (5)). Originally drawn up as a part of WIFODATA, they are now an independent component of CARSA. The program basis is formed by packages developed by partner institutions and by CKB:
- FREE / WILSON - Analysis
- HANSCH - Analysis
- programs for calculations of connectivities, pattern- and cluster-analysis, autocorrelations, molar volumes etc.
- program system LABSWARE

Furthermore extensive data collections, e.g. substituent coefficients, were set up.

RDSS

The program package RDSS (Reaction Design by Synthon Substitution) is an expert system for synthesis planning in organic chemistry (about predecessors s. (3) and (4) and literature cited therein). It is tool for the chemist, to predict and to plan reactions (s. figure 4).

RDSS simulates the steps of synthesis by means of the model of synthon substitution (6). By it it is possible to derive the knowledge base from reaction equations and to shift the acquiring of knowledge to the computer.
RDSS offers three variants of application:
1. Synthesis planning of target structures
2. Reaction prediction for given starting materials
3. "Mixed planning" (both directions combined)
All possibilities work with the same knowledge base, consisting of synthons and context descriptors. The synthon transformations describe reaction types and form the "production rules" of synthesis chemistry. The context descriptors depict the dependence of them on the structural context.
In RDSS it is not tried to gain knowledge for the problem-solving components from the chemist, like it is typical for expert systems of the second generation, but to derive the knowledge base automatically from reaction equations. For this task special analysis- and learning-procedures are utilized.
The user works with RDSS in the following manner: data are inputted as reaction equations, either interactive or separated from SPRESI-files. The system extracted automatically the chemical knowledge from the data in form of synthon equations and matching context descriptors. The system furnishes problem solutions depending upon the option of the user:
a) as precursors of target molecules
b) as products of starting materials
c) bi-directional as synthesis ways from educts to products.
The quality and quantity of the suggestions depend on the content of the knowledge base and the optional connection with the appropriate context descriptors. The principles of automatic knowledge acquiring from reaction equations and the synthesis simulation by synthon substitution are not realized in any other system up to now.
The program comprises the following components:
- I/O-component (input and storing of structures, reaction equations, and additional information and visualization in textual or graphical manner)
- topological analysis (canonization, ring recognition, fragmentation)
- data handling (collecting, separating, mixing, selecting etc. of data)
- transformation data handling, including the extension of the knowledge base and the forming of synthon library. An important part is the automatic recognition of synthon transformations in respect to the automatic knowledge acquiring.
- simulation of synthesis steps in both directions and combination of them

- 2 D-visualization

In a practical run RDSS works like follows:

- it awaits the input of equations or structures (interactive or from SPRESI-files)
- handles the data
- finds out the synthon equations (s. above)
- forms the knowledge base

On this basis it supplies in dialogue with the user:

- synthesis predictions (from starting molecules)
- synthesis plans (for target molecules)
- bi-directional synthesis plans (for targets from given starting structures)
- 2 D - visualization of the results (s. figure 5)

RDSS is written in C and portable to almost all hardware systems. The requirements are 16 or 32 bit PC or mainframe, 60 MByte harddisc and color graphics monitor (EGA). It is a product of VEB Chemiekombinat Bitterfeld and Central Institute of Cybernetics and Information Processes (Academy of Sciences of GDR, Berlin).

Acknowledgement: Thanks are due to the CARSA project group at CKB (P. Kemter, U. Lindner, D. Schönfelder, R.-D. Werner) for their very engaged work and the assistance at the preparation of demonstrations and texts.

References:

1 Weise A, Scharnow HG (1979) Z Chem 19:49
2 Moll R, Kemter P, Lindner U, Schönfelder D, Weise A (1988) Chem Techn (Leipzig) 40:33
3 Moll R (1989) in: Tosi C (ed) Strategies in Computer Chemistry. Kluwer, Dordrecht
4 Moll R (1989, publ. 1990) Anal Chim Acta, in press
5 Schönfelder D, Moll R, Mühlstädt M (1986) Z Chem 26:154
  Moll R (1989, publ. 1990) Gazz Chim It, in press
6 Weise A (1980) J prakt Chem 322:761

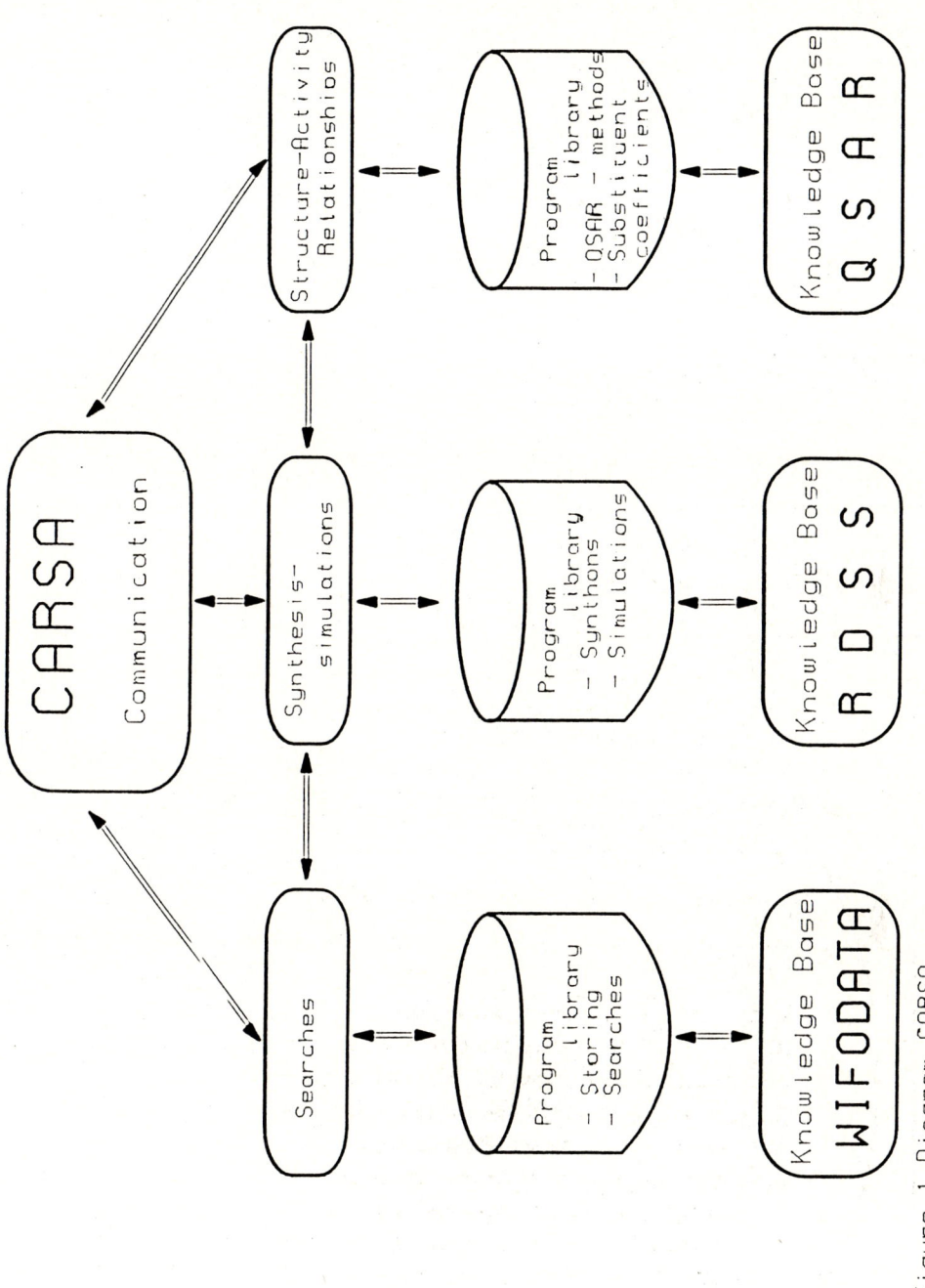

Figure 1 Diagram CARSA

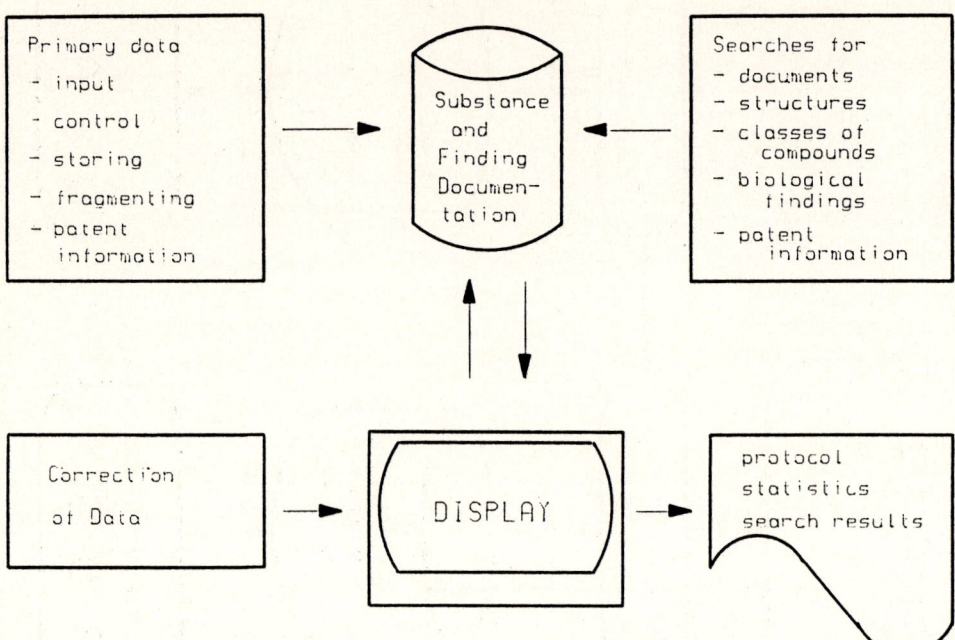

Figure 2 Diagram WIFODATA

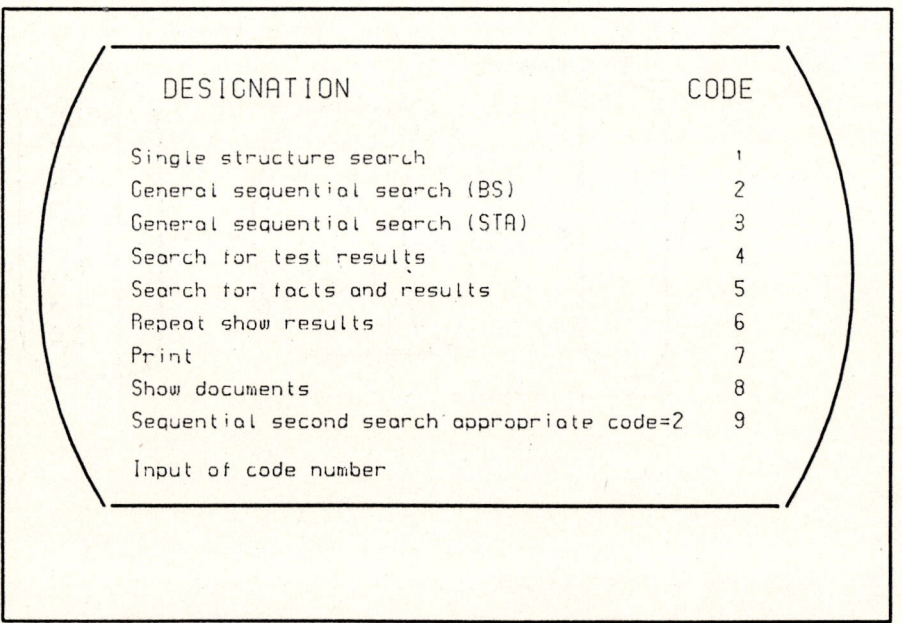

Figure 3 Menu WIFODATA

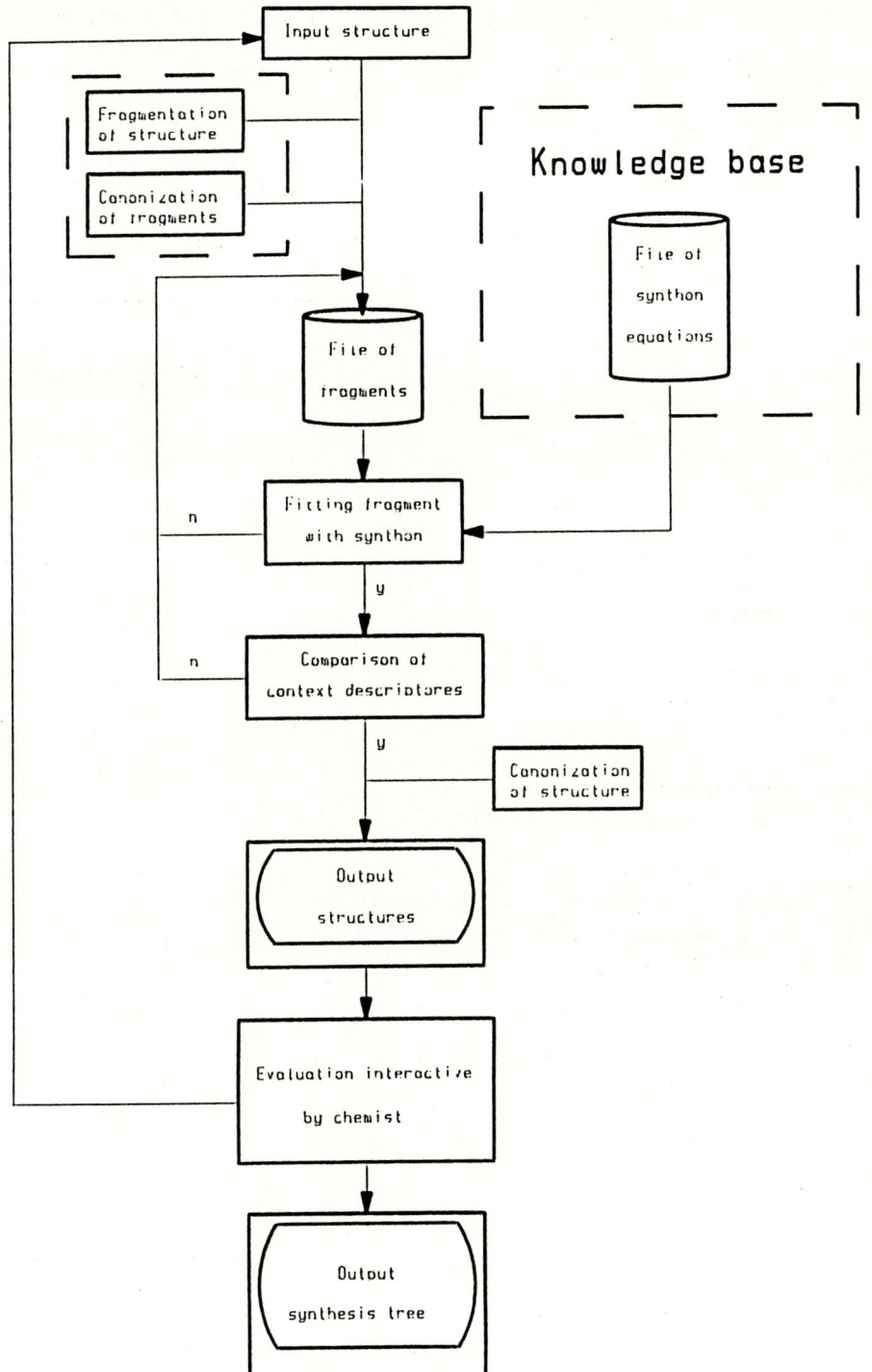

Figure 4 Diagram RDSS

Target Structure   Precursor Structure

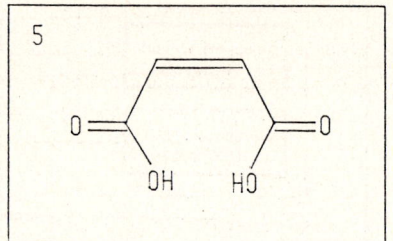

Oriented Fragment   Corresponding Synthon

O = C - O - C = O   ⇐   O = C - OH   O = C - OH

Synthesis Tree

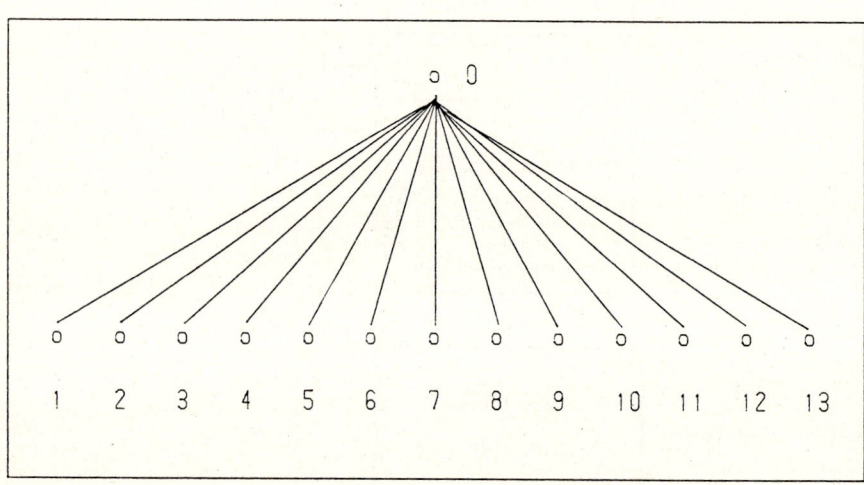

Figure 5 Partial results of a synthesis planning

Modelling of Polymer Gel Formation and Gel Reactions with Monte Carlo Methods for 3-dimensional Networks

Johannes Kinkel, Hanns J. Ederer and Klaus H. Ebert

Institut für Heiße Chemie, Kernforschungszentrum Karlsruhe, Postfach 3640, D-7500 Karlsruhe and SFB 123 der Universität Heidelberg, Im Neuenheimer Feld 293, D-6900 Heidelberg.

Abstract

The thermal treatment of poly-(p-methyl-styrene) was investigated experimentally. In contrast to polystyrene, gels are formed and decomposed again, depending on temperature. Based on the experimental results a reaction model for polymer decomposition, polymer gel formation and polymer gel degradation was developed. A Monte Carlo simulation program using methods from percolation theory is presented which covers thermal gel formation, gel formation caused by radiation, gel formation during polymerization, thermal, mechanical and ultrasonic decomposition of polymer gels and highly branched polymers.

Introduction

Mathematical modelling has become an important tool for the understanding of complex chemical reactions, for numerical predictions and comparisons with experimental results and for testing theories and assumptions [1].

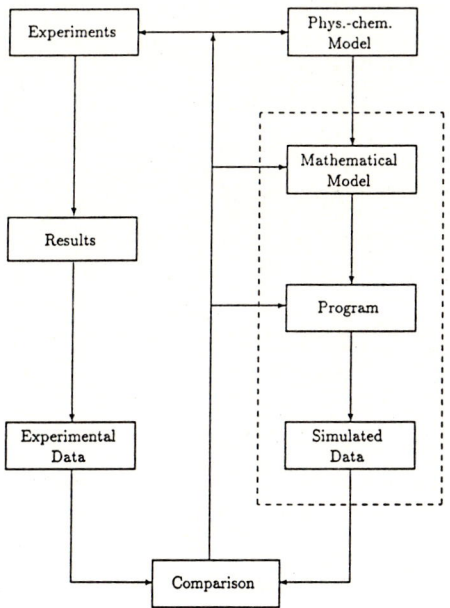

Fig. 1: Interdependence of experimental work and mathematical modelling

In Fig. 1 the connections and relations between experimental work on the one hand and modelling and simulations on the other hand are shown. On the left side following the arrows, the experiments produce results (e. g. spectra), from which, mathematically transformed, experimental data (e. g. concentration vs. reaction time) are obtained. From the physical-chemical model or the theory (top right side), a mathematical model is derived. This mathematical model may contain some approximations and assumptions. In the next step the mathematical model is converted into a numerical mathematical model, which is computable and forms the basis for a computer code. The computer program produces numerical results which, after some conversion, are compared with the experimental data. For the arrows on the right hand side it is true, that appropriate approximations are introduced to reduce the complexity of the model. The comparision of the experimental data with the calculated data may result in a feedback to either the theoretical model or to improvements of the experimental equipment or to alterations in the mathematical modelling process. As suggested by the scheme, this is an iterative process.

Experiments

The substances used for our studies of thermal behaviour — poly-styrene and poly-(p-methyl-styrene) — are shown schematically in Fig 2.

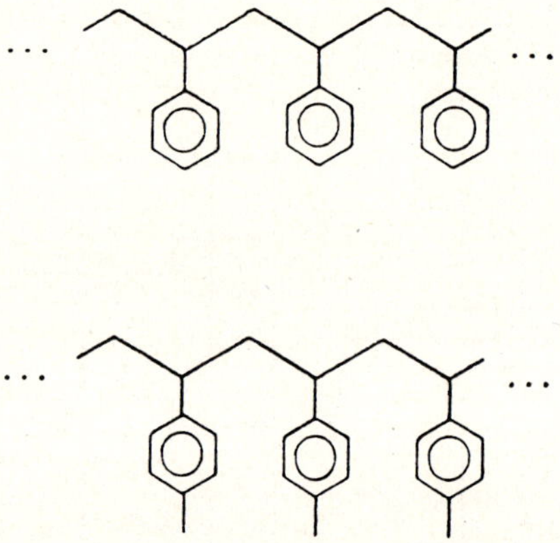

Fig. 2: Polymers used for studies of thermal reaction behaviour and gel formation

Thermal, radical and anionic initialized polymerization were applied for the preparation of both polymers. The rather simple equipments used for the experiments are shown in Fig. 3 [2,3]. The polymer sample is sealed in a glass ampulla. The ampulla is kept at constant temperature by a thermostat. In a tube reactor a polymer film is kept at a constant temperature, the volatile substances are pumped off and collected in a cooling trap.

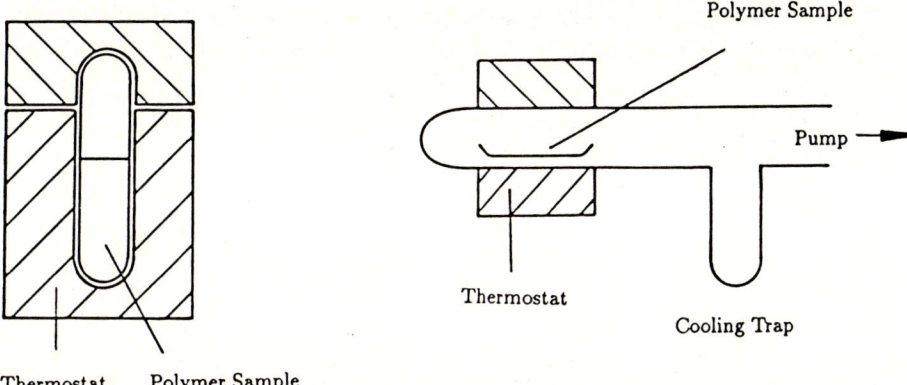

Fig. 3: Experimental setup for thermal treatment of the polymer samples. Left: batch setup; right: tube reactor connected to a vacuum pump for the removal of volatile substances.

The polymers and the volatile fraction were analyzed qualitatively and quantitatively using gas chromatography, gel permeation chromatography and mass spectroscopy. The experimental results obtained are the concentrations of the volatile substances, molecular weight distributions of the remaining polymers and the amount of non-soluble gel as a function of reaction time.

Reaction model

Based on experimental results, reaction models for the thermal treatment of poly-styrene and poly-(p-methyl-styrene) were developed consisting of elementary radical reactions. Some of them are shown in Fig. 4 representing the different types of radical reactions [4].

In the initiation step two free radicals are formed; in the radical recombination reactions two radicals are 'consumed' and the radical reaction chain is terminated. Another type of reaction is the transfer of hydrogen (and therefore also the transfer of the radical character) which is called metathesis reaction. The β-decomposition is the next type of chemical reaction shown. The reverse of this radical decomposition reaction is the reaction type called radical addition. In intramolecular additions rings are formed. Chemical bonds are ruptured by the initiation step and the radical decompositon reaction, therefore these are the most important types of reactions in polymer decomposition. New bonds are formed in radical additions to olefinic groups (being the most important reaction in radical polymerization) and by the radical recombination reactions (termination reactions). These two types of bond forming reactions are most effective for the thermal gel formation process. The hydrogen transfer reaction distributes the reactive radical sites more or less randomly within the polymer sample. There are different subtypes for each type of reaction, because of chemically different hydrogen and carbon atoms within the monomeric unit. On the other hand there are topological different reactions when the same type of chemical reaction takes place at different positions of the polymer chain. This complicates the 'chemistry' of the reaction considerably.

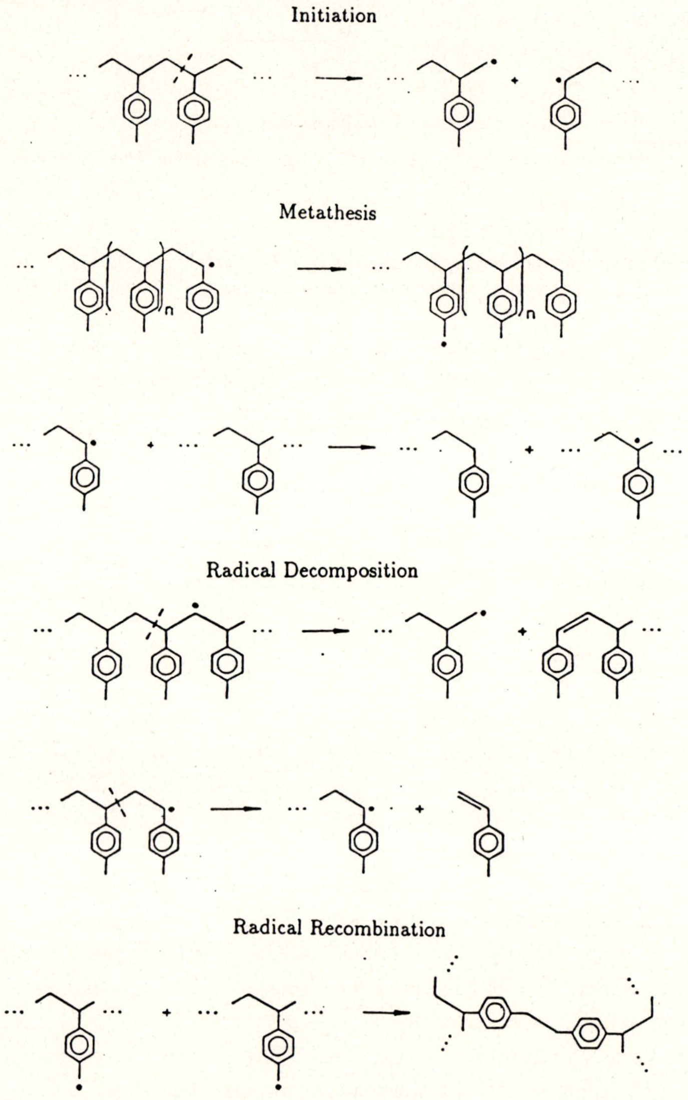

Initiation

Metathesis

Radical Decomposition

Radical Recombination

Fig. 4: Types of elementary reactions important for thermal decomposition and thermal gel formation of polymers

Mathematical modelling

In order to 'translate' a reaction model into a mathematical model, one has to answer three questions. The first one is, which ansatz fits the problem best.

The mathematical description of homogeneous complex chemical reactions is usually a system of stiff differential equations whereby for each chemical species involved a differential equation has to be established [5] — if the reaction model consists of elementary reactions, the system of differential equations can be generated by a computer. A different way of modelling is via Monte Carlo methods [6,7], where one tries to imitate the stochastic way chemical reactions proceed in reality. Monte Carlo methods usually avoid the formulation of a mathematical model as differential equations; rather the theoretical conceptions and ideas of the chemical or physical problems are translated directly to a computer program. For the application of this 'apparently ingenious' method one has to pay a price: very long computation times and a very large memory size is required and the accuracy of the computed results is not very high.

For highly branched polymers and polymer gels, a deterministic ansatz with one differential equation for each radical species will lead to an exploding system with more than 10^{12} equations. Therefore the Monte Carlo simulation is not just the method of choice for the pending problem but a must.

The next question is how to represent the polymer chains, the branched polymers and the polymer gels. This was solved in the following way.

- The monomeric (or pseudo-momomeric) units are points arranged in a matrix. To each point of this matrix a vector is attributed, in which the coordinates of the neighbour points are contained. In addition, a color is entered, which is important for the representation of the network. The maximum functionality of the points is three.
- Linear polymers are simulated by appropriate connections between the points. This connection is done by entering the coordinates of the neighbours into the connectivity-vectors. This leads to the following interpretation for the functionality of the connectivity-vector. If there is no neighbour entered, the point stands for a monomeric unit. One entered neighbour represents an end of a chain, two neighbours stand for a linear member of the chain, and a point connected to three neighbours represents a branching point. Because arbitrary points may be connected and not only neighbouring ones, the resulting polymers can be considered as to have a three dimensional structure. If a very dense gel is generated by a lot of points with three connections, there may result a topological problem in transfering the computer representation in a real three-dimensional polymer model.

The third question is how the reaction types shown in Fig. 4 should be simulated using this connected points concept of polymers?
- The removal of a connection constitutes an initiation reaction or a radical decom-

position reaction. It is executed by deleting the coordinates of neighbour points from the connectivity-vector.
- A new bond is formed by a radical recombination reaction or a radical addition reaction.
- In radical chemistry the number of initiation reactions and the number of termination reactions must be equal. Therefore — if no radical additions are considered — the number of removals of connections must be greater or equal than the number of new connections; the difference then constitutes the radical decomposition reactions.
- The random selection of a bond, which is to be disrupted in the initiation reaction, is equivalent to the chemical assumption that each bond connecting monomers has the same probability to be thermally broken. The metathesis reaction as a precursor reaction for either a radical decomposition reaction or a recombination reaction transfers the radical character randomly within the polymer sample. This is modelled by the random choice of bonds which are removed or the random choice of points which are connected by a bond.
- The unzipping depolymerization reaction has shown to be important in thermal polymer decomposition. This is a special reaction of the β-decomposition type in case the radical is situated at the end of a polymer chain and a monomer is produced. The end of a polymer chain is described by points with one connection and a monomer is an isolated point without any connection. Therefore, the unzipping depolymerization is represented by the removal of the connection of an end point leading to an isolated point.
- Physical reaction conditions like the removal of monomers by pumping off can be described by characterizing the isolated points to be excluded from the Monte-Carlo 'game'.

Computer program

In Fig. 5 a flow sheet of the computer program is given. The program is written in VS-Pascal and runs on an IBM 3090. A typical run needs up to 45 minutes CPU time. We used up to 75000 points representing pseudo-monomers. The initial condition is established by constructing an appropriate network of the points. Then a loop of n reaction steps follows, where n is a measure for the reaction time. Within this loop a certain number of random connection removals, bond formations and depolymerization reactions occurs.

After the n reaction steps — which is the reaction time increment — the resulting network is analyzed. The aim of this analysis is to obtain the molecular weight distribution together with the different molecular weight averages, the amount of gel, defined as the largest molecule within the polymer sample, the amount of volatile substances, the compactness of the gel from the ratio of points with two and three connections within the largest molecule.

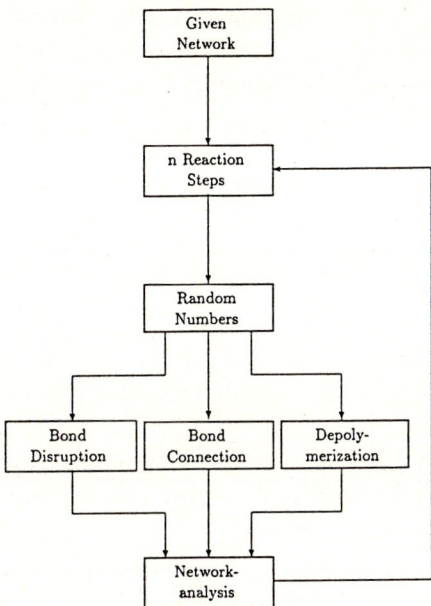

Fig. 5: Flow sheet of the computer program used to simulate thermal polymer reactions.

The analysis is realized by a coloring technique — connected points have the same color — known from percolation theory [8].

The three random number generators used modulo algorithm; one of them is used to extract numbers generated by the other two, a method proposed by Press et al. [9].

Ultrasonic decomposition of polymer gels [10, 11]

A preformed gel network is used as initial condition for the simulation of gel decomposition. In the simulation calculation only bond breaking reaction steps are considered.

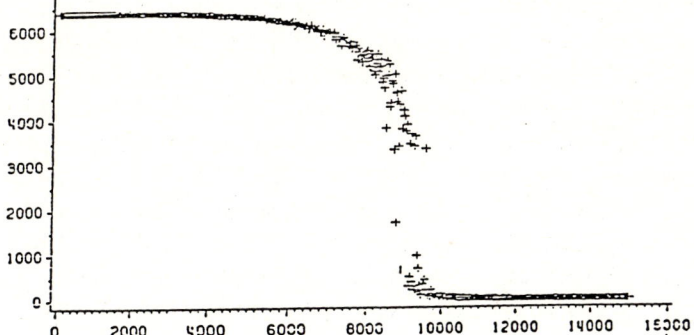

Fig. 6: Median of molecular weight vs. reaction time during the simulation of ultrasonic decomposition of a given polymer gel.

In Fig. 6 the course of the median of the molecular weight with reaction time is shown. The median means the molecular weight where the molecular weight distribution can be separated into two halfs with equal mass. The gel status discontinues, if more than one fourth of the bonds with connectivity three are broken. Then the gel decomposes totally very quickly.

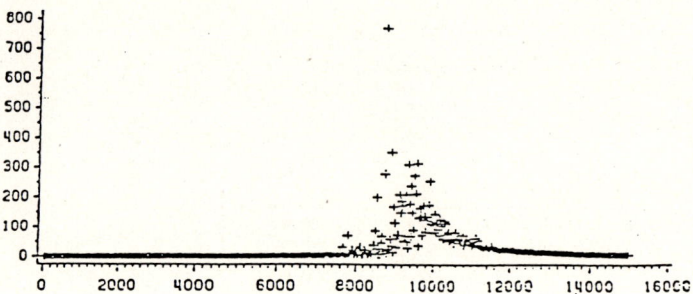

Fig. 7: The weight average of the polymer formed with reaction time during the ultrasonic decomposition of a given gel.

In Fig. 7 the weight average of the polymer (without the gel molecule) shows a maximum at the same time the gel decomposes rapidly. At the beginning of the reaction only a few small molecules are removed from the gel and at the time of decomposition of the gel, larger parts of the gel are transferred into polymer molecules which then decompose with time to smaller molecules. This behaviour is in accordance with the experimental results of ultrasonic decomposition of polymer gels. This is somehow surprising because the ultrasonic decomposition of linear polymers is not a random process but breaks polymer molecules near the middle of the polymer chain.

Thermal treatment of poly-(p-methyl-styrene)

One of the basic ideas within the mechanism of thermal gel formation considers a disruption of a bond followed be a reconnection. The breaking of a bond is the thermal initiation reaction producing two reactive free radicals. Hydrogen transfer reactions prefer the hydrogen of the p-methyl-group leading to a stabilized, less reactive free radical with benzylic structure. The only reactions these radicals can perform are recombinations. For such a model it is obvious that the number of polymer molecules (number average molecular weight) remains constant. The weight average, however, increases with reaction time, which is caused by the simultaneous production of smaller and larger molecules. Computer simulations of this model shows clearly that a gel is formed very rapidly in case the number of bond breakings and bond formations is about equal to the number of polymer molecules in the polymer sample considered.

If, as at higher temperatures, the radical decomposition reactions become important, because the metathesis reactions are less selective and the higher activation energies of these reactions are overcome by the higher temperatures, then the Monte Carlo model has to consider more bond disruptions than bond reconnections. In Fig. 8 the amount of gel with reaction time is shown for a ratio of two for the bonds broken to the bonds formed, and in Fig. 9 this ratio is three.

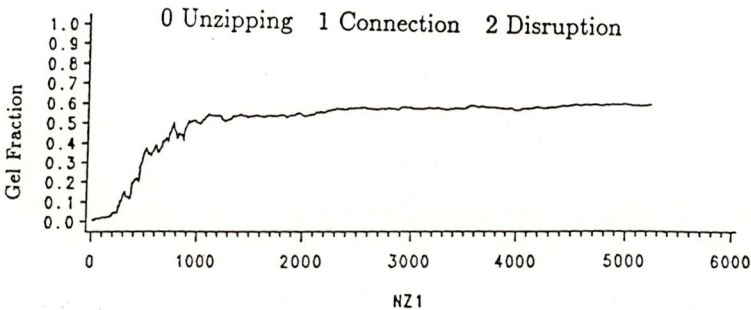

Fig. 8: Polymer gel formed using **two** bond disruptions and one bond formation in the elementary time step. No depolymerization reaction is considered

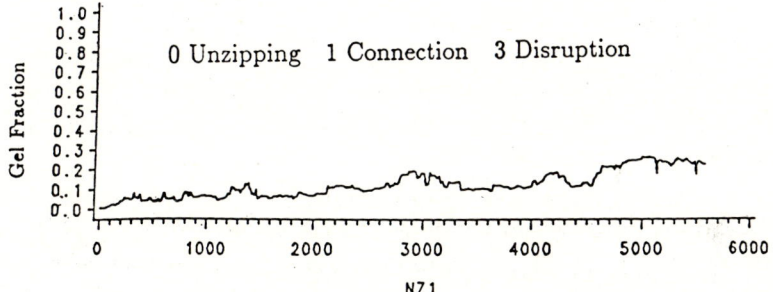

Fig. 9: Polymer gel formed using two **three** disruptions and one bond formation in the elementary time step. No depolymerization reaction is considered

Fig. 8 shows that polymer networks and gels are still formed, although twice as much bonds are broken as new ones formed. The question which immediately arises is: where is the upper limit of this ratio? Computer experiments show that there is only little or no gel formation, if more than three times as much bonds are disrupted than formed. The amount of gel formed if the ratio is exactly three to one is shown in Fig. 9. The gel formation is much smaller than in the case of Fig. 8, but still present, although considerable more chemical bonds are broken than new bonds are formed. But because the number of molecules is increasing with the course of the reaction, gel formation is accompanied by the production of very small molecules, oligomers and monomers (Fig. 10).

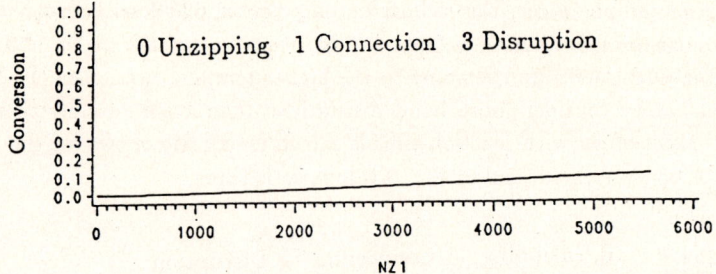

Fig. 10: Volatile fraction formed using **three** disruptions and one bond formation in the elementary time step. No depolymerization reaction is considered.

The volatile fraction formed in these computer simulations is too low compared with the experimental results. The amount of volatile molecules can be easily increased if depolymerization reactions (unzipping steps) are 'switched on' in the model calculations.

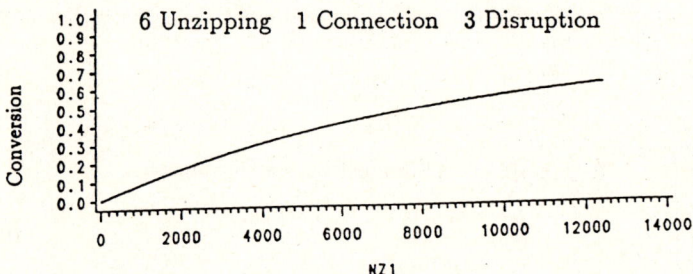

Fig. 11: Volatile fraction formed using **three** disruptions and one bond formation in the elementary time step together with **six** depolymerization steps.

An inspection of the volatile portion in the experiments of thermal polymer treatment shows, that depolymerization is the most effective type of reaction in polymer decomposition. The experiments in the closed reaction vessel lead to a much higher formation of gels and only a very small amount of low molecular weight substances is formed, whereas in experiments, where the volatile portion is pumped off, the formation of low molecular weight species is higher. This result can — in the context with computer simulations — be explained easily by the radical addition reaction of the monomers formed in depolymerization. There may exist a sort of 'dynamic equilibrium' between free radical depolymerization and free radical polymerization reactions. This unzipping and immediate readdition, which only can occur in the closed vessel experiments, leads faster to branching and therefore to gel formation within the polymer sample.

In the following experimental results of thermal treatment at 623 K are compared with computer simulations based on an unzipping-formation-disrupture ratio of 16-1-2.

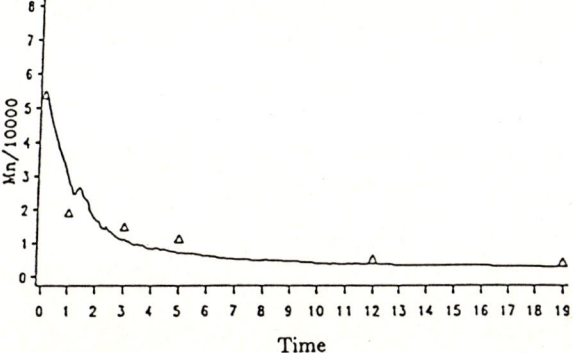

Fig. 12: Number average of the soluble polymer versus reaction time in hours at 623 K. The points are experimental data and the line computer simulations.

Fig. 12 shows the variation of the number average molecular weight of a soluble polymer is shown versus reaction time. The agreement is surprisingly well.

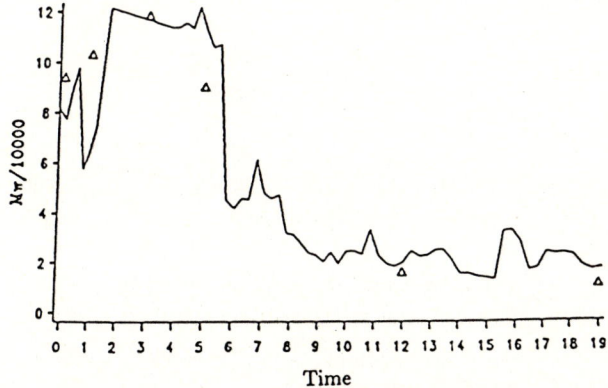

Fig. 13: Weight average of the soluble polymer versus reaction time in hours at 623 K. The points are experimental data and the line computer simulations.

Fig. 13 shows the weight average molecular weight of the soluble polymer versus reaction time, which reaches a maximum value after about 3 hours reaction time and then decreases steadily. The scattered shape of the simulations are due to the Monte Carlo method. It will become smoother the more monomer points are used in the calculation or the more simulations with different random numbers starting with the same initial conditions are calculated.

The quantity of the gel fraction (reference is the total initial polymer sample) with reaction time is shown in Fig. 14 and the respective volatile fraction (which also is called conversion) in Fig. 15. After six hours, both the gel fraction and the conversion become nearly constant and from the values it can be concluded that the

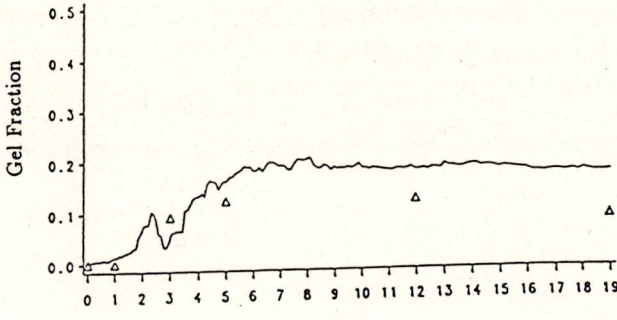

Fig. 14: Gel fraction versus reaction time. The points are experimental data and the line computer simulations.

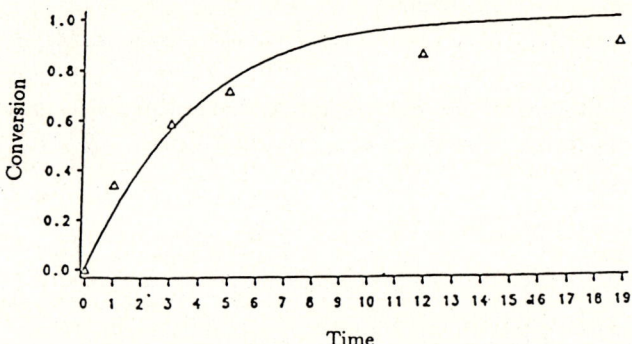

Fig. 15: Conversion or fraction of volatile compounds versus reaction time. The points are experimental data and the line computer simulations.

remaining (non volatile) polymer is nearly totally a non-soluble gel.

Outlook

For future research in this field the Monte-Carlo program should be rewritten in FORTRAN to be able to run it on a super computer, which is necessary, if extended simulations should be carried out. Radical addition reactions should be included into the simulation model as an additional feature to increase its validity. Also the mobility of the polymer molecules or parts of the molecules should be considered. That may lead to a deviation from the uniform randomization of the reaction points for bond disruption and bond formation. This might be considered as a problem of minor importance for the model simulations, but it may contribute to avoid the forming of polymer networks with forbidden topology. It should also be considered to develop a Monte Carlo program for the simulation of the thermal treatment of polymers which includes deterministic model calculations in the form of differential equations for the chemistry of the low molecular weight substances.

Literature

[1] Ebert K.H., Deuflhard P., Jäger W. (Editors)
Modelling of Chemical Reaction Systems
Springer Verlag (1980)

[2] Schröder U.
Thesis, Universität Heidelberg (1985)

[3] Kinkel J.
Thesis, Universität Heidelberg (1989)

[4] Ebert K.H., Ederer H.J., Schröder U.K.O.
Macromol. Chem., **183**, 1207 (1982)

[5] Isbarn G., Ederer H.J., Ebert K.H. in
Ebert K.H., Deuflhard P., Jäger W. (Editors)
Modelling of Chemical Reaction Systems
Springer Verlag (1980)

[6] Binder K.
Monte Carlo Methods in Statistical Physics
Springer Verlag (1986)

[7] Binder K.
Applications of the Monte Carlo Method in Statistical Physics
Springer Verlag (1987)

[8] Essam J.W.
Percolation theory
Rep. Prog. Phys., **43(7)**, 833-912 (1980)

[9] Press W.H., Flannery B.P., Teukolsky S.A., Vetterling W.T.
Numerical Recipes
Cambridge University Press (1986)

[10] Basedow A. M.
Habilitation thesis, Universität Heidelberg (1979)

[11] Mutschler H.
Thesis, Universität Heidelberg (1989)

MOLECULAR DYNAMIC SIMULATION OF THE INTERFACE AQUEOUS IONIC SOLUTION / LIPID MEMBRANE

K. Nicklas, J. Böcker, M. Schlenkrich, P. Bopp, J. Brickmann
Institut für Physikalische Chemie
Technische Hochschule Darmstadt
Petersenstrasse 20,
6100 DARMSTADT (Federal Republic of Germany)

Abstract: The interface between an ionic solution and a membrane, modelled by an ensemble of COO^- groups with translational and rotational degrees of freedom, is studied by molecular dynamics (MD) computer simulations. The charged membrane leads to a layering of the ions and the water molecules. Several water layers can be distinguished with structural properties very different from those found in the bulk phase.

1. INTRODUCTION

A detailed knowledge of the microscopic structure of ionic solutions and the particle dynamics in the interfacial region between an aqueous phase on one side, and a biomembrane on the other, is an essential prerequisite for the understanding of many elementary biological processes. An example is the mechanism of ion transport through narrow transmembrane channels where the overall transport rate is influenced by the ion association to the channel mouth, the stripping-off process of the ions hydration shell and the intra channel migration step (1).
In the present paper molecular dynamics (MD) simulation results on a model system are presented, which contains cations and water molecules between two charged model membrane surfaces. It is not the intention of this work to generate quantitative results for real membrane systems but to understand how model parameter systematically influence the structure and the dynamics of ionic solutions close to biological surfaces. In section 2. the model is described and some detailes of the simulation are summarized. The results of the simulations are shown in section 3. and in the final section some conclusions are drawn.

2. THE MODEL AND DETAILS OF THE SIMULATION

The system studied consists of 859 TIP4P water molecules (2), 60 Na^+ ions and 60 COO^- groups representing the two membrane surfaces, one on each side of a central liquid lamina. The number of degrees of freedom for the COO^- groups was reduced to translations of the center of mass within the surface plane and to rotations around the axes of symmetry. The carbon were given a mass of 54 atomic units to mimic the first elements of the lipid alkyl chain. The density of the headgroups in the membrane surface is 0.05 $Å^{-2}$ as was found in experiments (3). The model system is schematically shown in figure 1.

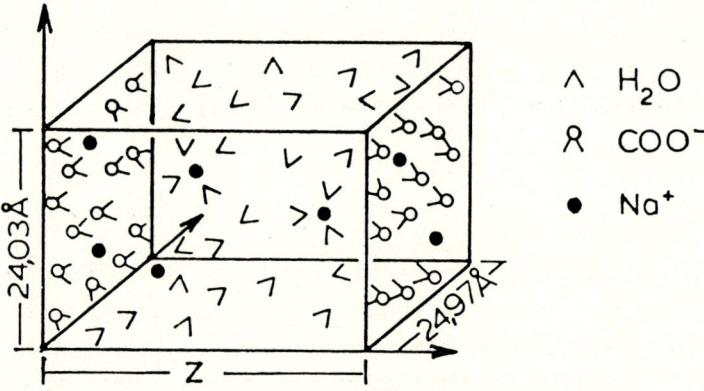

Fig.1 Sketch of the simulated system

As additional degrees of freedom both membranes were free to move along the Z coordinate. Simulations showed the system to be stable, as an equilibrium system volume was reached corresponding to a water density of 1 g cm^{-3} in the center of the lamina. The mass of each membrane was given by the sum of the COO^- group masses. Both membrane surfaces then performed an oscillation with an frequency of about 4 cm^{-1}, which is well outside the range of typical frequencies found in water or aqueous ionic solutions (4). The total potential energy of the system can be partitioned into the following contributions:

$$V_{tot} = V_{W-W} + V_{W-S} + V_{S-S} + V_{W-I} + V_{S-I} + V_{I-I} \quad [1]$$

where W stands for a water molecule, S for a headgroup in the membrane (COO⁻ groups) and I for a Na⁺ ion. The TIP4P model potential (2) is used for the intermolecular water-water interaction V_{W-W}. The other (S-S, S-I and W-S) interaction potentials were developed accordingly along the same lines as sums of Coulomb and Lennard Jones (12,6) site to site pair potentials, using the Lorentz-Berthelot combination rules and models by Jorgensen and Gao (5) and Heinzinger (6) for the COO⁻ group and the Na⁺ ion, respectively. A model potential by Bounds (7) was used for the water-ion van der Waals interaction. The range of interaction was limited by the shifted force method (8) to 7 Å and 11 Å for Lennard Jones and Coulomb interaction, respectively. Thus there was no direct interaction between the two membrane surfaces. All simulations were carried out at constant number of particles and constant total energy. Within the simulation time of 50 ps the energy drift $\Delta E/E$ was found to be less than 10^{-5}. Consequently, no rescaling of velocities had to be performed. This is important for the analysis of the dynamics of the system since a velocity scaling may lead to systematic errors.

3. STRUCTURAL RESULTS

A strong influence of the charged groups in the membrane on the adjacent liquid is to be expected from the work on ionic hydration (9) and on charged surfaces (10). In the interfacial region three structuring effects are in competition, namely that from the bulk phase, from the ionic hydration and from the membrane. Figure 2. shows the density of the sodium ions as a function of their distance from the membrane. (The positions of the carbon atoms of the COO⁻ groups are taken as a reference.) The results obtained for the two membrane surfaces limiting the model solution are superimposed in order to obtain an estimate of the statistical reliability.

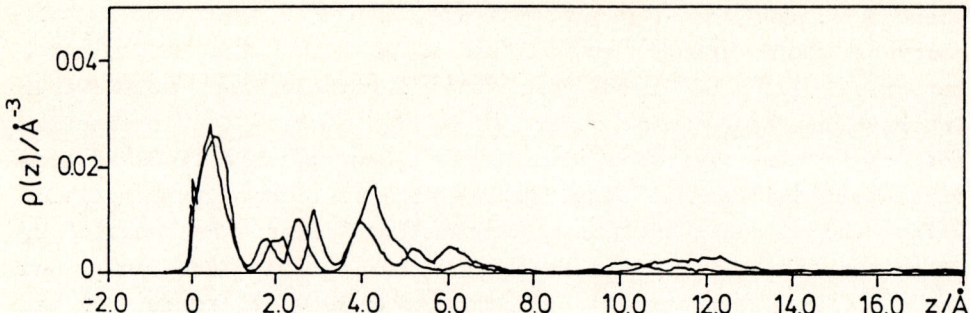

Fig.2 Density profiles for sodium ions as functions of the distance from the membrane surface. Results from both membranes are superimposed.

The most interesting feature of the simulations is the saturation of the first layer of adsorbed ions in the vincinity of the membrane with 12 to 13 ions vs. 30 COO⁻ groups. The fact that only about 40 % of the negative membrane charge is neutralized by these adsorbed ions is due to a mutual influence of the liquid and the membrane. The ions and the water molecules influence the structure of the membrane in such a way that favorable sites for the adsorption of ions are created. Figure 3 shows a sketch of a typical arrangement of model particles.

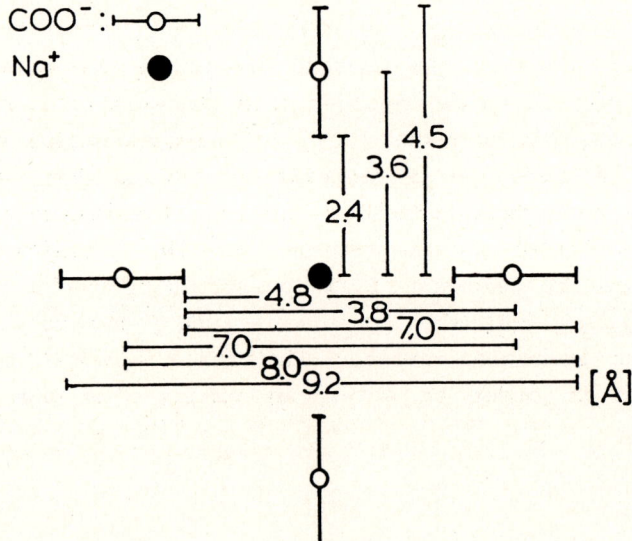

Fig.3 Typical arrangement of COO⁻ groups and Na⁺ ions on the membrane surface.

A detailed study of the distance distribution functions of the membrane groups, the water molecules (vide infra) and the ions in the vincinity of the membrane confirms this interpretation. At around 4 Å from the membrane a less pronounced second layer of ions is formed (Fig. 2). It contains about 8 ions and is separated from the first layer by an unstructured region with about 5 ions. Nearly 85 % of the membrane charge is neutralized within 5 Å from the surface.

It is essential to consider both the effect of electric charges and the packing effects in order to understand the resulting structure. In a study of Torrie (11) it was found that about 80 % of the surface charge is neutralized by the first adsorbed layer of ions if an unstructured uniformly charged wall was used for the membrane surface.

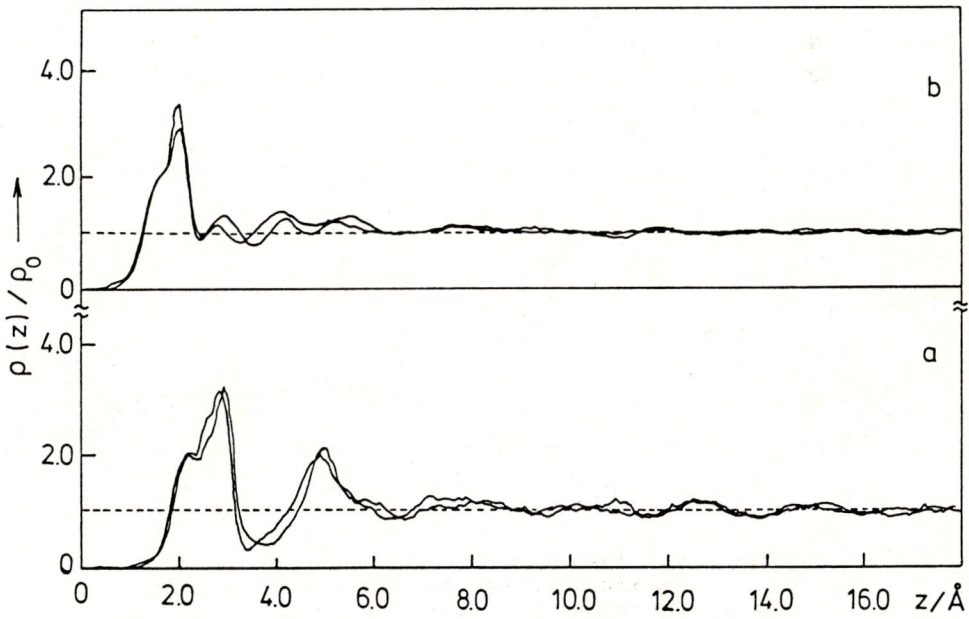

Fig.4 Density profiles for water molecule oxygens (a) and hydrogens (b) as a function of their distance from the membrane surface. Results from both membranes are superimposed.

Figure 4 shows, analogue to figure 2., the density of the water molecule oxygens and hydrogens as a function of their distances from the membrane. Density fluctuations as seen in this figure are an indication of a layering of the water molecules parallel to the membranes. In the case of the first layer this oberservation suggests the interpretation of this phenomenon as an adsorption. We shall make use of this concept of "adsorbed" layers of water in order to further characterize the structure of the water adjacent to the membranes. The shift between the oxygen and the hydrogen profiles is an indication of the average orientation of the water molecules with respect to the membrane. This orientation can be seen in figure 5a,b.

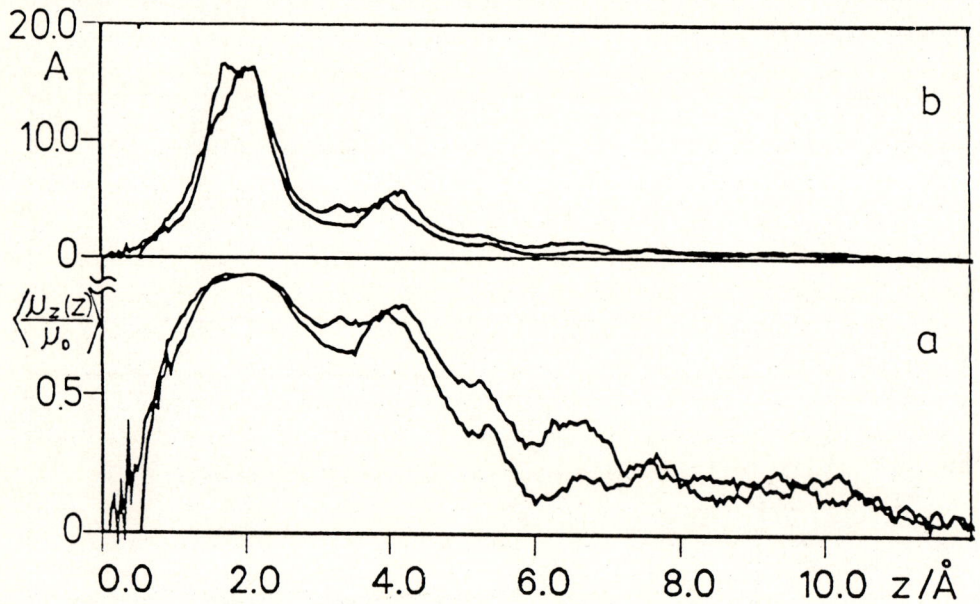

Fig.5 a: Average normalized water dipole moment component in direction of the membrane surface normal as a function of the membrane-oxygen distance.
   b: Resulting parameter A in Langevin function, as defined in the text, as a function of the membrane-oxygen distance. Results from both membranes are superimposed.

Figure 5a gives the values of the average normalized water dipol moment ($<\mu_z>/\mu_0$) parallel to the membrane surface-normal as a function of the oxygen-membrane distance. Figure 5b shows the value of the parameter $A=\mu_0 E_z/k_B T$ computed from curve 5a by means of Langevin equation:

$$<\mu_z>/\mu_0 = \coth(A) - 1/A \qquad [2]$$

where E_z is an effective electric field responsible for the orientation of the water molecules. Again the results from both interfaces are superimposed.

Since only very few water molecules are located within a distance of 1 Å from the membrane, the two curves in figure 5 show larger deviations in this region. The water molecules may penetrate into the interior of the membrane. In the present simplified model they attach to the positive carbon atom of the COO^- groups and sometimes reenter the liquid lamina. Almost complete orientational order is found for water molecules within a distance of 2.5 Å from the membrane. Due to the negative membrane charge the positive ends of the water dipols point towards the membrane surface. Beyond 2.5 Å the curves show an oscillating decay of orientational order followed by an continous increase of disorder beginning at about 5 Å membrane distance. The oscillations in orientational order correspond to the density oscillations seen in figure 3. A number of simulations (12) using different cut off radii for the Coulomb interaction in the range of 7 to 12 Å showed that the water dipole orientation as well as the oxygen and hydrogen density distribution were only weakly affected by this parameter.

For a detailed analysis, the liquid phase has been subdivided into layers parallel to the membranes. They are defined in terms of the oxygen distances from the membrane surfaces (Tab. 1). Oxygen-oxygen distribution functions (Fig. 6) have been evaluated independently for the various layers. These distribution functions have been normalized to unity at large distances. They are the analogue to the usual radial distribution function (RDF) in homogeneous systems. The results for the two membranes are again superimposed. The carbon-carbon distribution function of the COO^- groups are also plotted for comparison in this figure (curve a).

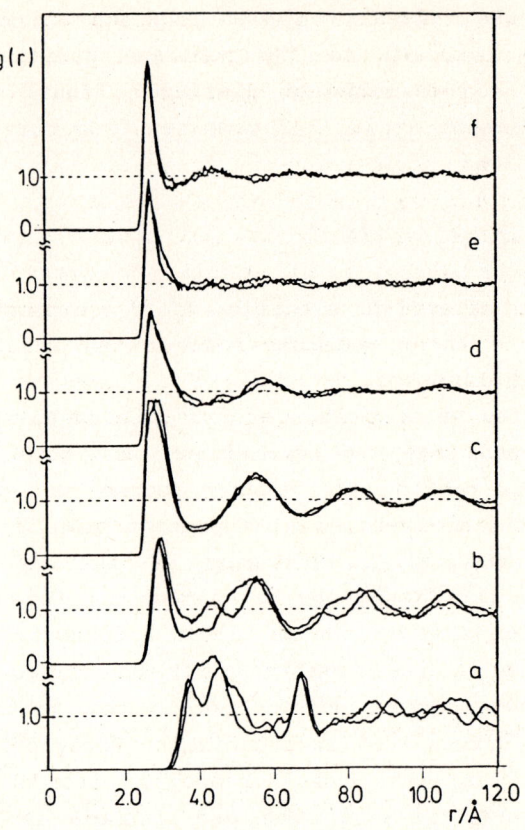

Fig.6 a: Carbon-carbon distribution function obtained for the membrane surface.
b-f: Oxygen-oxygen distribution functions obtained for the water layers parallel to the membrane surface defined in table 1. Results from both membranes are superimposed

Tab.1 Definition of water layers parallel to the membrane surface

Layer	b	c	d	e	f
Distance from membrane surface [Å]	1.7 - 2.5	2.5 - 3.3	4.4 - 5.4	6.9 - 7.9	15.0 - 16.0

In the first water layer (curve b), "adsorbed" to the membrane, the oxygen-oxygen distribution function displays a structure ranging up to 10 Å. This range is similar to that found for the carbon-carbon distribution function in the membrane, and larger than the typical ones found in water or aqueous solutions. There is a negative correlation between the C-C distances and the O-O distances in the first adsorbed water layer. The visualization of the particle trajectories (12) showed that this correlation corresponds to the predominant residence of the water molecules between (in terms of their x/y coordinates) the COO$^-$ groups. Going to larger membrane distances a continous decrease of the membrane influence on the oxygen-oxygen distribution functions is found. Curve f, corresponding to the central part of the lamina, shows the typical structure of pure bulk water. In contrast to simulations of an ion free membrane-water system, a structureless transition region was not found between the adsorbed water and the bulk water (12).

SUMMARY AND CONCLUSION

We have investigated the interface aqueous ionic solution/charged membrane with the aid of molecular dynamics simulation techniques and analyzed the structural consequences on the liquid and the membrane. The results of our investigations can be summarized as follows. The negative membrane charge is only partially neutralized by sodium ions directly adsorbed at the membrane surface. This corresponds to the limited density of favorable adsorbtion sites. It was attempted to make the system sufficiently large to include the complete transition region between bulk water and water adsorbed to the membrane. The density of the water is increased by a factor of about 2 next to the membrane. The overall influence of the membrane on the water density, the dipole orientation and the oxygen-oxygen distribution function of water extends about 6 Å into the liquid. These results show, that the structural properties of an ionic solution in the vincinity of a lipid membrane are substantially changed.

ACKNOWLEDGEMENT

A grant of computer time by the Höchstleitungsrechenzentrum (HLRZ) at the Kernforschungsanlage Jülich is accknowkedged. Financial support was given from the Deutsche Forschungsgemeinschaft and from the Fonds der Chemischen Industrie, Frankfurt. P.B. thanks the Deutsche Forschungsgemeinschaft for a Heisenberg fellowship

REFERENCES

1   Skerra A.,Brickmann J. (1987) Biophys.J. 51:969, and references therin
    Schlenkrich M., Bopp P.,Knoblauch M.,Skerra A., Brickmann J. (1988) Advances in Biotechnology of Membrane Ion Transport, edited by Jorgensen P.L., Verna A., Raven Press New York
2   Jorgensen W.L.,Chandrasekhar J.,Madura J.D., Impey R.W., Klein M.L. (1983) J.Chem.Phys. 79:926
3   Jain M. (1988) Biological membranes. John Wiley & Sons New York
4   Jancsó G., Bopp.P, Heinzinger K. (1984) Chem.Phys. 85:377
5   Jorgensen W.L., Gao J. (1986) J.Phys.Chem. 90:2174
6   Heinzinger K. (1985) Pure & Appl.Chem. 57:1031
7   Bounds D.G. (1985) Mol. Phys. 554:1335
8   Street W.B., Tildesley D.J., Saville G. (1978) ACS Symp.Ser. 86:144
9   Heinzinger K., Pálinkás G. (1987) Interaction of Water in Ionic and Non-Ionic Hydrates, edited by H.Kleeberg, Springer Verlag Berlin
10 Lee S.H., Rasaijah J.C., Hubbard J.B. (1987) J.Chem.Phys. 86:2383
11 Torrie G.M., Kusalik P.G., Patey G.N. (1989) J.Chem.Phys. 90:4513
12 Schlenkrich M., Nicklas K.,Brickmann J., Bopp P. Ber.Bunsen.Phys.Chem. 1990 in press

ELECTROANALYTICAL SIMULATIONS. 10.[1] THE SIMULATION OF FAST SECOND ORDER REACTIONS IN ELECTROCHEMICAL SYSTEMS

Bernd Speiser

Institut für Organische Chemie, Auf der Morgenstelle 18, D-7400 Tübingen 1, FRG

Abstract: Two methods for the simulation of fast chemical second order reactions coupled to an electron transfer at an electrode are used in conjunction with orthogonal collocation to calculate theoretical cyclic voltammograms: spline collocation and the technique of the heterogeneous equivalent. The techniques prove to produce reliable numerical results for the model of an electrochemically induced irreversible dimerization reaction after the application of a non-linear space coordinate transformation. Also, simulations of the ECE/DISP model (two electron transfers coupled by a chemical equilibrium and a disproportionation reaction) with the spline collocation method are discussed. It is shown in this case that the assumption about the DISP equilibrium usually required for the solution of the model equations can be dropped and the spline algorithm can be used throughout the parameter space.

INTRODUCTION

Processes at an interface between an electrode and an electrolyte can be investigated by means of electroanalytical techniques. In particular, cyclic voltammetry [1,2,3] has been shown to be a flexible tool for kinetic analyses due to the fact that the time scale of the experiment is easily varied by changing the potential scan rate v.

For a quantitative analysis of cyclic voltammetric data — e. g. the determination of reaction parameters such as rate or equilibrium constants, formal potentials of the electron transfer reaction or transfer coefficients — numerical techniques to compare model simulations and experimental results are essential. We have developed a system of computer programs (EASIEST) which supports this analysis [4,5,6].

---

[1] Part 9 of this series: Hertl P, Speiser B (1988), J Electroanal Chem Interfac Electrochem 250:237

One of the main problems in electroanalytical simulations is the treatment of fast chemical reactions coupled to the electron transfer. In these cases, the model consists of *stiff* differential equations. Their solution leads to an inacceptable increase of computer memory and *cpu* time if the implicit finite difference method is used for integration (see e.g. [7]).

Furthermore, within the diffusion layer adjacent to the electrode, for fast chemical reactions a thin *reaction layer* has to be modelled. The technique of orthogonal collocation [8] yields erroneous results in this case since information about the reaction layer is lost due to the spatial discretization performed [9].

In the simulation programs CVSIM and CASIM, which are part of EASIEST, we have implemented two techniques to overcome these problems: *spline collocation* [10], where the reaction and the diffusion layer are discretized separately and the resulting concentration profiles are splined together, and the *technique of the heterogeneous equivalent* [9,11] which originally has been developed for explicit finite difference schemes [12,13,14] and which removes the stiffness of the differential equation system by a quasi-steady state treatment.

These techniques will be used here to simulate cyclic voltammograms of very fast second order chemical reactions coupled to charge transfer steps at the electrode surface. Two reaction schemes which frequently occur in organic and inorganic electrochemistry will be considered: the ED mechanism, where an irreversible dimerization is initiated electrochemically, and the ECE model where two electron transfers are coupled by a chemical equilibrium. As a complication in the latter case, an additional disproportionation equilibrium is taken into account (ECE/DISP mechanism).

THEORY

General: A general dimensionless model equation for a species l diffusing to and from the electrode and reacting in homogeneous solution may be given as

$$\frac{\partial c_l^*}{\partial T'} = \beta \frac{\partial^2 c_l^*}{\partial X^2} + \rho^*(c^*) \tag{1}$$

where the concentration c_l^* is a function of space X and time T' and β is a dimensionless diffusion coefficient (for the definition of the dimensionless quantities, see e.g. [15]). The reaction term $\rho^*(c^*)$ is comprised of all rate law expressions containing the concentration c_l^*

$$\rho^*(c^*) = \sum_i \kappa_i \left(\prod_j c_j^* \right) \tag{2}$$

where the κ_i are dimensionless rate constants. The initial and boundary conditions for this partial differential equation depend on the mechanism chosen. The latter are usually derived from flux conditions at the surface and NERNST type equations.

Orthogonal collocation provides matrix elements $A_{i,j}$ and $B_{i,j}$, which only depend on the type and order N of a polynomial used to approximate the exact relationship $c_l^* = f(X, T')$ [16], such that

$$\left.\frac{\partial c_l^*}{\partial X}\right|_{X_i} = \sum_{j=1}^{N+2} A_{i,j} c_l^*(X_j, T') \qquad i = 2, \ldots, N+1 \tag{3}$$

and

$$\left.\frac{\partial^2 c_l^*}{\partial X^2}\right|_{X_i} = \sum_{j=1}^{N+2} B_{i,j} c_l^*(X_j, T') \qquad i = 2, \ldots, N+1 \tag{4}$$

Substituting (4) in (1) gives a system of ordinary differential equations, which can be solved for the $c_l^*(X_i, T')$ at the *collocation points* X_i.

Spline collocation: With spline collocation each equation of type (1) yields two partial differential equations, one for the reaction layer (index "1"):

$$\frac{\partial c_{l1}^*}{\partial T'} = \beta' \frac{\partial^2 c_{l1}^*}{\partial X_1^2} + \rho^*(c_1^*) \tag{5}$$

the other one for the diffusion layer (index "2"):

$$\frac{\partial c_{l2}^*}{\partial T'} = \beta' \frac{\partial^2 c_{l2}^*}{\partial X_2^2} + \rho^*(c_2^*) \tag{6}$$

Detailed derivations are given in [10].

Heterogeneous equivalent technique: When using the heterogeneous equivalent technique, it is assumed for each species m reacting very fast that

$$0 = \beta \frac{d^2 c_m^*}{dX^2} + \rho^*(c^*) \tag{7}$$

While RUŽIĆ and FELDBERGs approach [12,13,14] relied on an explicit solution of this ordinary differential equation, it has been shown, that application of orthogonal collocation to (7) yields a system of algebraic equations [11]. Usually, this system includes the boundary conditions. If all reactions in $\rho^*(c^*)$ are first order and the boundary conditions are linear, a system of linear equations results with the $c_m^*(X_i', T')$ in the reaction layer as unknowns (X_i' are the coordinates of the collocation points in the reaction layer). This system is easily solved, e.g. by GAUSS elimination and yields the concentration of c_m^* at the electrode surface ($X' = 0$) [11]. This value in turn is needed as a boundary concentration for the integration of the remaining partial differential equations (1) with $l \neq m$.

So far, the combination of orthogonal collocation and heterogeneous equivalent has not been applied to models with second order terms in $\rho^*(c^*)$. In this case, from (7) a system of non-linear equations arises, which can be solved iteratively.

Non-linear transformation of the space coordinate: Due to the second order reaction, in the case of the ED reaction the concentration profile is expected to be extremely steep in the reaction layer and to extend further into the solution than for a first order reaction. To account for this feature, a non-linear space coordinate transformation

$$U = \frac{X}{1+X} = \frac{x/x_R}{1+x/x_R} = \frac{x}{x_R + x} \tag{8}$$

is introduced, where x_R is an expression for the extension of the reaction layer. Both, the spline collocation equations of type (5) or (6) and equation (7) may be transformed by (8). Details of the resulting model equation system will be given elsewhere [17].

The effect of the transformation (8) is to expand the reaction layer in the U space, thereby decreasing the steepness of the concentration profile and to extend the simulated space from a finite part of the x coordinate to infinity.

Model mechanisms: The ED mechanism

$$A \underset{}{\overset{\pm e^-}{\rightleftharpoons}} B \quad E^0 \tag{9}$$

$$2\,B \overset{k_1}{\longrightarrow} C \quad k_1 \tag{10}$$

can be modelled by

$$\frac{\partial c_A^*}{\partial T'} = \beta \frac{\partial^2 c_A^*}{\partial X^2} \tag{11}$$

$$\frac{\partial c_B^*}{\partial T'} = \beta \frac{\partial^2 c_B^*}{\partial X^2} - \kappa_1 c_B^{*\,2} \tag{12}$$

$$\frac{\partial c_C^*}{\partial T'} = \beta \frac{\partial^2 c_C^*}{\partial X^2} + \frac{\kappa_1}{2} c_B^{*\,2} \tag{13}$$

with $\kappa_1 = k_1 / \frac{F}{RT} v$. The initial conditions are

$$T' = 0, 0 \leq X \leq 1 : c_A^*(X,0) = 1,\ c_B^*(X,0) = 0,\ c_C^*(X,0) = 0 \tag{14}$$

At a large distance from the electrode

$$T' > 0, X = 1 \ :\ c_A^*(1,T') = 1,\ c_B^*(1,T') = 0,\ c_C^*(1,T') = 0 \tag{15}$$

holds, whereas at the electrode surface ($X = 0$)

$$T' > 0, X = 0 \quad : \quad \frac{c_A^*(0, T')}{c_B^*(0, T')} = \theta_{A/B} S(T') \tag{16}$$

$$\left(\frac{\partial c_A^*}{\partial X}\right)_{X=0} = -\left(\frac{\partial c_B^*}{\partial X}\right)_{X=0} \tag{17}$$

$$\left(\frac{\partial c_C^*}{\partial X}\right)_{X=0} = 0 \tag{18}$$

Analogous formulations have been given by other authors [18,19,20] for use with finite difference algorithms or integration in LAPLACE space.

By application of spline collocation or collocation in combination with the heterogeneous equivalent with or without the use of transformation U four sets of model equations have been derived [17].

The ECE/DISP mechanism is given by [21]

$$A \underset{}{\overset{\pm e^-}{\rightleftharpoons}} B \qquad E_{A/B}^0 \tag{19}$$

$$B \underset{k_{-1}}{\overset{k_1}{\rightleftharpoons}} C \qquad K_1 = k_1/k_{-1} \tag{20}$$

$$C \underset{}{\overset{\pm e^-}{\rightleftharpoons}} D \qquad E_{C/D}^0 \tag{21}$$

$$B + C \underset{k_{-2}}{\overset{k_2}{\rightleftharpoons}} A + D \qquad K_2 = k_2/k_{-2} = \exp[\pm \frac{F}{RT}(E_{C/D}^0 - E_{A/B}^0)] \tag{22}$$

with $\kappa_1 = k_1/\frac{F}{RT}v$, $\kappa_{-1} = k_{-1}/\frac{F}{RT}v$ and $\kappa_2 = k_2 c_A^0/\frac{F}{RT}v$. In equation (22) the "+" sign corresponds to a reduction, the "−" sign to an oxidation process. In this case, the partial differential equation model is

$$\frac{\partial c_A^*}{\partial T'} = \beta \frac{\partial^2 c_A^*}{\partial X^2} + \kappa_2 c_B^* c_C^* - \frac{\kappa_2}{K_2} c_A^* c_D^* \tag{23}$$

$$\frac{\partial c_B^*}{\partial T'} = \beta \frac{\partial^2 c_B^*}{\partial X^2} - \kappa_1 c_B^* + \kappa_{-1} c_C^* - \kappa_2 c_B^* c_C^* + \frac{\kappa_2}{K_2} c_A^* c_D^* \tag{24}$$

$$\frac{\partial c_C^*}{\partial T'} = \beta \frac{\partial^2 c_C^*}{\partial X^2} + \kappa_1 c_B^* - \kappa_{-1} c_C^* + \kappa_2 c_B^* c_C^* - \frac{\kappa_2}{K_2} c_A^* c_D^* \tag{25}$$

$$\frac{\partial c_D^*}{\partial T'} = \beta \frac{\partial^2 c_D^*}{\partial X^2} + \kappa_2 c_B^* c_C^* - \frac{\kappa_2}{K_2} c_A^* c_D^* \tag{26}$$

The initial conditions are given by

$$T' = 0, 0 \leq X \leq 1 \; : \; c_A^*(X,0) = 1, \, c_B^*(X,0) = 0, \, c_C^*(X,0) = 0,$$

$$c_D^*(X,0) = 0 \tag{27}$$

They remain constant at $X = 1$ throughout the simulation

$$T' > 0, X = 1 \; : \; c_A^*(1,T') = 1, \, c_B^*(1,T') = 0, \, c_C^*(1,T') = 0,$$

$$c_D^*(1,T') = 0 \tag{28}$$

while we formulate at the electrode surface

$$T' > 0, X = 0 \; : \; \left(\frac{\partial c_A^*}{\partial X}\right)_{X=0} = -\left(\frac{\partial c_B^*}{\partial X}\right)_{X=0} \tag{29}$$

$$\left(\frac{\partial c_C^*}{\partial X}\right)_{X=0} = -\left(\frac{\partial c_D^*}{\partial X}\right)_{X=0} \tag{30}$$

$$\frac{c_A^*(0,T')}{c_B^*(0,T')} = \theta_{A/B} S(T') \tag{31}$$

$$\frac{c_C^*(0,T')}{c_D^*(0,T')} = \theta_{C/D} S(T') \tag{32}$$

Equations have been derived for the spline collocation technique [17].

All model formulations for the ED and the ECE/DISP mechanisms have been implemented into the framework of EASIEST. Simulations were performed on a CONVEX C2 using vectorized FORTRAN code and DDEBDF as an integrator [22]. Systems of non-linear equations were solved with the MINPACK routine HYBRJ1 [23].

RESULTS and DISCUSSION

ED model: Results for numerical simulations of all four formulations of the ED model are compared to limiting case and literature values in Table 1. For a small value of κ_1 the ED model shows a transition to the simple reversible electron transfer: the influence of the dimerization reaction becomes negligible. For large values of κ_1 SAVÉANT and VIANELLO [18] as well as NICHOLSON [19] have found a limiting case, where $\sqrt{\pi}\chi_p = 0.526$ and the peak potential shifts 30 mV for a ten fold increase of κ_1. In the medium range of rate constants, OLMSTEAD et al. give some numerical results of simulations [20].

Both spline collocation and the heterogeneous equivalent approach without a non-linear transformation give results that differ from the literature values, especially for fast rate constants. If we apply transformation U, however, the maximum deviation is 1 mV for $E_p - E^0$ and $\approx 0.5\%$ for $\sqrt{\pi}\chi_p$. These are also the estimated errors of the simulated data.

Table 1: Numerical simulation results for the ED model in comparison with the reversible limiting case and literatur values.

κ_1	reference values		spline collocation			
			without transformation		with transformation U	
	$\sqrt{\pi}\chi_p$	$E_p - E^0$ mV	$\sqrt{\pi}\chi_p$	$E_p - E^0$ mV	$\sqrt{\pi}\chi_p$	$E_p - E^0$ mV
10^{-6}	0.4463	+28.5 [a]	0.4463	+28	0.4451	+28
1.5	0.476	+20 [b]	0.4760	+21	0.4750	+21
10^6	0.526	−92 [c]	0.5356	−89	0.5289	−91
κ_1	reference values[c]		heterogeneous equivalent			
			without transformation		with transformation U	
	$\sqrt{\pi}\chi_p$	$E_p - E^0$ mV	$\sqrt{\pi}\chi_p$	$E_p - E^0$ mV	$\sqrt{\pi}\chi_p$	$E_p - E^0$ mV
10^2	0.526	−13	0.5253	−13	0.5276	−13
10^4	0.526	−52	0.5155	−54	0.5276	−52
10^6	0.526	−92	0.5021	−101	0.5276	−92

[a] reversible electron transfer; [b] values from [20]; [c] values from [19].

Thus, we conclude that our model formulations for the ED mechanism yield correct values if the non-linear space coordinate transformation U is used.

Only an increase of about 30% of the *cpu* time is observed if κ_1 is changed from 10^{-6} to 10^6 in the case of the spline algorithm: if the simulation extends over 500 mV and N=6 is used, *cpu* times vary between 22.6 and 29.9 s for the simulation of a single voltammogramm. For $\kappa_1 = 10^6$ spline collocation and heterogeneous equivalent use comparable amounts of *cpu* time.

It is interesting to note that the heterogeneous equivalent formulation of the ED mechanism given here fully retains the second order characteristics of the dimerization reaction. MAGNO et al. have given a similiar solution for the ED model [24], which, however, assumes a linear concentration profile of species B in the reaction layer.

ECE/DISP model: The ECE/DISP simulation results with CVSIM compare well with literature and limiting case values even without the use of transformation U [17]. Obviously, in this case the stretching of the concentration profiles is not as critical for a correct solution as in the case of the ED reaction. This may be explained by the fact, that the disappearance of B in the ECE/DISP mechanism is always controlled by the first order rate constant κ_1.

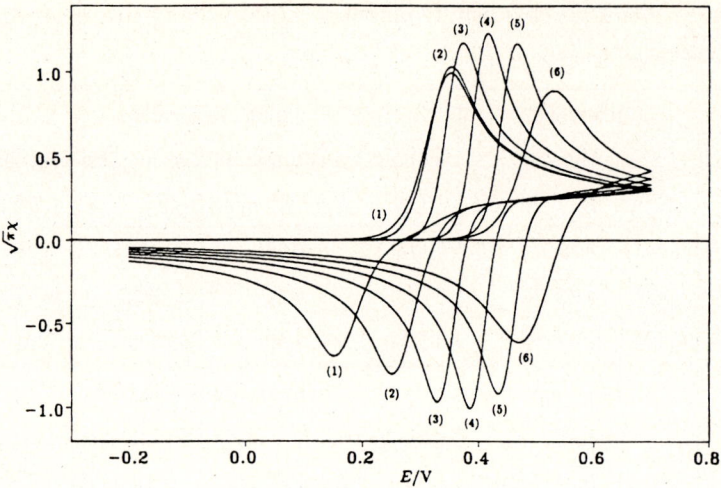

Figure 1: Simulated cyclic voltammograms for the ECE/DISP model with $E^0_{A/B} = +500$ mV, $\kappa_1 = \kappa_{-1} = \kappa_2 = 10^6$, $E^0_{C/D}$: (1) = 0. mV, (2) = +100 mV, (3) = +200 mV, (4) = +300 mV, (5) = +400 mV and (6) = +500 mV; oxidation currents are shown with a positive sign.

If the rate constants κ_1, κ_{-1} and κ_2 are varied between 10^{-6} and 10^6 the *cpu* time used for the simulation changes only by about 25%.

This is in contrast to AMATORE and SAVÉANTs observation [21] for the finite difference technique. Figure 1 gives a graphical representation of some simulated cyclic voltammograms of the ECE/DISP model. The three dimensionless rate constants κ_1, κ_{-1} and κ_2 were set to values of 10^6. While the formal potential $E^0_{A/B}$ was fixed at +500 mV, the corresponding value for the redox couple C/D was varied from 0 mV in steps of 100 mV to a value of +500 mV. This corresponds to values of 2.86×10^8, 5.82×10^6, 1.19×10^5, 2.41×10^3, 49.1 and 1.00, respectively, for the equilibrium constant K_2. With increasing $E^0_{C/D}$, the disproportionation equilibrium is situated less and less to the side of the product D. Simulation (1) with $E^0_{C/D} - E^0_{A/B} = -500$ mV corresponds to AMATORE and SAVÉANTs limiting case.

If the difference between the formal potentials decreases, several changes in the cyclic voltammograms occur: the peaks shift along the potential axis to more positive potentials. This is particularly true for the peak on the second part of the voltammogram (negative current function in Figure 1). The peak shape becomes sharper in curves (3) – (5) and then broader again [curve (6)]. The peak currents of both the oxidation and the reduction peaks strongly increase for intermediate values of $E^0_{C/D} - E^0_{A/B}$ and decrease again if the difference of the formal potentials approaches zero. If one considers only the first part of the voltammetric cycle, changes in the curves can be observed as soon as $E^0_{C/D} - E^0_{A/B}$ increases above -400 mV.

These changes can not be attributed to the change of the equilibrium constant K_2 since a similiar behaviour is observed in the ECE model, where only reactions (19) – (21) operate. Thus, the effect is asociated with the difference between the formal potentials of the two redox processes. From the results given here, it follows that experimental cyclic voltammograms of ECE/DISP systems may only be analysed by comparison to the limiting case curves if $E^0_{C/D}$ is at least 400 mV less positive (for oxidations) or less negative (for reductions) than $E^0_{A/B}$.

Systematic investigations elucidating the behaviour of the ECE/DISP model in the entire parameter space are currently under way.

CONCLUSION

Orthogonal collocation in its spline variant and in combination with the heterogeneous equivalent technique can be used to simulate fast second order reactions coupled to electron transfers at an electrode. In contrast to the finite difference methods only a small increase of *cpu* time with the model rate constants is observed. Thus, the algorithms on the basis of orthogonal collocation seem to overcome the problem of exceeding *cpu* times for simulations with very fast chemical reactions coupled to the electron transfer. The use of various formulations of a model ("general case", "limiting case", see e.g. [25]) is no longer needed if one wants to perform simulations in extended regions of the parameter space.

In the case of the ED model mechanism no linearization of the concentration profiles has to be assumed in the heterogeneous equivalent approach. In the case of the ECE/DISP mechanism the influence of the difference of the formal potentials can be investigated. Here, the applicability of the usual approximation $E^0_{C/D} - E^0_{A/B} \rightarrow \infty$ can be shown to be limited to $E^0_{C/D} - E^0_{A/B} < -400$ mV under the conditions used in the calculations.

ACKNOWLEDGEMENT

The author thanks the Deutsche Forschungsgemeinschaft, Bonn–Bad Godesberg for a fellowship and Dr. Peter Urban for several important discussions.

REFERENCES

[1] Speiser B (1981) Chem in uns Zeit 15:62

[2] Mabbott G A (1983) J Chem Educ 60:697

[3] Heinze J (1984) Angew Chem 96:823

[4] Speiser B (1985) Anal Chem 57:1390

[5] Speiser B (1989) In: Gauglitz G (ed), *Software-Entwicklung in der Chemie 3*, Proceedings of the workshop "Computer in der Chemie", Tübingen, 1988, Springer, Berlin, 333

[6] Speiser B, Comput Chem, accepted for publication

[7] Savéant J M, Xu F (1986) J Electroanal Chem Interfac Electrochem 208:197

[8] Pons S (1984) In: Bard A J (ed) Electroanal Chem 13:115

[9] Speiser B (1984) J Electroanal Chem Interfac Electrochem 171:95

[10] Hertl P, Speiser B (1987) J Electroanal Chem Interfac Electrochem 217:225

[11] Hertl P, Speiser B (1987) J Electroanal Chem Interfac Electrochem 235:57

[12] Ručić I, Feldberg S W (1974) J Electroanal Chem Interfac Electrochem 50:153

[13] Ručić I (1983) J Electroanal Chem Interfac Electrochem 144:433

[14] Ručić I (1985) J Electroanal Chem Interfac Electrochem 189:221

[15] Speiser B, Rieker A (1979) J Electroanal Chem Interfac Electrochem 102:1

[16] Villadsen J, Michelsen M L (1978) *Solution of differential equation models by polynomial approximation*, Prentice Hall, Englewood Cliffs

[17] Speiser B, Habilitationsschrift, Universität Tübingen, in preparation

[18] Savéant J M, Vianello E (1963) C R Acad Sc Paris 2597

[19] Nicholson R S (1964) Anal Chem 37:667

[20] Olmstead M L, Hamilton R G, Nicholson R S (1969) Anal Chem 41:260

[21] Amatore C, Saveant J M (1978) J Electroanal Chem Interfac Electrochem 86:227

[22] Shampine L F, Watts H A, DEPAC package overview, subroutine DDEBDF, SLATEC program library

[23] Moré J J, Garbow B S and Hillstrom K E (1980) *User Guide for* MINPACK-1, Argonne National Laboratory, Argonne

[24] Magno F, Perosa D, Bontempelli G (1985) Anal Chim Acta 173:211

[25] Amatore C, Saveant J M (1979) J Electroanal Chem Interfac Electrochem 102:21

THE RASHEVSKY-TURING SYSTEM: TWO COUPLED OSCILLATORS AS A GENERIC REACTION-DIFFUSION MODEL

M. KLEIN, G. BAIER AND O.E. RÖSSLER

Department of Theoretical Chemistry, University of Tübingen,
Auf der Morgenstelle 8, D-7400 Tübingen, FRG

A two-compartment reaction-diffusion system is investigated both analytically and numerically. Owing to the system's symmetry, a complete characterization of the three fixed points and their respective stability is achieved. Dependent on the coupling strength, different types of chaos can be observed in computer simulations. Some implications of the model are pointed out.

INTRODUCTION

Following the ideas of Rashevsky [1] who proposed a simple reaction-diffusion-transport equation to explain the occurrence of spatial polarity in a symmetrical cell, and Turing [2] who described a breakdown of symmetry in a multi-cellular morphogenetic system, Rössler and Seelig [3] developed a simplified version of the original model. They discussed a two-compartment, generic reaction-diffusion Rashevsky-Turing (RT)-system. It is based on two symmetrically built, homogeneous reaction oscillators with only one Michaelis-Menten-type nonlinearity. They are diffusively coupled with the effect of a cross-inhibition. The system represents a general dynamical model for the qualitative understanding of spontaneous differentiation, breakdown of symmetry, bistable flip-flop behavior and chaotic turbulent dynamical states [4]. Sporns and Seelig [5] used the RT-system as an enzyme-induction, substrate-inhibition model to simulate the biochemical coordination of cells which exchange molecules through communicating junctions ('gap junctions'). They added further nonlinear coupling terms which, nevertheless, do not significantly change the dynamical behavior of the basic system. Röhricht et al. [6] gave a first concrete physical interpretation of the RT-prototypic model. They presented experimental evidence that the spatio-temporal nonlinear behavior (impurity-impact-ionization-induced avalanche breakdown) in semiconducting germanium can be described qualitatively by the present four-variable RT-model. They took the morphogen A to represent the number density of moving charge carriers and B the mean energy per carrier.

THE SYSTEM

The reaction scheme of the simple two cellular RT-system is sketched in Figure 1.

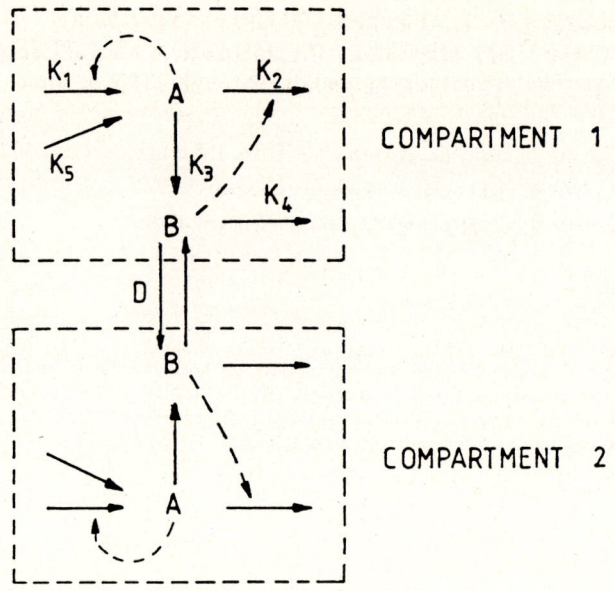

Fig. 1 Reaction scheme of the RT-System. (Constant pools were omitted from the sketch, catalytic rate control is indicated by dashed arrows.)

If we assume isothermy and homogeneity in either compartment, approximate identity of activity and concentration and fast relaxations of intermediate products, the four-variable ordinary differential equation for the reaction scheme reads:

$$\dot{x}_1 = (k_1 - k_3)x_1 - k_2 y_1 \frac{x_1}{x_1 + K} + k_5$$
$$\dot{y}_1 = k_3 x_1 - k_4 y_1 + D(y_2 - y_1)$$

$$\dot{y}_2 = k_3 x_2 - k_4 y_2 + D(y_1 - y_2)$$
$$\dot{x}_2 = (k_1 - k_3)x_2 - k_2 y_2 \frac{x_2}{x_2 + K} + k_5 \quad , \tag{1}$$

where $x_i, y_i (i = 1, 2)$ are the concentrations of the two morphogens A and B in compartment 1 and 2, respectively, the components $k_1, \ldots, k_5$ represent constant reaction rates, with $k_1 > k_3$, and K is a phenomenological Michaelis-Menten constant. The nonlinear Michaelis-Menten-type kinetic simulates a 'steady-state' approximation of some fast decaying intermediate products.

D is the diffusion coefficient of the cross-inhibiting coupling of the two cells between the two morphogens B. The excess of self-inhibition of the morphogens A via B within each cell, over the cross-inhibition generated by the other cell, is compensated by an autocatalysis of the morphogen A, within the cell itself so that in between the cells a net cross-inhibition survives.

STABILITY ANALYSIS OF THE FIXED POINTS

Dimensionless Variables

To simplify the mathematical treatment, we introduce dimensionless variables and parameters (with $i = 1, 2$):

$$a_i = x_i(k_1 - k_3)/k_5$$
$$b_i = y_i k_2/k_5$$
$$\tau = t(k_1 - k_3)$$
$$P_1 = K(k_1 - k_3)/k_5$$
$$P_2 = k_2 k_3/(k_1 - k_3)^2$$
$$P_3 = k_4/(k_1 - k_3)$$
$$P_4 = D/(k_1 - k_3).$$

The system now reads

$$\dot{a}_1 = a_1 - b_1 \frac{a_1}{a_1 + P_1} + 1.0$$
$$\dot{b}_1 = P_2 a_1 - P_3 b_1 + P_4(b_2 - b_1)$$
$$\dot{b}_2 = P_2 a_2 - P_3 b_2 + P_4(b_1 - b_2)$$
$$\dot{a}_2 = a_2 - b_2 \frac{a_2}{a_2 + P_1} + 1.0 \quad . \tag{2}$$

Fixed Point in the Uncoupled Case

The model is built from two potentially oscillating symmetrical subsystems. With $P_4 = 0$ we geometrically find the fixed point (a_u, b_u) within each compartment as the intersection of the line given by $b_u = (a_u P_2)/P_3$ with the hyperbola described as $b_u = a_u + P_1/a_u + (P_1 + 1)$. We obtain a quadratic equation in b_u. Physically relevant ($a_u > 0$, $b_u > 0$) are only those points with $J = P_2/P_3 > 1$ or, equivalently, $P_2 > P_3$. This yields

$$b_u = J a_u = \frac{1}{2} \frac{J}{J-1} \left((P_1 + 1) + \sqrt{(P_1 - 1)^2 + 4 P_1 J} \right) \quad . \tag{3}$$

Linearizing the ODEs at these points gives one the local Jacobian matrix

$$\begin{pmatrix} f & g \\ P_2 & -P_3 \end{pmatrix}$$

with

$$f = 1 - \frac{P_1 b_u}{(a_u + P_1)^2} \quad , \quad g = -\frac{a_u}{a_u + P_1} \quad ,$$

and finally the eigenvalues

$$\lambda_{1,2} = \frac{1}{2}\left((f - P_3) \pm \sqrt{(P_3 + f)^2 + 4P_2 g}\right) \quad . \tag{4}$$

The local stability depends on these eigenvalues. The transition from real to complex-conjugate eigenvalues occurs along the line

$$P_3 = -f + 2Jg \pm \sqrt{4Jgf + 4J^2 g^2} \quad .$$

A Hopf-bifurcation of the formerly stable fixed point leads to limit-cycle oscillations within each compartment if the real part of the complex-conjugate eigenvalues crosses the imaginary axis which is the case when $P_3 = f$.

Fixed Points in the Coupled Case

We look for the fixed points (a_{ie}, b_{ie}) with $(i = 1, 2)$ of the four-variable system, Eq.(2). In the following we use:

$$K = P_3/P_4$$
$$J = P_2/P_3$$
$$P = P_1 \quad .$$

Since, in Eq.(2), the b_{1e}, b_{2e} depend linearly on the a_{1e}, a_{2e}, we may solve for b_{1e} and b_{2e}. Starting out from

$$a_{1e} = \frac{P_3}{P_2} b_{1e} - \frac{P_4}{P_2}(b_{2e} - b_{1e})$$

$$a_{2e} = \frac{P_3}{P_2} b_{2e} - \frac{P_4}{P_2}(b_{1e} - b_{2e})$$

we get

$$b_{ie} = a_{je} \frac{J}{K+2} + a_{ie} \frac{(K+1)}{(K+2)} \quad ,$$

whereby $i, j = 1, 2$ and $i \neq j$. After insertion we finally end with

$$a_{ie} = \frac{(K+2)}{J}\left(a_{je}\left(1 - \frac{J(K+1)}{(K+2)}\right) + \frac{P}{a_{je}} + (P+1)\right) \quad .$$

The last equation is of the form $a_{ie} = F(a_{je})$. Therefore, after inserting a_{je}, so that $a_{ie} = F(F(a_{ie}))$, we get a 4th-order polynomial in a_{ie}. From the symmetry of the RT-system we expect this polynomial to decay into two of quadratic order. One is essentially the already known fixed-point equation, Eq.(3), in the form

$$\left(a_{ie}^2 - \frac{P+1}{J-1} a_{ie} - \frac{P}{J-1} \right) \; .$$

It contains the fixed points of the uncoupled case, which we will now call the "symmetric" fixed points, $a_{ie} = a_u$ and $b_{ie} = b_u$. The other quadratic polynomial is

$$\left(a_{ie}^2 + \frac{(P+1)(K+2)}{(K+2)-J(K+1)} a_{ie} - \frac{P(K+2)}{JK-(K+2)} \right) \; .$$

It describes two new fixed points:

$$a_{ie} = \frac{1}{2} \left(\frac{(P+1)(K+2)}{J(K+1)-(K+2)} \right.$$
$$\left. \pm \sqrt{\left(\frac{(P+1)(K+2)}{(K+2)-J(K+1)} \right)^2 + \frac{4P(K+2)}{JK-(K+2)}} \right) \; .$$

It remaines now to find the conditions for the physically relevant fixed points, $a_{ie} > 0$.

i) The first term will be positive if:

$$J > \frac{K+2}{K+1} \; .$$

ii) In case of a real root, even the smaller a_{ie} remains positive, if:

$$J < \frac{K+2}{K} \; .$$

iii) Even if $J = (K+2)/(K+1)$, the root remains positive.

Therefore two positive solutions of a_{ie} exist whenever

$$\frac{K+2}{K+1} < J < \frac{K+2}{K} \; .$$

The b_{ie} follow from the a_{ie}, specifically

$$b_{ie} = a_{ie} + \frac{P}{a_{ie}} + (P+1) = F(a_{ie}) \; .$$

Thus, the symmetry of the RT-system is responsible for two more pairs of fixed points arising. These "asymmetric" fixed points are

$$\begin{aligned} a_{1e} &= a_e^+ & \text{with} && b_{1e} &= F(a_e^+) & \text{Compartment 1} \\ a_{2e} &= a_e^- & \text{with} && b_{2e} &= F(a_e^-) & \text{Compartment 2} \\ a_{1e} &= a_e^- & \text{with} && b_{1e} &= F(a_e^-) & \text{Compartment 1} \\ a_{2e} &= a_e^+ & \text{with} && b_{2e} &= F(a_e^+) & \text{Compartment 2} \end{aligned}$$

Linearization gives the Jacobian matrix of all fixed points of the four-variable RT-system:

$$\begin{pmatrix} f_1 & g_1 & 0 & 0 \\ P_2 & -(P_3+P_4) & P_4 & 0 \\ 0 & P_4 & -(P_3+P_4) & P_2 \\ 0 & 0 & g_2 & f_2 \end{pmatrix} \quad (5)$$

with

$$f_i = 1 - \frac{P b_{ie}}{(a_{ie}+P)^2} \quad , \quad g_i = -\frac{a_{ie}}{a_{ie}+P} \quad .$$

Stability of the Symmetric Fixed Point

In the case of the common symmetric fixed point $a_{1e} = a_{2e} = a_u$, $b_{1e} = b_{2e} = b_u$, the f_i and the g_i are identical: $f_i = f$, $g_i = g$. We take advantage of the symmetry of both cells and introduce new variables:

$$a_{\text{Sum}} = \frac{1}{\sqrt{2}} (a_{1e} + a_{2e})$$

$$b_{\text{Sum}} = \frac{1}{\sqrt{2}} (b_{1e} + b_{2e})$$

$$a_{\text{Diff}} = \frac{1}{\sqrt{2}} (a_{1e} - a_{2e})$$

$$b_{\text{Diff}} = \frac{1}{\sqrt{2}} (b_{1e} - b_{2e}) .$$

The Jacobian matrix then reads

$$\begin{pmatrix} f & g & 0 & 0 \\ P_2 & -P_3 & 0 & 0 \\ 0 & 0 & f & g \\ 0 & 0 & P_2 & -(P_3+2P_4) \end{pmatrix} , \quad (6)$$

yielding two separated characteristic equations

$$(\lambda_{1,2}^2 + \lambda_{1,2}(P_3 - f) - P_3 f - P_2 g) \quad , \quad (7)$$

and

$$(\lambda_{3,4}^2 + \lambda_{3,4}(P_3 + 2P_4 - f) - P_2 g - f(P_3 + 2P_4)) \quad . \quad (8)$$

Eq.(7) is identical to Eq.(4), with the above eigenvalues. The eigenvalues from Eq.(8) are

$$\lambda_{3,4} = \frac{1}{2}\left(f - (P_3 + 2P_4) \pm \sqrt{(f + P_3 + 2P_4)^2 + 4P_2 g}\right) \quad . \tag{9}$$

Conditions in parameter space, for the transition from real to complex-conjugate eigenvalues, are

$$P_4 = -0.5(f + P_3) \pm \sqrt{-P_2 g} \quad ,$$

for the stability line of real eigenvalues,

$$P_4 = \frac{P_3 f + P_2 g}{2(P_3 - f)} \quad ,$$

and for the Hopf-bifurcation,

$$P_4 = 0.5(f - P_3) \quad .$$

Stability of the Two Asymmetric Fixed Points

In case of the asymmetric fixed points, no separation of the local matrix, Eq.(5), could be found. However, all coefficients of the polynomial of the characteristic equation are given explicitly. In such a case the Routh-Hurwitz criterion [7] is the tool that allows one to find at least the stability of those fixed points. The equation for the eigenvalues of the local matrix, Eq.(5), is a polynomial of 4th order,

$$\lambda^4 + c_1 \lambda^3 + c_2 \lambda^2 + c_3 \lambda + c_4 = 0 \quad .$$

The coefficients $c_k (k = 1 \ldots 4)$ are

$$c_1 = -f_1 - f_2 + 2(P_3 + P_4)$$
$$c_2 = -2(P_3 + P_4)(f_1 + f_2) - P_2(g_1 + g_2) + 2P_3 P_4 + P_3^2$$
$$c_3 = -(P_3^2 + 2P_3 P_4)(f_1 + f_2) + 2(P_3 + P_4)f_1 f_2$$
$$\quad - P_2(P_3 + P_4)(g_1 + g_2) + P_2(f_1 g_2 + f_2 g_1)$$
$$c_4 = (P_3^2 + 2P_3 P_4)f_1 f_2 + P_2(P_3 + P_4)(f_1 g_2 + f_2 g_1) + P_2^2 g_1 g_2 \quad .$$

The Routh-Hurwitz criterion requires for the stability of the fixed points

$$c_1 > 0$$

$$\det \begin{vmatrix} c_1 & c_3 \\ 1 & c_2 \end{vmatrix} > 0$$

$$\det \begin{vmatrix} c_1 & c_3 & 0 \\ 1 & c_2 & c_4 \\ 0 & c_1 & c_3 \end{vmatrix} > 0$$

$$\det \begin{vmatrix} c_1 & c_3 & 0 & 0 \\ 1 & c_2 & c_4 & 0 \\ 0 & c_1 & c_3 & 0 \\ 0 & 1 & c_2 & c_4 \end{vmatrix} > 0 \quad .$$

The conditions for stable fixed points are

$$c_k > 0 \quad \text{and} \quad c_3(c_1c_2 - c_3) > c_1^2 c_4 \quad .$$

Using the above criteria, now a map of parameter space can be indicated which, however, is not given here.

DYNAMICAL BEHAVIOR OF THE RT-SYSTEM

The Rashevsky-Turing model is a four-variable system of two coupled, potentially oscillating subsystems and therefore reveals many features from the variety of dynamical behaviors of nonlinear science. Two different cases dependent on the strength of the diffusive coupling are of special interest. With <u>strong</u> cross-inhibition between the compartments present ($D = 12$ in Eq.(1)), we find "bichaotic" behavior with spatio-temporal polarity of the two cells, a Feigenbaum scenario as a route to chaos, and a transition in phase space as an 'interior crisis.' With <u>weak</u> coupling present ($D = 1$ in Eq.(1)), a different kind of chaotic state, "symmetrical chaos," becomes possible. This dynamical state appears to be reached via locking on a two-torus.

Bi-Chaoticity

From the above stability analysis of the 'symmetric' and 'asymmetric' fixed points it follows that there may exist a dynamical state for the two strongly coupled oscillating subsystems in which all fixed points are unstable. The bifurcation diagram Figure 2 gives an overview.

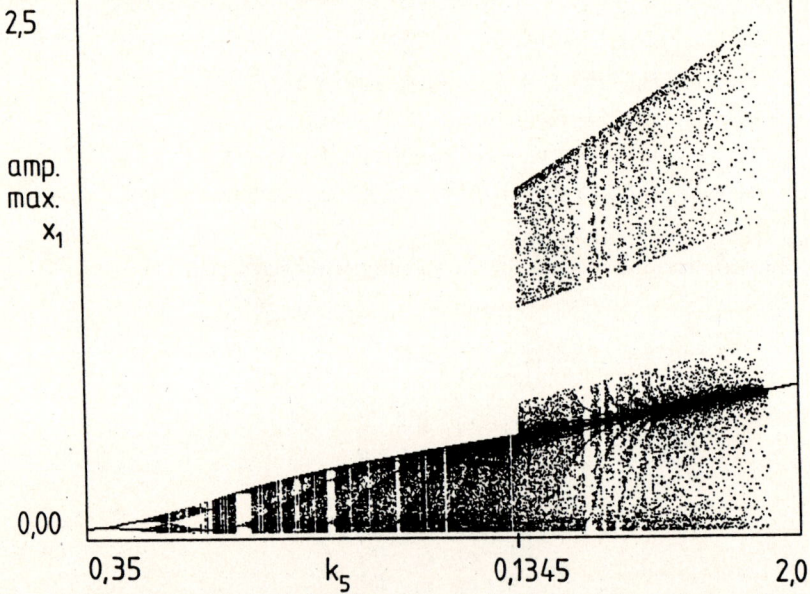

Fig. 2 Feigenbaum scenario of the 'bichaotic' state in Eq.(1). 100 transient amplitudes were omitted, 200 maximal amplitudes of x_1 are plotted vs. the evocation parameter k_5. (Note the two types of maximal amplitudes beyond $k_5 > 1.345$.)

At low values of k_5 there is limit-cycle oscillation present. Increasing the parameter toward 1.344, we find a cascade of period-doubling bifurcations, chaotic subbands, as well as windows of periodic motion- the typical Feigenbaum scenario. As one can see from Figure 3a and Figure 4a, there exists strong polarity between the two subsystems. The four-dimensional state space contains two bounded regions which do not overlap. As a correspondance of the asymmetry of the initial conditions, one cell always dominates the other and polarity never changes. This corresponds to a spatially ordered situation. At a value of k_5 around 1.3445, the attractor extension in both cells abruptly increases so that the two cells become indistinguishable. The trajectory in phase space is injected from one subdomain into the other. Thus a change in polarity of the formerly spatially ordered system occurs at random time intervals. Such a situation is called 'interior crisis' [8,9]. The resulting new attractor again has windows of periodic motion, separating chaotic subbands, until eventually a reversed bifurcation cascade occurs in the parameter interval about $k_5 = 1.9$, leading to a stable period-one limit-cycle in the now synchronized suboscillators. The whole Feigenbaum scenario may be regarded as a symmetry-breaking phase transition from the 'phase-locked' oscillation at $k_5 = 2.0$ and the 'phase-lagged' situation of nearly 90^0 phase difference between the variables of both suboscillators at $k_5 = 0.4$ (cf. [6]).

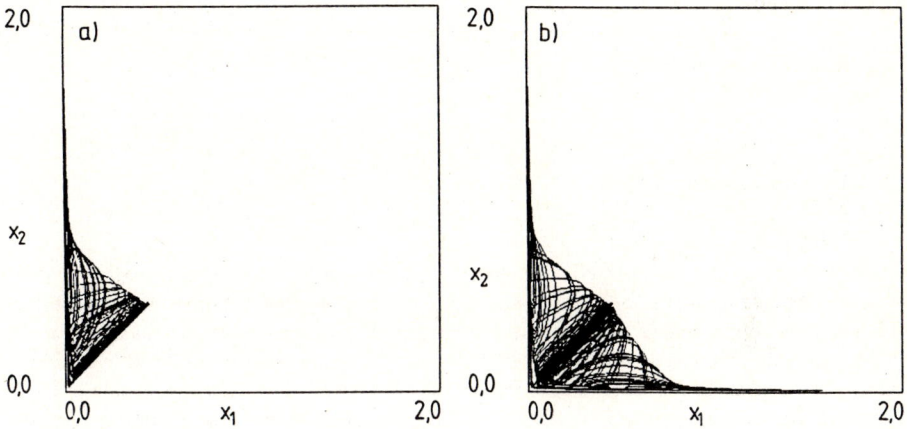

Fig. 3 Projection of the 4D flow in phase space of Eq.(1) onto the x_1, x_2 plane. Initial conditions: $x_1 = x_2 = 0.75$, $y_1 = 0.5$, $y_2 = 0.6$. Parameters: $k_1 = 10.8$, $k_2 = 6.0$, $k_3 = 6.0$, $k_4 = 3.0$, $K = 0.03$, $D = 12.0$. a) Polarized, spatially ordered, situation, $k_5 = 1.344$. b) Overshooting case, $k_5 = 1.345$.

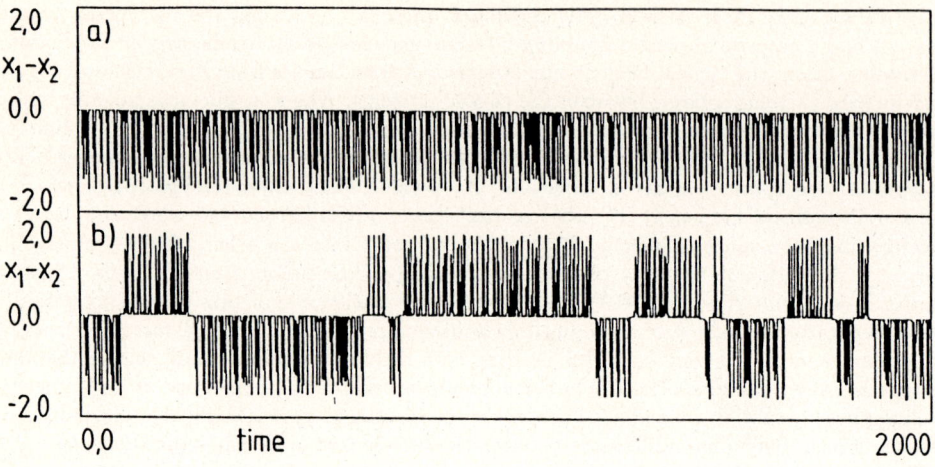

Fig. 4 Time evolution of the difference variable (x_1-x_2). Initial conditions and parameters as in Figure 3. a) Spatially ordered situation, $k_5= 1.344$. b) Overshooting case, $k_5= 1.345$.

Symmetric Dynamics

The 'symmetrical chaotic' behavior in contrast, occurs at weak couplings between the subsystems. Unlike those of the 'bi-chaotic' state the 'asymmetric' fixed points are now stable nodes and the 'symmetric' one is an unstable focus. There are three attractor basins present. The chaotic attractor's basin is embedded in between the basins of the two stable nodes. Figure 5 shows the symmetric, non-polarized flow in phase space. The Poincaré cross section, Figure 6, reveals the underlying toroidal structure. (The circle appears to be closed eventually.) The bifurcation diagram, Figure 7, reveals limit-cycle oscillations of period two at low values of k_4. There next is a sharp transition to a period-6 cycle. With increasing k_4, the situation gets more complicated. We have a mixing between very long-lasting transients and chaotic behavior. For k_4 less than 4.3275, there is chaos present, until, for k_4 larger than 4.3275, the chaotic attractor ceases to exist and only the stable nodes survive. The scenario resembles a route to chaos via a locked mode on the surface of a two-torus [10].

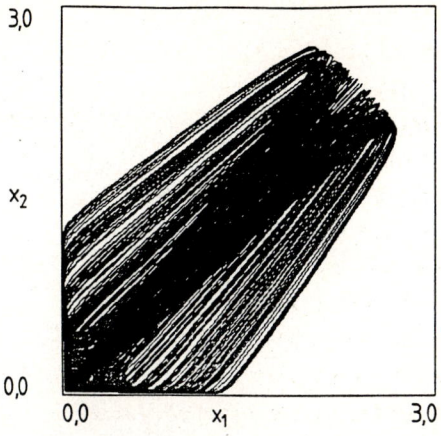

Fig. 5 Projection of the 4D flow in phase space of Eq.(1) onto the x_1,x_2 plane. Initial conditions: $x_1=x_2=0.75$, $y_1=0.5$, $y_2=0.6$. Parameters: $k_1=10.8$, $k_2=4.5$, $k_3=6.0$, $k_4=4.264$, $k_5=1.5$, $K=0.03$, $D=1.0$.

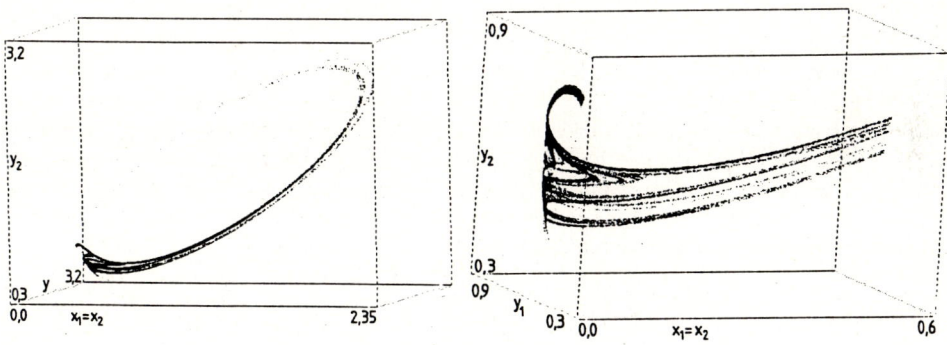

Fig. 6 Poincaré cross section of the 4D flow of Eq.(1), obtained in the $x_1=x_2$ hyperplane. Initial conditions and parameters as in Figure 5. Left side: Full view, right side: Close up.

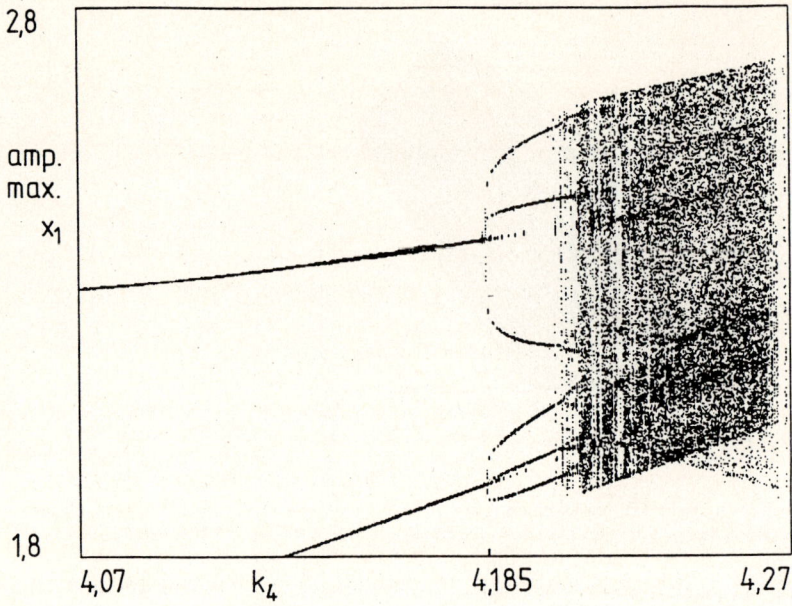

Fig. 7 Bifurcation scenario of the 'symmetric' state of Eq.(1). 200 transient amplitudes were omitted, 100 maximal amplitudes of x_1 were plotted vs. k_4. Parameters as in Figure 5.

OUTLOOK

The RT-system provides a suitable model for demonstrating general features of reaction-diffusion and activator-inhibitor systems. Due to the symmetry between coupled oscillators and the resulting simplicity of the pertinent ordinary differential equations, an analytic treatment of the steady-state behavior is possible. The visualization of the global behavior of a 4D-system is a challenge for the simulation by computer. In the case of attractive chaotic solutions, numerical integration is the only tool to obtain time evolution series, phase space portraits of attractors, cross sections of flows and bifurcation scenarios, respectively.

Recent studies on basin boundaries of attractors provide new insight in the interplay amongst multiple coexisting basins of attraction (self-similarity, fractality). These studies frequently focus on discrete mappings. Their extension toward generic continuous systems will yield similar structures in differential equations (e.g. final state sensitivity in state and parameter space). First results pointing to the existence of self-similar basin boundaries in continuous systems were recently found in another two cellular four-variable system [11]. The RT-system is a good candidate for the future study of the behavior of the boundary separating different coexisting chaotic attractors.

ACKNOWLEDGEMENT

We thank B. Röhricht, J. Parisi and J. Peinke for stimulating discussions.

REFERENCES

[1] N. Rashevsky, Bull. Math Biophys. 2, pp. 15-25, 65ff. and 109-121 (1940)
[2] A.M. Turing, Phil. Trans. R. Soc. London B 237, 37 (1957)
[3] O.E. Rössler, F.F. Seelig, Z. Naturforsch. 27b, 1444 (1972)
[4] O.E. Rössler, Z. Naturforsch. 31a, 1168 (1976);
 C.R. Kennedy, R. Aris, in New Approaches to Nonlinear Problems in Dynamics, P.J. Holmes ed., Philadelphia SIAM, 211 (1980)
[5] O. Sporns, F.F. Seelig, Biosystems 19, 83 and 237 (1986);
 O. Sporns, F.F. Seelig, Physica 26D, 215 (1987)
[6] B. Röhricht, J. Parisi, J. Peinke, O.E. Rössler, Z. Phys. B Condensed Matter 65, 259 (1986)
[7] A. Hurwitz, Math. Ann. 46, 273 (1895);
 L. Cesari, Asymptotic Behavior and Stability Problems in ODEs, Springer Verlag, Berlin (1959)
[8] C. Grebogi, E. Ott, J.A. Yorke, Physica 7D, 181 (1983)
[9] J.M. Thompson, H.B. Stewart, Nonlinear Dynamics and Chaos, Wiley, Chichester (1986)
[10] P.Bergé, Y. Pomeau, C. Vidal, Order Within Chaos, Hermann, Paris (1984)
[11] O.E. Rössler, J.L. Hudson, M. Klein, *Self-similar basin boundaries in a continuos system*, in: Nonlinear Dynamics in Engineering, W. Schiehlen, ed., Springer Verlag, Berlin (1989).

PHOTOREACTIONS IN SOLIDS
EXPERIMENT AND SIMULATION

D. Fröhlich and G. Gauglitz

Institut für Physikalische und Theoretische Chemie
Universität Tübingen
Auf der Morgenstelle 8
D-7400 Tübingen

Abstract

To reduce the problems in evalution of photokinetic rate equations, the liquid reaction system has to be stirred in order to get closed solutions. In the case of the examination of photoreactions in solids, the evaluation requires the application of numerical methods. In the special case of azobenzene in silgel 604A a further problem shows up, the photolysis product does not stay fixed at its site of production, but exhibits diffusion. It is demonstrated that even in this case an evalution is possible using modified rate equation, which take into consideration time and space dependencies as well.

The evaluation of photoreactions in solution cause nowadays few problems. Taking into account some basic assumptions, it is possible to find closed solutions for the differential equations which describe the problem. The most important requirement is that the reaction products always are homogeneously distributed during the irradiation. This can be achieved either by using very diluted solutions, in which only a small part of the light passing the solutions is absorbed, or by continuously stirring to have always a homogeneous solution. The practical importance of the first approach is very small since in highly diluted systems the reaction turnover is rather poor. For this reason in common, the solution is continuously stirred to achieve the above mentioned homogeneity. Under these conditions, one is restricted to solvents with low viscosity and low irradiation intensities. On the other hand, the differential equations have to be solved numerically in high viscous medium, transparent solids, and especially at high intensities used in pulsed lasers.

As an example, azobenzene is photokinetically examined in a silicon rubber polymer. The observed photoisomerization is uniform:

$$\text{trans Azobenzene} \underset{h\nu}{\overset{h\nu}{\rightleftarrows}} \text{cis Azobenzene}$$

Both the photoisomers have bands of absorption in the ultraviolet and the blue wavelength region. Therefore, the progress of the reaction can be observed spectroscopically. The silicon-rubber[1] is a polymethylsiloxane. Its chains are crosslinked by vinyl groups. It is transparent down to wavelengths below 250 nm.

Theory

Both the following equations describe the change of the concentration during irradiation in any volume element of the matter:

$$\frac{da}{dt} = \dot{a} = -\varphi_1^A \cdot I_{abs}^A + \varphi_2^B \cdot I_{abs}^B \tag{1}$$

$$\frac{db}{dt} = \dot{b} = -\dot{a} \tag{2}$$

By assuming monochromatic irradiation and an intensity I_o independent of time, the amount of light absorbed I_{abs}^i of any component i can be calculated if the absorptivities ε_i are known:

$$I_{abs}^A(x) = 1000 \cdot \overline{\varepsilon}_i' \cdot a \cdot I_o \cdot e^{-\int_0^x \sum_i^n (\overline{\varepsilon}_i' \cdot c_i) \cdot d} \tag{3}$$

The local rate of reaction at any distance x from the front surface is given by the following expression:

$$\frac{\partial a}{\partial t} = I \cdot \left(-\varphi_1^A \cdot \overline{\varepsilon}_A' \cdot a + \varphi_2^B \cdot \overline{\varepsilon}_B' \cdot b\right) \cdot e^{-\int_0^x (\overline{\varepsilon}_A' \cdot a + \overline{\varepsilon}_B' \cdot b) \cdot dx} \tag{4}$$

It turns out that the concentrations depend on the radiation time t as well as on the geometric element in the polymer. To avoid this geometric dependence, one transforms the equation into a pseudotime $\Theta(x,t)$, which is defined as follows:

$$\Theta = I \cdot \int_0^t e^{-\int_0^x (\overline{\varepsilon}_A' \cdot a(x,t) + \overline{\varepsilon}_B' \cdot b(x,t)) \cdot dx} \tag{5}$$

Furthermore, the following definitions are used:

$$\overline{R}_1 = \overline{\varepsilon}'_A \cdot \varphi_1^A \qquad (6)$$

$$\overline{R}_2 = \overline{\varepsilon}'_B \cdot \varphi_2^B \qquad (7)$$

$$\overline{Q} = \overline{R}_1 + \overline{R}_2 \qquad (8)$$

$$I = 1000 \cdot I_0 \qquad (9)$$

Two equations result which look very similar to the relationship obtained in thermal reactions. The integration supplies relationships for the concentrations a and b, which depend on the pseudotime Θ:

$$a(\Theta) = \frac{a_0}{\overline{Q}} \cdot \left(\overline{R}_2 + \overline{R}_1 \cdot e^{-\overline{Q}\Theta} \right) \qquad (10)$$

$$b(\Theta) = \frac{a_0}{\overline{Q}} \cdot \overline{R}_1 \cdot \left(1 - e^{-\overline{Q}\Theta} \right) \qquad (11)$$

The substitution of these equations in the integrand of eq. (5) the following relationship is obtained:

$$\overline{\varepsilon}'_A \cdot a(\Theta) + \overline{\varepsilon}'_B \cdot b(\Theta) = \overline{C}'_1 + \overline{C}'_2 \cdot e^{-\overline{Q}\Theta}, \qquad (12)$$

whereby the constants $\overline{C}'_1$ and $\overline{C}'_2$ are defined as following:

$$\overline{C}'_1 = \frac{a_0}{\overline{Q}} \cdot \left(\overline{\varepsilon}'_A \cdot \overline{R}_1 + \overline{\varepsilon}'_B \cdot \overline{R}_2 \right) \qquad (13)$$

$$\overline{C}'_2 = \frac{a_0}{\overline{Q}} \cdot \overline{R}_1 \cdot \left(\overline{\varepsilon}'_A - \overline{\varepsilon}'_B \right) \qquad (14)$$

In a two step sequence this equation is differentiated first with respect to time t and furtheron with respect to the geometric point x.

$$\frac{\partial^2 \Theta}{\partial t \partial x} = - \left(\overline{C}'_1 + \overline{C}'_2 \cdot e^{-\overline{Q}\Theta} \right) \cdot \frac{\partial \Theta}{\partial t} \qquad (15)$$

In consequence, this equation is integrated with respect to time t. A formular is obtained which contains an integration constant f(x), which is independent of time. It is a fact, that at time t = 0 no gradient of concentration exists. Furthermore, the pseudotime $\Theta(x,0)$ equals zero. By this means, the integration constant can be calculated, too:

$$f(x) = - \frac{\overline{C}'_2}{\overline{Q}} \qquad (16)$$

Formal integration$^{(2)}$ up to the pseudotime yields

$$\int_0^\Theta \frac{d\Theta(x,t)}{\overline{C_1'}\cdot\Theta + \frac{C_2'}{\overline{Q}}\cdot\left(1 - e^{-\overline{Q}\Theta}\right)} = -x + g(t) . \qquad (17)$$

The integration constant g(t) is obtained from the starting conditions for x = 0.

$$\Theta(0,t) = I \cdot t . \qquad (18)$$

It is

$$g(t) = \int_0^{It} \frac{d\Theta(0,t)}{\overline{C_1'}\cdot\Theta + \frac{C_2'}{\overline{Q}}\cdot\left(1 - e^{-\overline{Q}\Theta}\right)} . \qquad (19)$$

Both the integrals given in eqs. (17) and (19) can be combined:

$$\int_{It}^\Theta \frac{d\Theta}{\overline{C_1'}\cdot\Theta + \frac{C_2'}{\overline{Q}}\cdot\left(1 - e^{-\overline{Q}\Theta}\right)} + x = 0 \qquad (20)$$

This integral has to be solved by a numerical procedure, the so-called Newton-Raphson iteration. It allows to determine the pseudotime Θ at any time t at any geometric point x. The Θ values can be used to calculate the concentrations of reactants a and b from eqs. (10) and (11). By this approach the dependence of the absorbances on time for the total photoreaction can be calculated. The aim is to determine the quantum yields φ_1^A and φ_2^B. These parameters are varied as long as the calculated absorbance time curve does not fit to the measured one.

Results and Discussion

The examination of the azobenzene/Silgel system showed some systematic deviations between measured and calculated absorbance data. These deviations cannot be attributed to a consecutive reaction. The photoisomerization at the beginning of the reaction is supposed to result a photostationary state. It is characterized by no further change in absorption. But in practise, the absorption has slowly changed without any appearance of a consecutive photoproduct observable in the reaction spectra. Further experiments show that azobenzene has a very high rate of diffusion in the silicon polymer. This fact can be used to explain the deviation between the absorption time curves observed.

Without any diffusion the sum of the concentration of trans- and cis-azobenzene is constant at every geometric point in the polymer. The radiation builds up a gradient of concentration which is the driving force for the diffusion. Cis-azobenzene is formed just in the front of the irradiated polymer. In consequence it diffuses along the light path away from the source of radiation according to the gradient of concentration. In contrary, trans-azobenzene moves in the opposite direction into the volume elements where the concentration of trans-azobenzene is poor and which are closer to the source of radiation. Besides this gradient in the direction of the light path there is another gradient between irradiated and not irradiated volume which is even more distinct. It turns out that diffusion along the light path has no influence on the measured absorbances. The two reasons are that first the path of irradiation and of measurement are the same and second the measurement of absorption is an integrating method. It summarizes all the cis- and trans-azobenzene molecules within this light path according to the law of conservation of mass. Therefore diffusion doesn't alter the concentration regarding only this direction. One finds that the only difference between diffusion and no diffusion is that the absorption time curve is slightly altered before the system reaches the photostationary state. In contrast to this very small deviation, the diffusion perpendicular to the direction of the radiation causes more differences. Cis-azobenzene permanently moves out of the irradiation volume and trans-azobenzene will move in. There will be an increase or decrease of the number of molecules in the light path of measurement, if the diffusion coefficient of both the isomers differs significantly.

The experimental set-up is chosen such, that the irradiated volume is much smaller than the non-irradiated. This is demonstrated in fig. 1.

Fig. 1
Experimental set-up

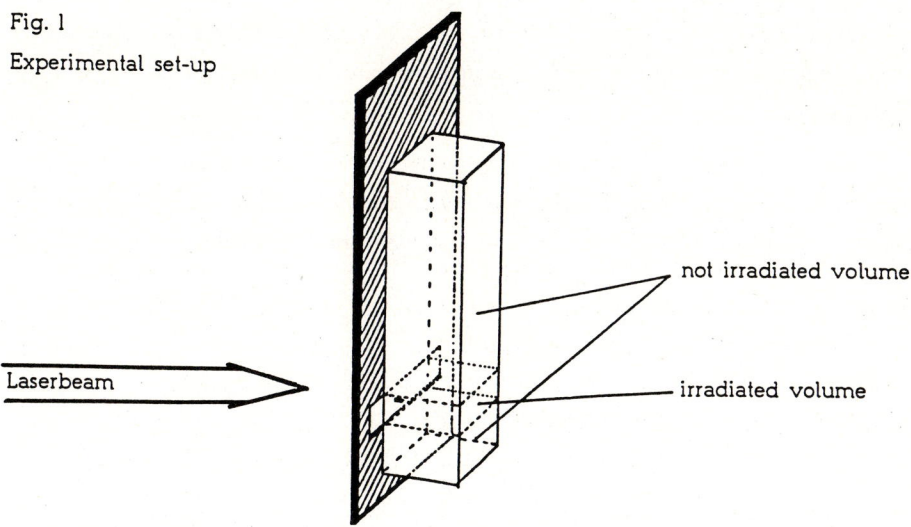

For this reason, the change in absorption in the "photostationary state" can last a very long time. Finally, to consider the diffusion, the equations for the rate of reaction have to be taken as follows

$$\frac{\partial a}{\partial t} = -\varphi_1^A \cdot I_{abs}^A + \varphi_2^B \cdot I_{abs}^B - D_A \cdot \frac{\partial^2 a}{\partial x^2} \qquad (21)$$

$$\frac{\partial b}{\partial t} = \varphi_1^A \cdot I_{abs}^A - \varphi_2^B \cdot I_{abs}^B - D_B \cdot \frac{\partial^2 b}{\partial x^2} \qquad (22)$$

Thereby, the amount of light absorbed at the geometric point x is defined by

$$I_{abs}^i(x) = \overline{\varepsilon}_i' \cdot c_i \cdot I(x) \cdot dV \qquad (23)$$

No closed solution can be derived for this system of differential equations. The procedure of evaluation is the following:

The polymer is devided into small volume elements. The changes of concentration are computed in these elements for small time steps. It turns out that neither space nor time discretisation can be selected independently. It is necessary to fulfil

$$\frac{1}{2} \geq D_i \cdot \frac{\Delta t}{\Delta x^2} \qquad (24)$$

to obtain a stable numerical procedure. If the term to the right is larger than 0.5 then the values of concentration become unstable. After a few steps of calculation, numerical overflow can occur.

Taking a pulsed laser as a light source the diffusion can be separated from the photoreaction because the time for the photoreaction is very short ($\leq$ 20 ns). In this period of irradiation, no diffusion can happen. Between the pulses and during the recording of the reaction spectrum, just diffusion has to be considered since azobenzene shows negligible thermal reaction.

As fig. 2 proves, neglection of diffusion shows deviation between simulated and measured absorption time curves. Taking this diffusion into account (calculated in 2 dimensions), the agreement between simulated and measured curve becomes drastically better as can be seen in fig. 3.

Literature:

(1) Silgel 604, Wacker Chemie, Burghausen, FRG
(2) H. Mauser, "Formale Integration", Bertelsmann Universitätsverlag Düsseldorf 1974
(3) J. Crank: The Mathematics of Diffusion, Clarendon Press Oxford, 2. Edition 1975
(4) H. C. Kessler, The Kinetics of Light Absorption in Photobleaching Media, J. Phys. Chem. $\underline{71}$ (8), 2736-7 (1967)

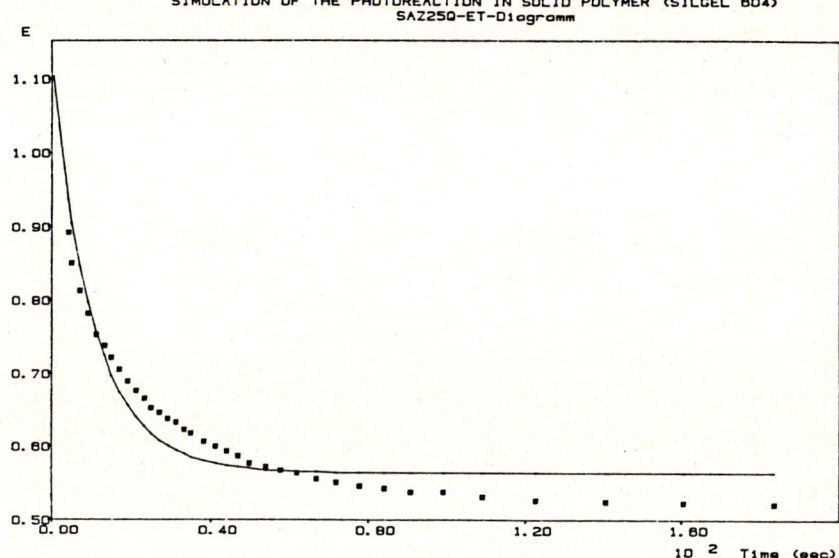

Fig. 2: Absorption of the reactands versus time. Diffusion is neglected, therefore only a bad coincidation of the simulation and the measurement is achieved.
Dotted line: measured values   solid line: simulation

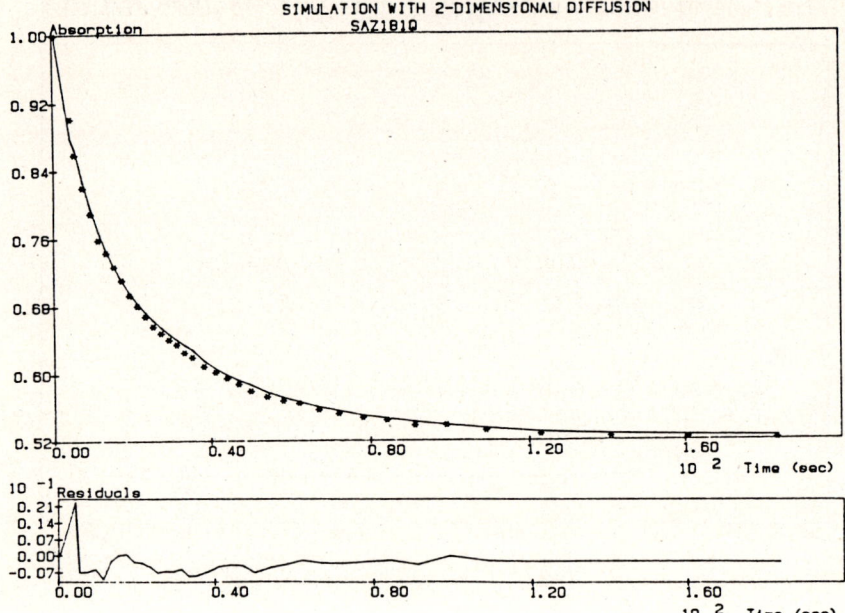

Fig. 3: Considering the diffusion in two dimensions leads to a better coincidation of the simulation and the measured values.

Dotted line: measured absorption/time curve of azobenzene in polymer (same as fig.2)
solid line: simulation regarding the diffusion

The Role of Machine Learning in Knowledge Acquisition

Kai Zercher [1,2]
Bernd Radig [1]

[1] TU München, Institut für Informatik
Orleanstr. 34, D-8000 München 80

[2] Siemens AG, ZFE IS INF 32
Otto-Hahn-Ring 6, D-8000 München 83

Abstract: Acquiring the knowledge for a knowledge-based system has proven to be a difficult task. Machine learning techniques are one possible approach to tackle this problem. Three case studies are described which show that machine learning techniques can produce superior results than more traditional knowledge acquisition techniques. Finally, some conclusion drawn from these examples are presented.

1. Introduction

Expert systems or - to use a more general term - knowledge-based systems are the most prominent result of artificial intelligence (AI) research and have found their ways to numerous industrial applications. However, developing such a system has many times proven to be more difficult than expected. Most often the biggest obstacle is the problem of knowledge acquisition, i.e., expressed very simplified, how to get the knowledge of the expert into the machine. More formally, knowledge acquisition is defined as the task of identifying, extracting, and formalizing the knowledge required by a knowledge-based system. Initially one thought that this task should not be too difficult if an expert is available, because he/she knows how to solve the problems. But it became clear that there is a tremendous difference between solving a problem, something an expert is good at, and explaining exactly how such a solution can be obtained by formalized means, a task most experts are not very good at. Knowledge acquisition proved to be "not just a problem of accessing and translating what is already known, but the familiar scientific and engineering problem of formalizing models for the first time" [8].

This observation prompted more and more researchers to investigate the problem of knowledge acquisition, which is now regarded as an established subfield of AI, having its own section in most general AI conferences. In 1986 the first AAAI Knowledge Acquisition Workshop was held in Banff, Canada and similar events - like the European workshops on knowledge acquisition - have followed. The international journal 'Knowledge Acquisition' was established in 1989 in response to the increased importance of the field.

Knowledge acquisition is concerned with the development of formal methods and strategies, which ease the task of knowledge acquisition by making it less time consuming and more reliable [3, 4, 19]. Most researchers distinguish three general approaches: manual, semi-automatic, and automatic. Manual techniques require that a knowledge engineer works together with an expert but without the use of computerized tools. A knowledge engineer is somebody who is familiar with both the various knowledge acquisition techniques and the computer technology needed to implement the knowledge-based system. In addition he/she should have good psychology and communication skills. The best known manual techniques are the various forms of interviews (unstructured, structured, ...), brainstorming, and protocol analysis. With semi-automatic techniques computerized tools come into play. An example are repertory grid methods [18]. Here, key terms of a domain are identified and the program systematically asks the expert for attributes which are common to two terms and which distinguish them from a third one. Thereby a domain can be structured and relevant properties can be identified. Still these techniques require that the expert is available for a sufficient amount of time and that he is either familiar with the tool or is assisted by a knowledge engineer. Finally, we have the automatic procedures, the learning techniques.

2. Machine Learning

Machine learning techniques transform available data from various sources into knowledge required by the problem solving component (Figure 1). The data can in principle be any type of knowledge. Machine learning is not just a simple transformation step, but usually a complex extraction process which goes along with a change in the abstraction level of the knowledge. A typical setting is that the expert provides examples of solved problems, e.g., in medical applications, a description of findings/symptoms of a patient along with the determined diagnosis. Then the program inductively finds a general concept description, e.g., a rule for diagnosing a specific disease. With this setting the task of knowledge acquisition is neatly divided into two subtasks, solving of specific problems by the experts (something which they are known to be good at) and systematically searching for general rules by a computer program (something at which they are usually superior to humans). This division of labor has proved to be very successful (see Section 3).

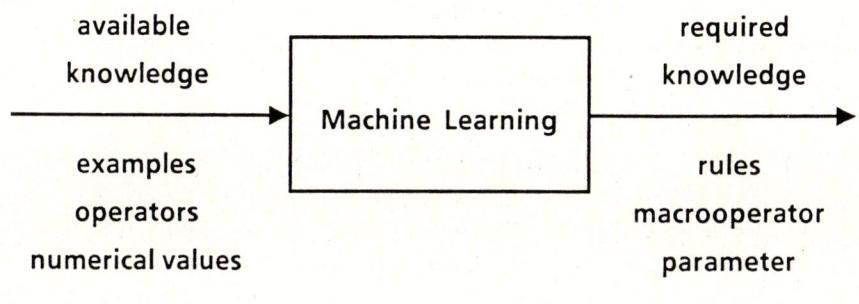

Figure 1

Machine learning is not a new research area [5]. Techniques of adaptive control theory (e.g., parameter learning) and pattern recognition (e.g., statistical parameter estimation, multilinear regression analysis, cluster techniques) have been known for a long time. In recent times neural networks have become popular. All these techniques deal primarily with numerical data. Although such data are important, many AI researchers felt that symbolic data, like attribute-value descriptions, graph representations or predicate calculus descriptions are at least as important. Machine learning as a subfield of AI deals primarily with this type of data. Some early works date back to the 1950s, but interest has increased a lot during the last decade [9, 12, 13]. In this paper we concentrate on this type of machine learning.

We have already presented a definition of machine learning. An alternative one states that "learning denotes changes in the system that are adaptive in the sense that they enable the system to do the same task or tasks drawn from the same population more efficiently and more effectively the next time" [20]. This definition stresses two important goals of machine learning. First, acquiring the ability to solve new kinds of problems or to provide a superior solution to old ones. Second, improving the speed in which a solution can be found.

We classify machine learning techniques into four main categories: rote learning, inductive learning, deductive learning, and learning by analogy. Rote learning (learning by heart) is the simplest form of learning. Provided data is stored without any modification. Still issues like organization of the stored information and questions of efficient retrieval can complicate that approach. A well known example is Samuel's checkers player [9]. Inductive learning derives general knowledge from specific examples. It is probably the most widely used approach. Like with any form of induction, whether done by humans or machines, there is no guarantee that the final result is correct. One must use so-called inductive leaps which are in principle not formally justified. An advantage of using programs for doing induction is that they are less biased than humans and search more systematically for the best suited generalization. Still inductive programs also need some form of bias, which tells them which generalizations are preferred or which structure the desired concept description should have. A differently motivated approach is deductive learning. Here a domain theory is already present, which enables the system to solve all relevant problems. Unfortunately the domain theory is too general or too large and most importantly, finding a solution takes too long. The task of deductive learning is now - guided by specific problem solving traces - to combine and to specialize parts of the domain theory in order to learn new, more efficiently usable knowledge. The best known technique is explanation-based learning [15, 10, 11]. Figure 2 compares inductive and deductive learning. The goal of both is to find knowledge at the right level of abstraction, i.e., a level that is appropriate for the problem solving task. Induction takes data (examples) which are too low level and which are then generalized. Deduction uses a domain theory which is too general and has to

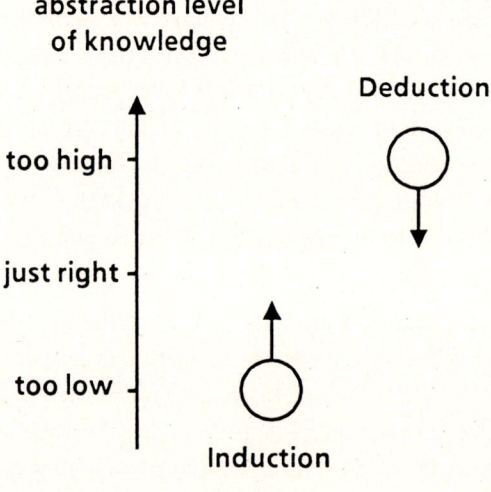

Figure 2

specialize this knowledge. In recent years various approaches have been proposed in which both techniques are combined [1, 16]. Learning by analogy tries to utilize the same principle which underlies large portions of human problem solving. To solve a new problem, one looks for a similar old problem and then tries to transform the old and known solution into a solution to the new problem [6, 7].

3. Case studies

3.1 Learning diagnosis rules for soybean diseases

A very interesting study was done by Michalski and Chilausky [14]. They compared two different knowledge acquisition approaches in the context of designing an expert system for soybean disease diagnosis. The first strategy was an interview technique. The experts - assisted by the knowledge engineers - directly developed the rules. The second strategy used the inductive learning algorithm AQ11. About 600 examples were available. Each example was a description of a diseased plant including general information. It was expressed in 35 attribute-value pairs, e.g., time of occurrence = July, precipitation = above normal, stem = abnormal, canker lesion color = brown, etc. Each description was also associated with the diagnosis of an expert, e.g., diagnosis = brown spots. The whole set of examples was evenly split into two sets, a training set used by the inductive learning program and a test set on which the learned rules as well as the rules proposed by the experts were evaluated. One of the learned rules was the following:

   IF   Leaf malformation = absent, Stem = abnormal,
        Stem cankers = below soil line, Canker lesion color = brown
   THEN Diagnosis = Rhizoctonia root rot

In the final comparison of both knowledge acquisition approaches, the inductive learning program was clearly superior. The table below shows a summary of the results:

	Induction	Interview
First choice correct	97.6 %	71.8 %
One choice correct	100 %	96.9 %

The expert system evaluated rules in such a way, that in some cases several diagnoses were proposed. 'First choice correct' lists the percentages where the highest ranked diagnosis was the correct one, 'One choice correct' lists the percentages where the correct diagnosis was among the proposed ones. These results are even more impressive when one takes into account that the rule formalism used by the experts was slightly more powerful then the one used by the inductive learning program.

3.2 Induction of diagnosis rules for a satellite power system

Another comparative study was done by Pearce [17]. Here, the problem was to design diagnosis rules for a satellite power system. Again one approach was a classical interview technique. The experts were directly asked for diagnosis rules. The second approach consisted of three steps: The experts were asked to define a qualitative model of the power system. Because of their engineering background and the existence of detailed satellite documentation this was a relatively easy task for them. Then qualitative simulation was used to generate an exhaustive set of examples, namely corresponding to all possible qualitative behaviors of the power system (assuming that at most one error was present). Finally, the AQR induction program was run with the example set and a set of diagnosis rules was generated. Since the example set was complete - a rather unusual condition - this process is also an instance of data compression. Both rule sets were evaluated on a small set of test examples generated by an already existing real time satellite simulator (not to be confused with the qualitative model and its simulator). In addition, the rule set integrity was checked by a specialized tool. The results are summarized below:

	Qualitative model + simulation + induction	Interview
Numbers of rules	75	110
correctness	100 %	72 %
Overall time effort	3 -4 MM	6 MM
Integrity problems	none	type errors, unreachable and dead end rules

A nice result of this study is that the approach which used inductive learning did not only produce a higher quality rule base but that it also took less time.

3.3 Learning rules for detecting chemical substructures from infrared spectra

As part of the system EXPERTISE [2] for evaluating infrared spectra, rules were needed which deduce the likely presence of substructures in a chemical substance from its given infrared spectrum. Although it is known for some important substructures how they manifest in a spectrum, acquiring rules for a large set of substructures either from books or by asking experts seemed to be too time consuming. Instead a learning system was designed whose overall structure is depicted in Figure 3.

It is assumed that a library of infrared spectra of known substances is given; the spectra are coded as line tables. A fragment generator is used to find substructures for which a

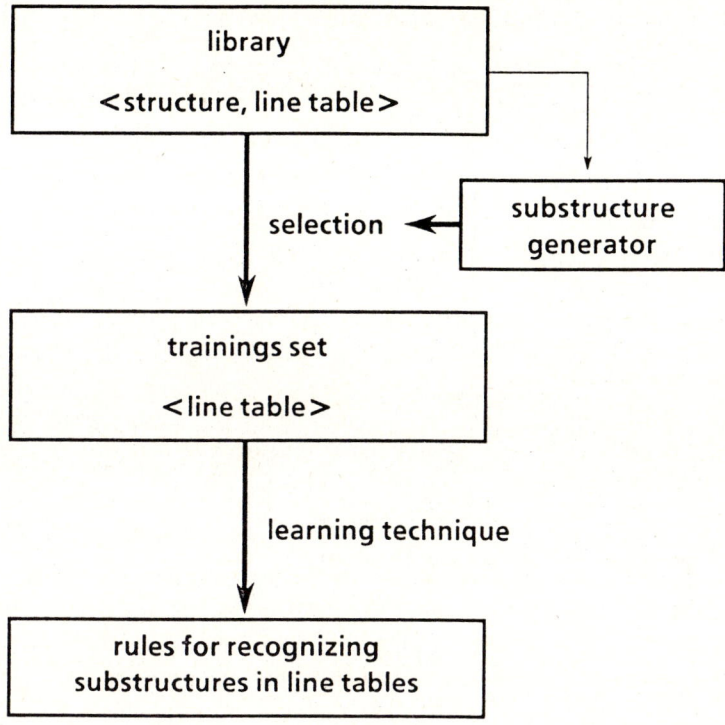

Figure 3

recognition rule should be learned. Once a substructure has been determined, all spectra of substances which include that substructure are retrieved from the library; they form the training set. A main assumption in infrared spectroscopy is that a substructure, even if present in different molecules, will produce peaks at similar places of the spectrum. The learning system uses this property the other way around. It looks for peaks which are present in all spectra of the training set at similar wavelengths and having similar intensities. It assumes that these peaks are characteristic for the selected substructure. For each such set of peaks a general condition is constructed. It requires that a peak has wavelength and intensity in specified intervals; projected on a spectrum, the condition requires that the peak is within a specified rectangle. A rule consists of several such conditions and is satisfied only if all conditions are fulfilled. The learning algorithm guarantees that the intervals are as small as possible in order to avoid a strong overgeneralization. It also guarantees that the learned rules are correct with respect to the training set, a property not shared by some statistical approaches. Another advantage is that the rules are readable by experts. For substructures for which the peaks they cause are known the learned rules were mostly consistent to what the experts believe. Obviously the system also learns rules for substructures for which no such expert knowledge exists.

The set of learned rules is utilized in the system EXPERTISE, which is successfully used by chemists at a Philips research laboratory.

4. Conclusions

What can we learn from these case studies? First, machine learning can be a very useful technique. It can help to find qualitatively better rules in a shorter time. It can utilize data which would otherwise be ignored. Second, machine learning is usually only one step in the overall knowledge acquisition process, but sometimes a crucial one. Third, one should ask experts for that kind of information which they can easily provide with a high quality. Experts are good in solving problems, but not necessarily in specifying formal rules. Experts might be able to construct a causal model, but not diagnosis rules. The process of transforming the provided information into the form required by the expert system can then often be left to a learning program.

What are the preconditions for a successful application of machine learning? First, a sufficient amount of examples and/or domain theory must be available. It is clear that one can not expect to learn a whole rule base from just one example. The amount of data required depends largely on its quality and the kind of knowledge one wants to learn. The more irrelevant information is contained in the data the more data one needs. The same holds if parts of the data are wrong (noise). In fact, symbolic learning algorithms are in general much less noise tolerant than most numerical ones. Finally, the more degrees of freedom one has in the form and contents of a rule to be learned, the more data are needed.

A second precondition is that there should be evidence that the needed knowledge is in fact implicitly contained in the data. Such could be that a novice on his way to become an expert can utilize the data. Finally one should make sure that there are no simpler ways to obtain the needed knowledge.

Machine learning has proved to be a very useful technique for knowledge acquisition. Most of the early applications were of purely academic nature, but one can now find an increasing number of real world applications. In times where more and more information is already digitally recorded it can be expected that machine learning applications will become increasingly important. Still, these techniques have not reached a stage where one can simply pick a program from a library and apply it without knowing anything about the underlying principles. Today an understanding of the field is still required. One has to select the right algorithm, to modify it if necessary, to prepare the data and to control the results, but the possible benefits seem worth this effort.

Acknowledgements

We wish to thank Angelika Hecht and Ruxandra Scheiterer for helpful comments on earlier drafts of this paper.
The first author thankfully acknowledges the support by a Ph.D. grant from Siemens AG.

References

1 Bergadano, F., Giordana, A., "A knowledge intensive approach to concept induction", p. 305 - 317, proceedings of the fifth international conference on machine learning, Ann Arbor, 1988.

2 Blaffert, T., "Ein Gesamtsystem zur Auswertung von Infrarotspektren durch Lernen und Erkennen spektraler Merkmale sowie Zerlegung und Synthese chemischer Strukturgraphen", Dissertation (in preparation), Universität Hamburg, 1989.

3 Boose, J.H., Gaines, B.R., "Knowledge acquisition for knowledge-based systems", AAAI 88, tutorial notes MP4, 1988.

4 Boose, J.H., "A survey of knowledge acquisition techniques and tools", Knowledge Acquisition, Vol. 1 Nr. 1, p. 3 - 37, 1989.

5 Buchanan, B.G. et al., "Models of learning systems", in "Encyclopedia of computer science and technology", p. 24 - 50, Marcel Dekker, New York, 1978.

6 Carbonell, J.G., "Learning by analogy: Formulating and generalizing plans from past experience", Chap. 5 [12].

7 Carbonell, J.G., "Derivational analogy: A theory of reconstructive problem solving and expertise acquisition", Chap. 14 [13].

8 Clancey, W., presentation at the first knowledge acquisition for knowledge-based systems workshop, Banff, Canada, 1986.

9 Cohen, P.R., Feigenbaum, E.A. (eds.), "The handbook of artificial intelligence", Vol. 3, chap. 14, William Kaufmann, 1982.

10 DeJong, G., Mooney, R., "Explanation-based learning: An alternative view", Machine Learning 1, p. 145 - 176, 1986.

11 Ellman, T., "Explanation-based learning: A survey of programs and perspectives", ACM computing surveys, Vol. 21, Nr. 2, p. 163 - 221, June 1989.

12 Michalski, R.S., Carbonell, J.G., Mitchell, T.M. (eds.), "Machine learning: An artificial intelligence approach", Vol. 1, Morgan Kaufmann, 1983 (also Springer Verlag, 1984).

13 Michalski, R.S., Carbonell, J.G., Mitchell, T.M. (eds.), "Machine learning: An artificial intelligence approach", Vol. 2, Morgan Kaufmann, 1986.

14 Michalski, R.S., Chilausky, R.L., "Learning by being told and learning from examples: An experimental comparison of the two methods of knowledge acquisition in the context of developing an expert system for soybean disease diagnosis.", International Journal of Policy Analysis and Information Science, Vol. 4, No. 2, p. 125 - 161, 1980.

15 Mitchell, T.M., Keller, R.M., Kedar-Cabelli, S.T., "Explanation-based learning: A unifying view", Machine Learning 1, p. 47 - 80, 1986.

16 Pazzani, M.J., "Integrating explanation-based and empirical learning methods in OCCAM", proceedings of the third European working session on learning (EWSL), 1988.

17 Pearce, D.A., "The induction of fault diagnosis systems from qualitative models", p. 353-357, proceedings AAAI 88, 1988.

18 Shaw, M.L.G., Gaines, B.R., "An interactive knowledge-elicitation technique using personal construct technology", in Kidd, A. (ed.), "Knowledge acquisition for expert systems. A practical handbook", Plenum Press, 1987

19 Shaw, M.L., Gaines, B. R., "Knowledge acquisition: Some foundations, manual methods and future trends", in Boose, J., et. al. (eds.), "Proceedings of the third European workshop on knowledge acquisition for knowledge-based systems" (EKAW 89), 1989.

20 Simon, H.A., "Why should machines learn?", Chap. 2, [12].

NEURAL NETWORKS

Marjan Tušar,
SRC Kemija, Ljubljana
Jure Zupan
'Boris Kidrič' Institute of Chemistry, Ljubljana

Abstract: The basic principles of neural networks (weight vector, layers, connections, feed-back corrections, etc.) are described and discussed. In particular, Hopfield and Hamming net, Kohonen learning, and back-propagation algorithm are described more in detail. Besides the basic architectures of these nets some of the problems that may arise in the actual applications are pointed out.

INTRODUCTION

The neural network is a multi-layer (single, two, or more layers) network of identical electronic circuits which resembles or mimics the neuron: a brain cell. Each computer simulated neuron 'i' consists of: a set of m input lines, a vector of adaptable weights W_j (w_{i1}, w_{i2}, .. w_{im}), a procedure for calculating the 'net' output Net_j on the basis of inputs and weights, and of a sigmoidal function yielding values between 0 and 1 the result of which is called neuron's true output Out_j. The mentioned basic elements (ref. 1) of a 'neuron' cell and its computer 'counterpart' are shown in Figure 1.

The neural networks are basically supposed to solve problems in one of the following three categories: clustering, content dependent retrieval, and reduction of representations. All three types of problems are very often encountered in various fields of chemistry (ref.2).

The neural network procedure starts with a set of N neurons (all or only those on the top layer) obtaining a m-channel signal X ($x_1, x_2, x_3, ... x_m$) simultaneously on their inputs. The components x_i of the input signal X can be binary (i.e. zeros/ones or +1/-1 in some applications) or real values. Although the neural network approach was first applied to problems where binary inputs are prevailing (such as image recognition (ref.3)), in chemistry the most attention is (or should be) focused to neural net architectures capable to handle continuous valued input. Each j-th neuron, according to its m-dimensional weight vector W_j with corresponding weights ($w_{j1}, w_{j2}, ... w_{ji}, ... w_{jm}$) yields a net output Net_j:

$$Net_j = \sum_{i=1}^{m} x_i w_{ji} \qquad /1/$$

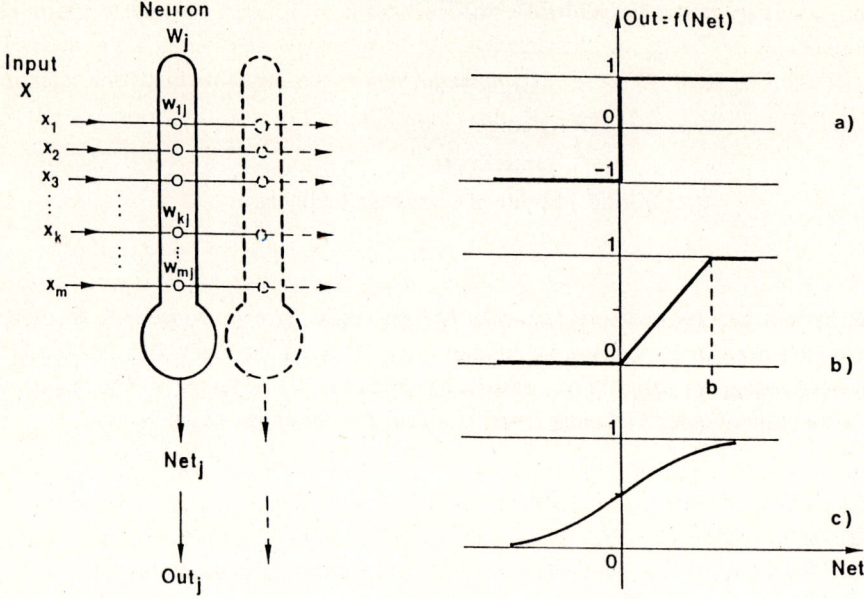

Fig. 1 Computer simulated neuron included into the neural net (left) and three types of squashing function: hard limiter (a), threshold logic (a), and sigmoidal function (c).

which is further on 'normalized' or 'squashed' into neuron's final output Out_j with a function $f(Net_j)$ that for any real input Net_j yields an output in a limited ($[-1, +1]$ or $[0,1]$) interval:

$$Out_j = \begin{cases} 1 & \text{for } (Net_j - \Theta_j) \geq 0 \\ -1 & \text{for } (Net_j - \Theta_j) < 0 \end{cases} \qquad /2a/$$

$$Out_j = \max\{b, \min\{0, (Net_j - \Theta_j)\}\} \qquad /2b/$$

$$Out_j = \frac{1}{1+e^{-(Net_j - \Theta_j)}} \qquad /2c/$$

The most frequently used 'squashing' is the sigmoidal function /2c/. The threshold Θ_j improves the adjustability of neurons in the same manner as in standard pattern recognition approaches (ref. 4) where it is handled as an extension of the dimension of the measurement vector **X** for one variable:

$$\mathbf{X}\,(x_1, x_2, x_3, \ldots x_m) \longrightarrow \mathbf{X'}\,(x_1, x_2, x_3, \ldots x_m, 1) \qquad /3/$$

Therefore, the js are regarded as an additional component in each weight vector $\mathbf{W_j}$ and can be trained exactly in the same way as all other components w_{ji}. In order to achieve this the $\mathbf{W_j}$s are iterated with one dimension more ($w_{j1}, w_{j2}, ... w_{j,m+1}$), while the input vectors are augmented with one additional input line conveying to neural network a permanent value of '1':

$$\text{Net}_j = \sum_{i=1}^{m} x_i w_{ji} + x_{m+1} w_{i,m+1} = \sum_{i=1}^{m+1} x_i w_{ji} \qquad /4/$$

where $x_{m+1} = 1$ and $w_{i,m+1} = \Theta_j$. The essential difference between the learning algorithms as implemented in standard pattern recognition methods and neural network approach, respectively, is the nonlinear transformation (eq. /2a-c/) of the dot product between the input vector $\mathbf{X'}$ and weight vector $\mathbf{W}$ in the later approach.

The output signals Out_j of the neurons can be connected to: inputs of other neurons on deeper levels, fed-back to neurons on the same level, inputs of neurons in certain constant or time-dependent neighborhood of the most excited neuron, or any combination of such inter-connections. The few most commonly used neural network approaches will be discussed in the next paragraphs.

In order to test and to show how some neural network designs perform on real (although small, i.e. up to 120-dimensional) input patterns an IBM PC computer based program package called NEURON was developed in our laboratory. Program NEURON can handle Hopfield, Hamming, Carpenter, Kohonen, and up to 5 layer back-propagation neural nets. The program has a built-in bit editor enabling the user to set up any desired bit pattern on a 6 x 20 bit matrix and a possibility to read patterns from a file, hence providing an adequate flexibility for handling a wide variety of different inputs. All examples presented in this paper were generated using program NEURON.

HOPFIELD NET

One of the simplest neural net designs is the Hopfield net (ref. 5-7) which handles binary patterns rather than the real ones. The two states in bit patterns for the Hopfield net discussed in this paper will be assigned as +1 and -1. The Hopfield net is designed to solve contents-dependent retrieval problems. It requires a selection of M prespecified m-dimensional exemplar patterns $\mathbf{R_j}$ from which a m x m-dimensional weight matrix W is calculated. On the basis of matrix W the unknown (modified, slightly different, or otherwise corrupted) inputs should be recognized as (associated with) one of the exemplar patterns $\mathbf{R_j}$.

The most significant feature of the Hopfield net are permanent weights w_{ij}, which are calculated in advance from all M exemplar patterns in the following manner:

$$w_{ij} = \sum_{s=1}^{M} x_i^s x_j^s \quad \text{for } i \neq j \quad \text{and} \quad w_{ij} = 0 \quad \text{for } i = j \qquad /5/$$

The output of each neuron Outj is calculated as:

$$\text{Out}_j = f\left(\sum_{i=1}^{m} w_{ij} x_i\right) \qquad /6/$$

where w_{ij} are permanent weights and x_i is the i-th component of the m-dimensional input pattern **X**. The 'squashing' function $f(x)$ is the hard-limiter shown in figure 1a and in eq. /2a/. The outputs Out_j are then connected back as the new inputs and the iteration is repeated until the outputs do not change anymore. The entire procedure is shown schematically in figure 2.

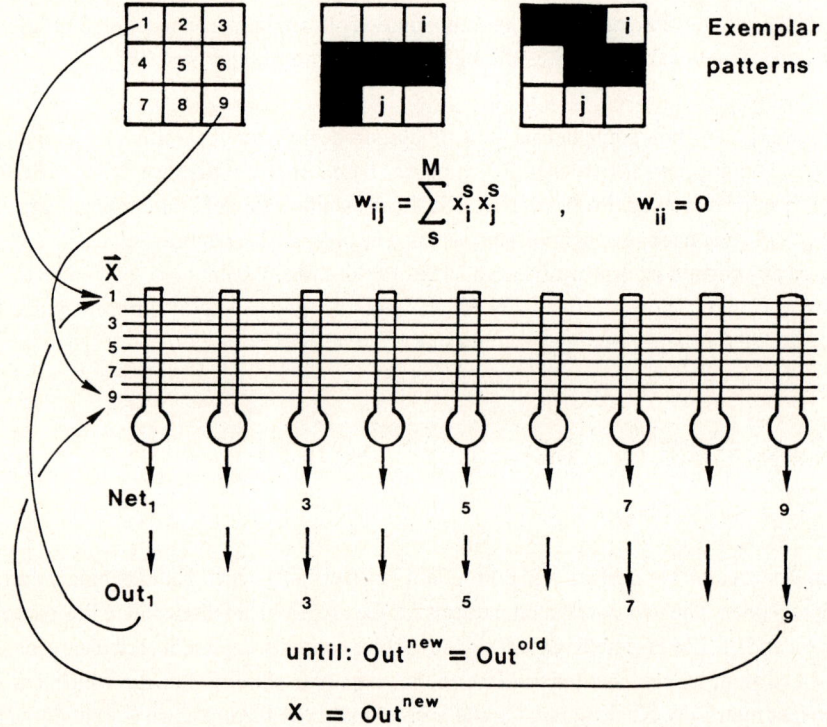

Fig. 2. The Hopfield net. There are M 9-dimensional exemplar patterns, hence the weight matrix contains 81 weights. The iteration is repeated until the outputs do not change anymore. The expected result is that after the input of a corrupted pattern and a number of iterations the neural net will produce the exemplar pattern most similar to the input.

In order to obtain a good prediction it is necessary to chose exemplar patterns to differ as much as possible. Due to the way the weight matrix is formed (eq. /5/), similar (i.e. heavily overlapping) exemplars will cause the unknown pattern to converge towards a kind of an 'average' exemplar, thus producing a 'no-match' result (ref. 8). The weak point of the Hopfield net is a large number of weights required for the recognition. The number of exemplar patterns cannot be increased without a considerable increase in resolution of the pattern matrix (vector **X**, Fig. 2), what in turn results in a quadratic increase of the weight matrix **W**.

Hopfield net can be used in a design of intelligent instruments. If, for example, a base line correction is to be applied automatically, the first task of the instrument is to recognize what type the actual base line corresponds to. Figure 3 shows 5 different base-line types and 5 corresponding patterns as simulated with the program NEURON. The pattern in the lower left part represents an input pattern (reduced recorded spectrum) which, after 3 iterations converges to the exemplar pattern no. 5. In a real-world application the image matrix should have at least $250 \times 20 = 5000$ bits what means that $2.5 \cdot 10^7$ matrix elements have to be scanned for each prediction. At the present, this cannot be economically carried out without a hardware implementation of parallel neural network architecture.

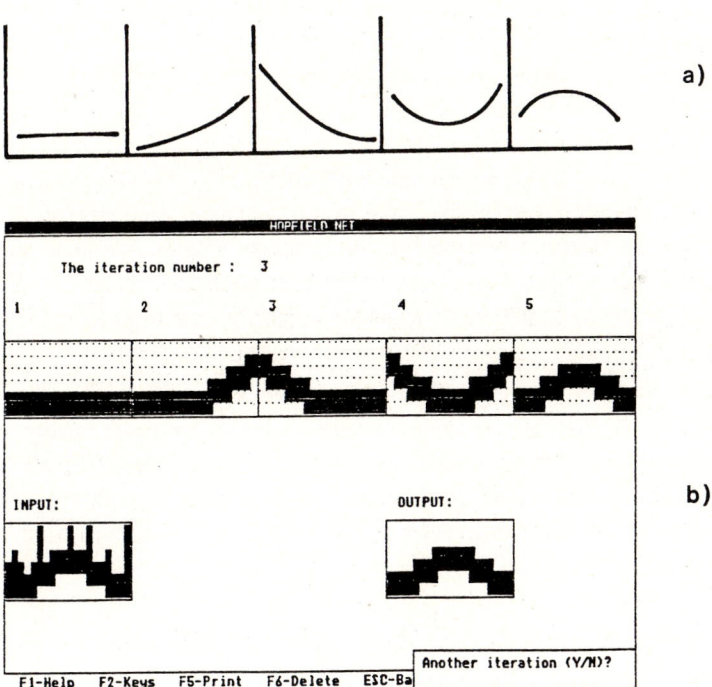

Fig. 3 Five different base-line types (a) and corresponding exemplar patterns as simulated on a neural net program NEURON (b). In the lower left corner as an input a reduced spectrum is simulated. The goal of the net is to recognize the type of the base line (an afterwards react accordingly) of any recorded spectrum which is considered as a 'corrupted' exemplar.

HAMMING NET

The Hamming net was suggested in order to avoid large weight matrices required by the Hopfield net. The recognition of (or association with) the exemplar patterns in the Hamming net is achieved in two steps. First, the net output Net$_j$ is calculated in a form of the Hamming distance (ref. 9) between the m-dimensional sample **Z** and all exemplar patterns:

$$\text{Net}_j = \sum_{i=1}^{m} w'_{ij} z_i \qquad \text{where } w'_{ij} = x^j_i / 2 \qquad \qquad /7/$$

Second, the Net$_j$ is transformed into Out$_j$ using the threshold logic (eq. /2b/) and carried forth to the next layer of neurons represented by a rectangular M x M matrix of weights w_{ik}:

$$w_{ik} = 1 \quad \text{for } i = k \qquad \text{and } w_{ik} = -\varepsilon \qquad \text{for } i \neq k \qquad \text{with} \quad \varepsilon < m/2 \qquad /8/$$

This second layer of constant weights (marked as w_{ik} to be distinguished from the previously used weights marked as w'_{ij}, eq. /7/) is used only to determine the exemplar most similar to the input pattern **Z**. At the first glance it might seem that employing an M x M-dimensional neural net matrix only to determine the smallest distance is an overkill, especially if all distances are already at hand. However, the point is that the 'max-net' (as this second net is usually referred to; ref. 8) does the job **automatically** with no programming and no instructions involved, and what is even more important - it can easily be hardwired. Compared to the Hopfield net, the Hamming approach needs only M m-dimensional vectors (arrays) and an M x M-dimensional weight matrix, both together of course, being orders of magnitude smaller than the m x m-dimensional weight matrix (eq. /5/) used by the Hopfield net.

All calculations on the first level: Hamming distances and Net-Out transformations can be made directly in memory and do not require large amount of storage. Due to the much smaller amount of weights taken into the account a much shorter time is needed to obtain the final result. Figure 4 schematically shows the flow of data in the Hamming net: first, calculation of weights w'_{ij}, j, and w_{ik}, then estimation of Hamming distances Net$_j$ between the input pattern **Z** and all exemplar patterns R$_j$, their transformation into Out$_j$, and finally, finding out the maximal Out$_j$ by applying the 'max-net'.

To show the difference between the Hopfield and Hamming net the same exemplar patterns as in the previous case were input to program NEURON selecting the Hamming option. Figure 5 shows the result of picking up the appropriate base-line similar to the procedure shown in figure 3. Here, the 'squashed' Hamming distances between the sample **Z** and all exemplar patterns serve as initial inputs to the 'max-net'. By repeatedly feeding-back the 'max-net' outputs to its inputs, only one output, the one associated with the most similar exemplar is finally obtained in 'on' or 'high' position.

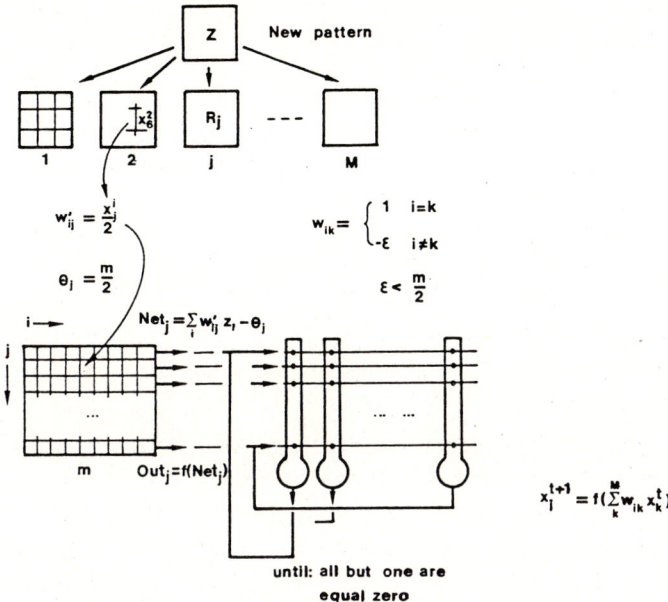

Fig. 4 Hamming net determines the exemplar pattern most similar to the input one. The 'max-net' is quadratic weight matrix having '1's on the diagonal and $-\varepsilon$ on all other positions.

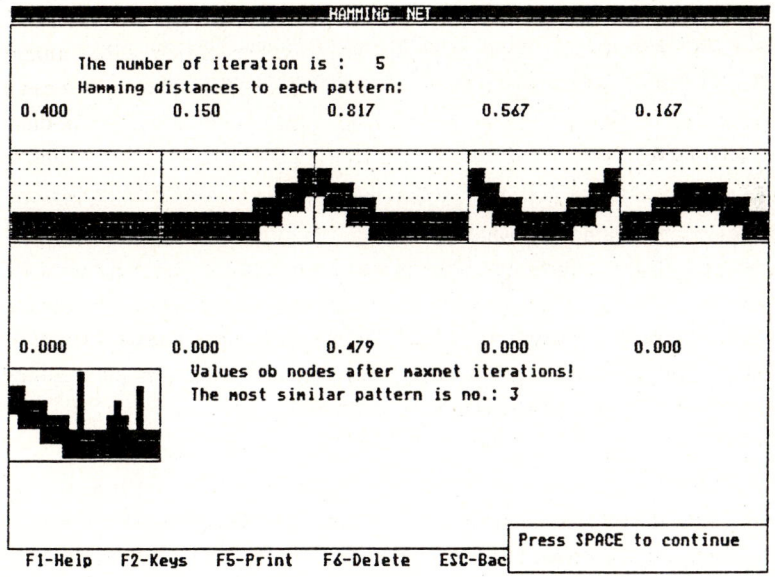

Fig. 5 Hamming net applied to the same example for determining the appropriate base line in fig. 3 (Hopfield net). The calculated Hamming distances between the input pattern and each exemplar pattern calculated in the first step are shown above each exemplar. The input pattern (reduced infrared spectrum) is shown in the lower left corner of the screen.

KOHONEN LEARNING

Kohonen type of net-work architecture (ref. 8,10,11) resembles the neural inter-connections, self-correcting, and feed-back of signals in the brain most closely, especially if the computer emulated neurons are ordered in a planar surface (Fig. 6). The idea of Kohonen neural net architecture is that, after a number of inputs is presented to the net, the weights will acquire such values that **similar** input patterns will tend to give the largest output signals in nodes that are **topologically close** together. This is achieved by introduction of the concept of the topological neighborhood of each neuron. The topology of the neighborhood (circular, rectangular, or hexagonal) depends on the application and is usually changing (shrinking concentrically towards the center) during the learning procedure.

For this type of neural net architecture it is characteristic that for any input X_i ($x_{i1}, x_{i2},...x_{im}$), only one neuron (and its neighborhood) is selected to be stimulated once (in some applications few times) more by changing its weights to enhance the output of the selected neuron. The selection of the neuron to be stimulated can be achieved in many different ways depending on the type of application: if the learning is unsupervised (i.e. no in advance or prescribed target is sought) the neuron with the largest output or the neuron with weights most similar to the components of the input pattern X_i can be selected; in the case of supervised learning, however, the neurons that must be stimulated are predefined for each class of inputs in advance.

In this approach the components of the input signal X_i are real values. The changing of weights in the particular neuron (and its neighborhood) is carried out in the following manner:

$$w_{ij}^{new} = w_{ij}^{old} + a (x_i - w_{ij}^{old}) \qquad /9/$$

where the parameter a is the 'gain term' which varies with the topological distance from the central neuron and decreases with the number of inputs presented to the net. The gain term in general decreases with the increasing topological distance, but may in certain applications have a kind of decreasing wave form (Mexican-hat-shape, ref. 10, p. 126). The correction of weights (eq. /9/) is made only once and then the net is ready for a new input. The planar lay-out of neurons and the learning procedure in the Kohonen net is schematically shown in figure 6.

In order to show the Kohonen learning, a slightly modified experiment described by Kohonen (ref. 10) and Lippmann (ref. 8) was repeated by the program NEURON. This experiment shows that the Kohonen net consisting of neurons (each having 2 inputs and 2 weights) in a rectangular array can be trained that the two weights of each neuron will show an arbitrary prespecified neuron's position. In our experiment each input X (x_1, x_2) in the learning consisted of two coordinates from an area within a circle. The development of the weight values as obtained by the program NEURON after each hundred different input vectors were presented to the net are shown in figure 7. In this example, the gain decreases linearly with increasing topological distance.

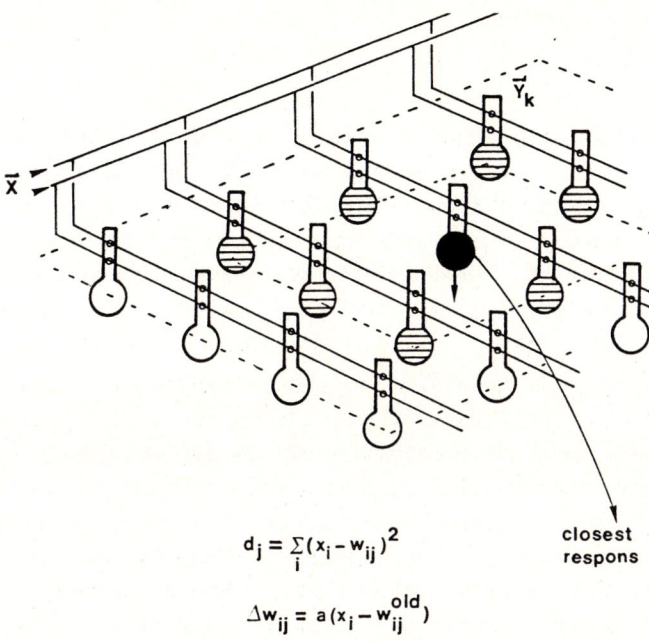

$$d_j = \sum_i (x_i - w_{ij})^2$$

$$\Delta w_{ij} = a(x_i - w_{ij}^{old})$$

closest respons

Fig. 6 In Kohonen learning the neurons are topologically ordered in a plane. The neighborhood of the neuron where the stimulation (correction of weights) is carried out and the gain term a (eq. /9/) must decrease with time.

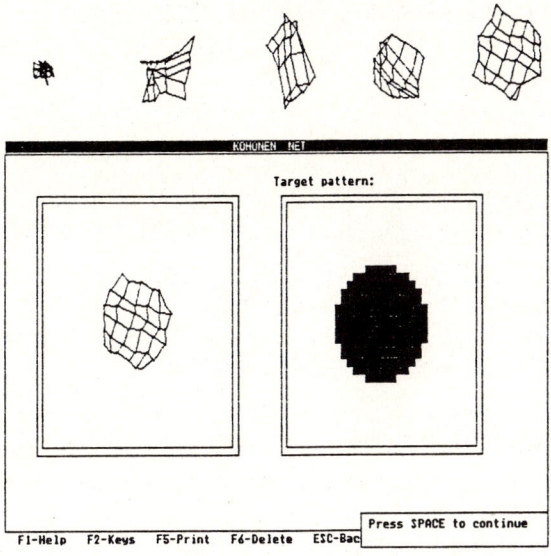

Fig. 7 Steps in the development of a 'circular' grid of neurons' weights from a 6 x 6 matrix of neurons as simulated by the program NEURON. Each grid shown is based on the weights of corresponding neurons taken as coordinates in the plane. The grids were drawn after 1, 100, 200, 300, 500, and 600 randomly selected points from the circle were input to the net.

BACK-PROPAGATION OF ERRORS

On of the most exciting and lately most popular neural net designs (ref. 12-14) is the multi-layer 'back-propagation' procedure initially developed by Rumelhart (ref. 15). The back-propagation algorithm does not reflect a particular similarity to real processes in brain, but it has a sound fundamental background (ref. 15). The net and transformed outputs are calculated in the same way as in the previously described nets, the only difference being in the way the weights in different layers are changed to yield the sought outputs.

The basic idea of the back-propagation algorithm is that the weights in all layers are changed in the direction to minimize the cost function C (difference between the final output and the desired target T; ref. 15). The errors appearing on outputs of each layer are determined and corrected in a sequence: starting with errors in the final layer and continuing backwards through the 'hidden' layers up to the input layer (Fig. 8b). The procedure of inputting the same signal and changing weights using the back-propagation of errors is repeated until an acceptable agreement between the output and target T is obtained. Next, a new input signal is applied to the net. The cost function C is expressed as:

$$C = \sum_{cases} \sum_j (T_j - Out_j)^2, \qquad Out_j = f(Net_j), \qquad Net_j = \sum_k w_{jk} x_k \qquad /10/$$

where function f is defined with eq. /2c/. By applying the gradient descent method the correction for each weight can be obtained:

$$w_{ij}^{new} = w_{ij}^{old} - a\, \partial C / \partial w_{ij} \qquad /11/$$

$$\partial C / \partial w_{ij} = (\partial C / \partial Out_j)(\partial Out_j / \partial Net_j)(\partial Net_j / \partial w_{ij}) \qquad /12/$$

The derivatives on the right side of eq. /12/ follows from the expressions /10/:

$$\partial C / \partial Out_j = -2(T_j - Out_j), \qquad \partial Out_j / \partial Net_i = f'(Net_j), \qquad \partial Net_j / \partial w_{ij} = x_i \qquad /13/$$

Inserting the derivatives /13/ into equation /12/ we obtain:

$$\partial C / \partial w_{ij} = -2 f'(Net_j)(T_j - Out_j) x_i \qquad /14/$$

where x_i is the input component to the weight w_{ij} and the middle part $f'(Net_j)(T_j - Out_j)$ is the back-propagated error DEL_j (ref. 14). In this form DEL_j can be calculated only for the output layer, while for hidden layers the $T_j - Out_j$ is not known. Therefore, the previously calculated errors have to be back-propagated through the particular hidden layer in order to obtain this part of eq. /14/:

$$\partial C/\partial w_{ij} = -2 f'(Net_j)(\sum_k DEL_k w_{jk}) x_i \qquad /15/$$

finally, the weights are corrected using the expression obtained by inserting eq. /14/ or /15/ into the eq. /11/:

$$w_{ij}^{new} = w_{ij}^{old} + a\, DEL_j\, x_i \qquad /16/$$

where the input x_i to a layer is actually the output Out_i from the layer above it. The back-propagation algorithm handles real inputs and is mainly implemented as a two or three layer network of neurons. It may, however, be extended to any number of layers, but the use of more than three layers is meaningful only if many dimensional inputs are employed. It was shown (ref. 8) that using the three layer back-propagation for two-dimensional inputs a correct recognition of a surface of any shape can be achieved. Because the back-propagation algorithm requires the targets for each class to be known in advance, the back-propagation always acts as a supervised learning.

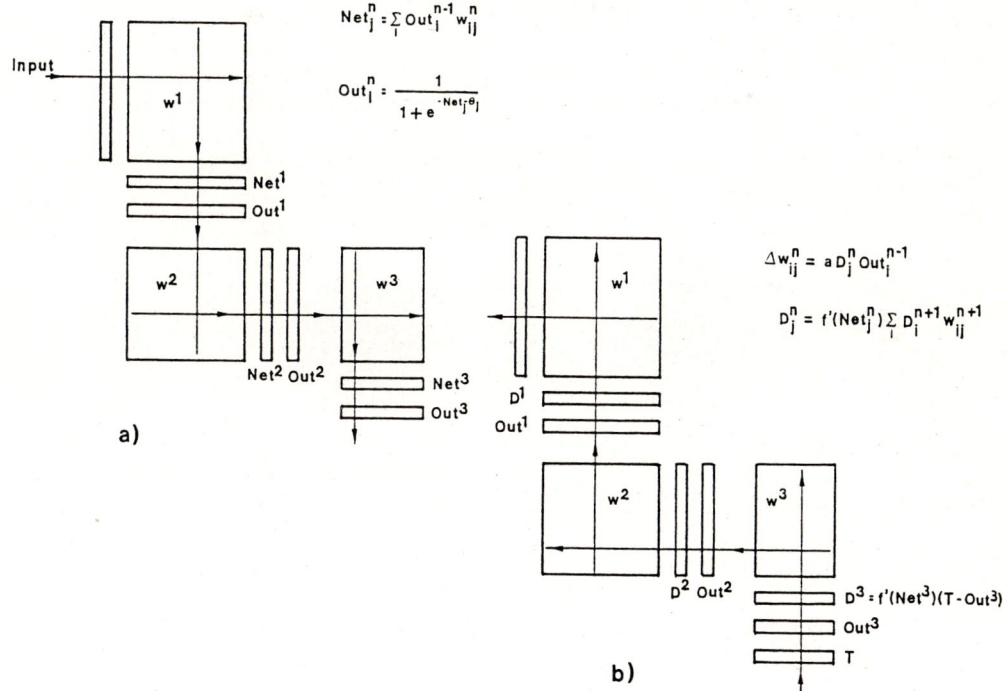

Fig. 8 Propagation of signals through three layer network (a) and back-propagation of errors through the same net(b). The parameter a in equation for changing weights is the gain term, usually around 0.5, but can slowly decrease during the training process.

The propagation of signals through three layers of neurons and the subsequent back-propagation of errors together with the correction of weights are schematically shown in figure 8, while six phases in the training for distinguishing between the points in the circle and the points outside it as performed using program NEURON are shown on figure 9. The shown example is the same as described by Lippmann (ref. 8). Back-propagation can be used in virtually all areas in chemistry where the standard pattern recognition methods were used until recently: from supervised clustering to associative memories and intelligent instruments (ref. 16,17).

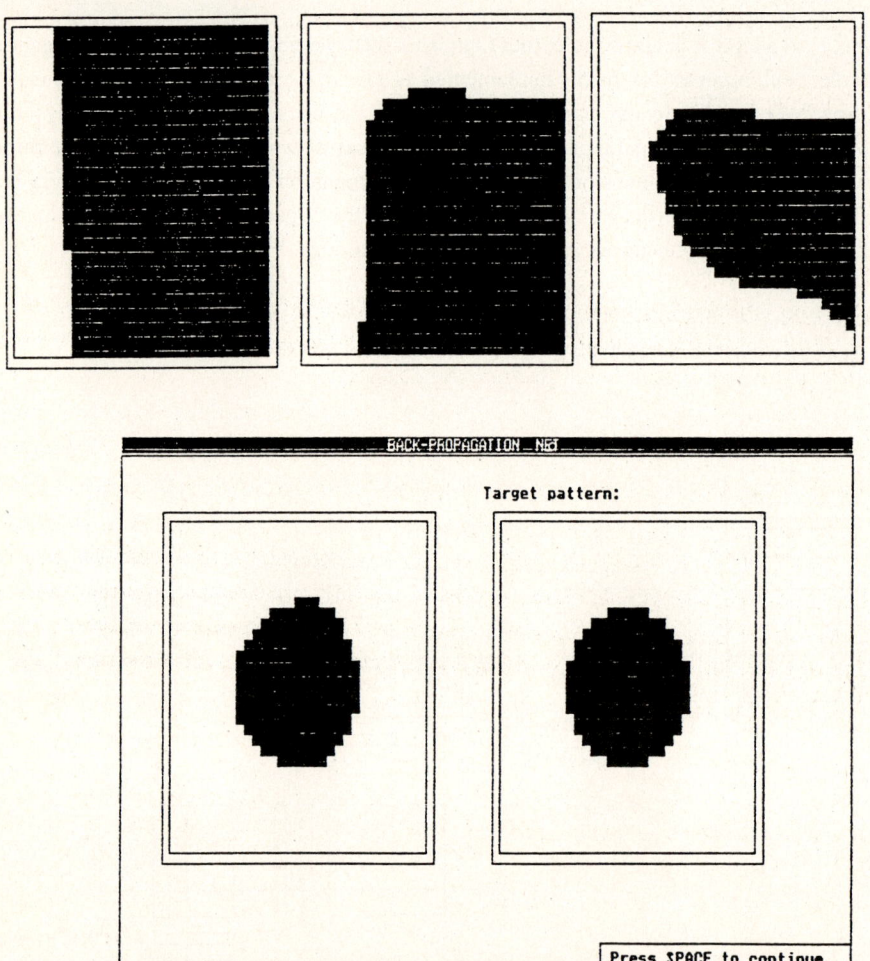

Fig. 9 The pattern to be learned (in the lower right corner) and the 'recognition' surface after 5, 20, 100, and finally 725 input points from both areas (outside and inside the 'circle') were presented to the two-layer network of 40 hidden neurons with two input and two output nodes.

CONCLUSION

As it can be seen from this short (and not at all exhaustive or complete) overview, the neural networks have many different applications. In chemistry, the applications range from clustering of multivariate data, structure and property predictions based on different kind of spectra, to content dependent retrievals on the basis of incomplete data. Even if some of the neural networks were developed strictly for handling bit (zero-one) type of inputs, they can easily be used for handling real values. An example for such a conversion is shown in figures 3 and 5 for Hopfield and Hamming nets, where the real values of each base-line intensities were converted to the bit images of the base-line.

The actual choice for the problem to be solved and the selection of the neural net design is, however, left entirely to the reader's imagination. However, it must be made clear that everything that can be achieved, at present stage, using neural networks can be achieved by some other 'classical' method as well. Additionally, neural network architectures are far from being 'robust', i.e. an exactly the same neural network design with the same set of parameters (number of neurons, number of layers, dimension of weight vectors, initial values of weight vectors, gain term, neighborhood, squashing function, threshold, etc.) cannot be used for another application. On the contrary, each application needs a number of trials and fine tuning to find out the most appropriate set of parameter that will actually perform in the way or achieve the goal as it was planned.

Neural networks are mainly very computation intensive. In order to achieve good results in the real applications large input vectors (input matrices - images) and correspondingly larger weight matrices are required what causes many iterations for convergence and consequently eats-up considerable amount of computer time. Reliable results are normally obtained after a number of trials in the search for the best set of parameters what Additionally takes a lot of time. On the bright side, however, the neural network offer some genuine possibilities to solve problems (especially in the domain of associative learning and content dependent retrieval) involving large amounts of data. The solutions yielded by the neural nets, once they are trained to solve specific problems, can be obtained much faster (for some applications like content dependent retrieval even orders of magnitude faster) compared with standard methods. It is hoped that once the special purpose parallel processors performing general purpose neural network applications will be available cheap enough, many instrumental and experimental problems in chemistry could be solved in a very convenient way.

REFERENCES

1. R Rosenblatt, Principles of Neurodynamics, Spartan Books, New York, 1959
2. J Zupan, Algorithms for Chemists, John Wiley, Chichester, 1989,
3. JY Jau, Y Fainman, SH Lee, Applied Optics **28**(2), (1989) 302-305
4. TL Isenhour, P Jurs, Analytical Chemistry, **43**, (1971), 20A-35A

5 JJ Hopfield, DW Tank, Biological Cybernetics, **52** (1985), 141-152
6 JJ Hopfield, DW Tank, Science, **233**, (1986), August, 625-633
7 RJ McEliece, EC Posner, ER Rodemich, SS Venkates, IEEE Trans. Inform. Theory, **IT-33** (4), (1987), 461-482
8 RP Lippmann, IEEE ASSP Magazine, April **(1987)**, 4-22
9 K Varmuza, Pattern Recognition in Chemistry, Springer Verlag, Berlin, 1980, p.26
10 T Kohonen, Self-Organization and Associative Memory, Springer-Verlag Berlin, 2-nd ed.,1988
11 T Kohonen, Neural Networks, **1**, (1988), 3-16
12 WP Jones, J Hoskins, BYTE, October (1987), 155-162
13 R Hecht-Nielsen, IEEE Spectrum, **March (1988)**, 36-41
14 PD Wasserman, T Schwartz, IEEE EXPERT, **Spring (1988)**, 10-15
15 DE Rumelhart, GE Hinton, RJ Williams, Learning Internal Representations by Error Propagation' in 'Parallel Distributed Processing: Explorations in the Microstructure of Cognition', DE. Rumelhart, JL. McClelland, Eds., MIT Press, Boston, 1986,
16 J Zupan, Anal. Chim. Acta, (1989) in press
17 M Bos, to be published in Anal. Chim. Acta

APPLICATION OF FUZZY NEURAL NETWORK TO SPECTRUM IDENTIFICATION

Matthias Otto and Uwe Hörchner

Department of Chemistry, Bergakademie Freiberg,
Akademiestr. 6, 9200 Freiberg, GDR

Abstract: The main concept of using neural network models for pattern association and pattern classification is outlined. Applications for handling spectroscopic patterns are demonstrated with storing UV-spectra recorded under different experimental conditions in a neural network and identifying unknown spectra by presenting them to the network. Essentially Kosko's Adaptive Bidirectional Associative Memory [2] is explored that enables to learn similar spectra of one chemical compound as well as spectra of different chemical identity. The final comparison of sample and reference spectra is carried out by fuzzy set operations.

INTRODUCTION

Neural networks or parallel distributed processing can be considered an alternative to symbolic knowledge manipulation used in classical artificial intelligence approaches. The idea of neural networks is based on parallel processing of information rather than to handle individual pieces of information sequentially.

The human brain as the biological analogy of neural networks consists of more than 10 billion neural cells with complicated interconnections. Although current computer elements are about a million times faster than neurons, neurons have a thousandfold greater connectivity. This high degree of connectivity enables huge amounts of information to be processed in many small pieces in a parallel fashion.

Neural models can be effectively simulated on essentially any computer and for the foreseeable future, the bulk of neural network research is expected to be carried out with simulations on conventional computers. However, the high connectivity may represent a major problem for uniprocessor computers used nowadays.

A neural network consists of many units (the neurons) with inputs and outputs. Each unit sends its numeric output modified by inhibitory or excitatory weights (transmission coefficients, synapses) to another unit - the unit fans out - so that the output of a unit is the input to another or to many other units. One single unit multiplies the inputs by the individual weights and sums them up (fan in vector). The resulting value is the activation value of the unit which is often modified by applying an appropriate transformation function. The activation value is than processed to other units by applying the appropriate weights.

The result is the transformation of a pattern vector of input values to a pattern of output values. The information is contained in the form of static knowledge in the weights or synapses (long term memory) and in the form of dynamic knowledge in the activations of the neurons (short term memory).

In contrast to classical AI-techniques where knowledge is represented explicitly in form of a rule or a static pattern, with neural networks knowledge is implicitly stored in the weights that enable certain patterns to be created or associated repeatedly.

The process of learning can be defined then as adjusting the connecting weights such, that by presenting a certain input pattern the correct output units will be activated. Thus, learning is the change or fine tuning of the synapses.

Different learning paradigms are feasible:

Autoassociate learning is based on presentation of patterns to the system which are associated with themselves. After adjusting the weights the neural network is able to correctly recognize patterns that are shown to the inputs of the system, even if parts of these patterns are missing.

Pattern association differs from autoassociative learning in a way that pairs of individual input and output patterns are presented to the system and the neural network learns to associate a certain output pattern with a certain input pattern.

Pattern classification is similar to the above mentioned learning paradigms, i.e. patterns are shown to the inputs together with their class membership as their output patterns (unsupervised learning). The system learns these patterns by tuning the weights and allows unknown patterns to be classified with respect to the correct class.

Regularity detection is used for evaluating important features of patterns that are presented only as input patterns. It is a non-associative learning strategy and can be classified as an unsupervised learning method. The learning algorithm recognizes the most important features and groups the patterns into classes. This kind of learning is used with competitive learning techniques.

From the many neural network models that have been proposed [1] in the present work the Adaptive Bidirectional Associate Memory (ABAM) [2] is investigated. This model has no hidden units but is based on non-linear propagation rules and adaptive resonance. Since the activations of the neurons can be considered as fuzzy sets common operations of fuzzy set theory [4] can be applied on the input and output patterns for deriving a quality index in order to compare, e.g., sample spectra with those from a library of candidate reference spectra. The method is demonstrated here for spectrum identification in the UV-spectral range.

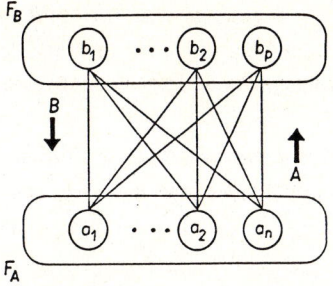

Fig. 1. A bidirectional Associative Memory (BAM) representing the neuron fields symmetrically connected by the synapses

Adaptive Bidirectional Associative Memory

The neural model of the Bidirectional Associative Memory (BAM) [2] is represented by the input field or bottom-up field, F_A, that consists of the n neurons a_i. The output (top-down) field F_B stands for the p neurons b_j. In Fig. 1 the BAM model is represented with the synaptic connections m_{ij}. The neurons indicate neuron names and neuron states at the same time. Interconnections between the m neuron pairs (A_i, B_i) is given by the correlation matrix M according to:

$$M = A_1^T B_1 + A_2^T B_2 + ... A_m^T B_m \qquad (1)$$

or for single synaptic connection
$m_{ij} = \Sigma a_i b_j$

This encoding scheme represents the simplest biologically plausible learning law discussed by Donald Hebb [3]. The main idea of Hebb correlation learning is that if the neuron a_i and b_j are activated simultaneously, the weight connecting these units is increased.
If a new input pattern A is presented to the BAM field F_A associative recall is possible by mutual calculation of B by:
$B = A M$ \qquad (2)
and feeding back the B-vector to A giving a new A'-vector as follows
$A' = B M^T$ \qquad (3)
This process is repeated until the system equilibrates on a fixed pair (A, B) as the final associative recall.
Since in the original BAM the activations of neurons are restricted to the values 0 or 1 summation of the weighted sums of a_i or b_j is thresholded to generate binary inputs or outputs [2].
If the activations a_i and b_j are transformed to the signals $S(a_i)$ and $S(b_j)$ then continuous BAM's can be simulated. Kosko proposed the following dynamic system [2]:

$$\dot{a}_i = -a_i + \Sigma_j S(b_j)m_{ij} + I_i \qquad (4)$$
$$\dot{b}_j = -b_j + \Sigma_i S(a_i)m_{ij} + J_j \qquad (5)$$
where a_i, b_j are passive decay terms and I_i, J_j represent constant external inputs.

Typically, the transform function is of sigmoid type, e.g. $S(x) = 2/(1+\exp(-cx))$, with c being a constant. For the time dependent change of the synapses the simple Hebb learning law can be formulated as:
$$\dot{m}_{ij} = -m_{ij} + S(a_i)S(b_j) \qquad (6)$$
The dynamic system of equations (4) to (6) forms now an adaptive BAM (ABAM) that can be used for storing and recalling spectral data.

Learning (Encoding) UV-Spectra

For input of spectra to the ABAM the digitized spectra have to be converted into vectors representing the fileds F_A and F_B. The easiest way is to overlay the spectra with a grid and then a value of 1 is assigned to pixel positions that are crossed by the spectrum and 0 otherwise. Fig. 2a shows an example for a very coarse grid of 10x10 pixels for the fields F_A and F_B each. In this case the pairs (A_i, B_i) are formed from the same spectrum, i.e. the ABAM associates different patterns from one spectral identity. For presenting the spectra to the ABAM the pictures are arranged in two vectors row by row.

a.

b.

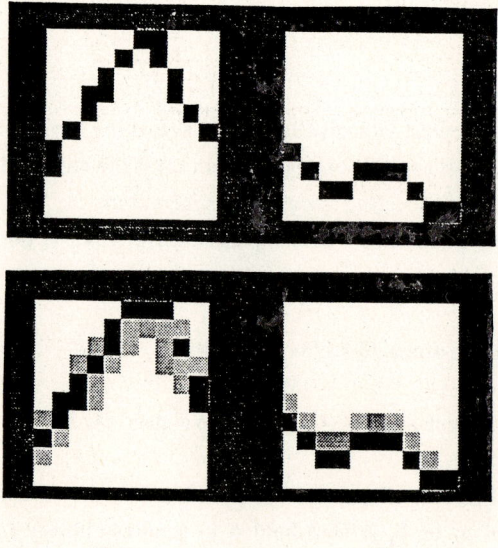

F_A F_B

Fig. 2. Demonstration for digitizing spectra arranged in the field F_A and F_B as a 10x10 pixel matrix

Spectra of the same chemical compound recorded under different experimental conditions can be encoded in the ABAM by presenting them for some time (several snap shots) to the network. Fig. 2b demonstrates the signals in A and B after having learned three different spectra. In dependence on performance of the network the three spectra are either stored as separable patterns or as in the present case as fuzzy spectra of a certain chemical identity. Spectra that are recognized as different patterns even with the very coarse grid in Fig. 2 are shown in Fig. 3.

Spectrum 4

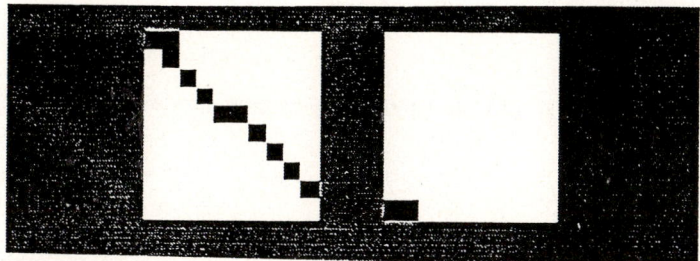

Spectrum 5

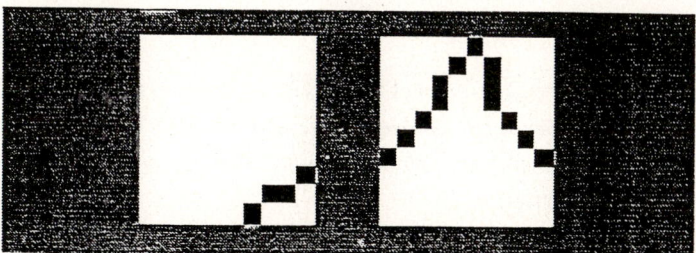

Fig. 3. Spectra of different identity as compared to the spectra in Fig.2.

The spectra of Fig. 3 have also been shown as inputs to the ABAM in order to recall the patterns in further trials. Care must be taken in the choice of scaling constants for the terms in equ. (4) to (6) since these constants determine the learning rate. If, for example, the forget term $-m_{ij}$ in equ. (6) is too high compared to the second term present learning would wash out past learning.

Decoding and Identification of Spectra

For association of an unknown sample pattern the coded spectrum is presented to the ABAM and convergence is awaited within some time.

In order to quantify the match between the associate pattern and the sample pattern, the spectra are compared by fuzzy set operations. This is possible because the signal vectors $S(A) = \{s(a_1), s(a_2),...s(a_n)\}$ and $S(B) = \{s(b_1), s(b_2),...s(b_p)\}$ can be considered as fuzzy sets over the universe A and B, respectively. Thus pattern comparison is performed by intersecting the recalled pattern S_r with the sample pattern S_s in the usual manner giving as the membership function for the resulting set C [2]:

$$m(x)_C = \min\{s_r(x), s_s(x)\} \qquad (7)$$

Normalization to the interval [0,1] is carried out by computing the relative cardinality according to:

$$rel_s card\ C = (card\ C)/(card\ S) \qquad (8)$$

where the cardinality of the standard set S [4] is here the joint universe of the sample set S_s. As the result values of relative cardinalities of 1 stand for perfect match of the sample spectrum to the recalled spectrum and values near to 0 represent no similarities between sample and the candidate reference spectrum.

The three similar spectra of Fig. 2 (1-3) and those of Fig. 3 (4,5) were recalled with the ABAM after having trained it with the three similar spectra as seen from Table 1. The previously learned spectra 1-3 are recognized as "known identities" and the unknown spectra 4 and 5 give very low relative cardinalities.

Table 1. Recall of spectra from Fig. 2 (1-3) and Fig. 3 (4,5) with the ABAM

Spectrum Number	Relative cardinality as similarity index for match of spectra after learning	
	Spectra 1 to 3	Spectra 1 to 5
1	0.893	0.95
2	0.816	0.947
3	0.978	0.953
4	0.077	0.953
5	0.059	0.967

After that the ABAM was allowed to learn all of the five spectra and again recall was intended. The results were less encouraging since all of the spectra were recalled with similarity values (relative cardinality) less than 0.2. Somewhat better results were obtained by increasing the number of learning steps from 2 to 5. But even then the spectra 4 and 5 were recalled with relative carinalities of about 0.5 only. The reason is to be attributed to the fact that the dynamic behaviour of the ABAM cannot be fully exploited if the spectral patterns are presented as crisp patterns, i.e. with signal values of 1 and 0 else.

However, by decreasing the original inhibitory and excitatory activations by a factor of 10 and by judicious choice of scaling constants for the whole dynamic system in equ. (4) to (6) all the 5 spectra can be recalled with nearly complete match as shown with the relative cardinalities in Table 1, third column.

Conclusion

Although the present work is aimed at demonstrating the principles of the adaptive neural network only other possibilities for associating spectral patterns can be already envisaged. The ABAM can be used to learn other pairs like spectra and chemical structures or chemical elements, spectra and chemical names or pictures, spectra and structural features or spectra and class membership.

The applicability is limited by the computer storage capacity for the correlation matrix as well as by the dimensionality of the correlation matrix, since theoretically the number of trainable patterns has to be less than the dimensionality of the lowest dimensioned field F_A or F_B [2].

The main advantages of using neural networks for pattern association or pattern classification derive from the facts that search time is independent on the number of stored patterns, correlations between spectra and chemical features can be learned rather than have to be explicitly declared, and fuzzy spectra of one chemical identity can be processed with these systems.

References
1 Kohonen T (1984) Self-Organization and Associative Memory, Springer-Verlag, New York
2 Kosko B (1987) Appl. Optics 26:4947
3 Hebb D O (1949) The organization of behaviour, New York, Wiley
4 Otto M (1988) Chemometrics Intell. Lab. Systems 4:101

TOOLS FOR AUTOMATIC PROGRAM GENERATION

H. Armitage, A. Khuen and D. Ziessow

Iwan-N.–Stranski-Institut für Physikalische
und Theoretische Chemie,
Straße des 17. Juni 112, D-1000 Berlin 12

Abstract: The efficiency of program development can be improved by employing appropriate software tools. In this paper we discuss the compiler generator LEX-YACC and the GENTRAN module of the computer algebra (C. A.) system REDUCE. Their use is demonstrated for examples from physical chemistry. The LEX-YACC program, a UNIX tool, is employed to generate the user interface of a spectral analysis program. REDUCE is used to modify REDUCE subroutines in a symbolic manner. The resulting code is then translated into "C" code by the GENTRAN module and the resulting subroutine inserted into a numerical program.

1.0 INTRODUCTION

The advance of computer and network technology demands an equivalent development with respect to software. The efficiency in programming as well as the quality of the programs must be improved. How can this be achieved in the case of chemical software?

Many programs employed in chemical research mainly make use of numerical algorithms, either to evaluate experimental data or to model physical phenomena. In writing such programs, a programmer adheres to a basic rule: The program is divided into small subprograms each handling a well defined task. There are subroutines for data input/output, various algorithms and the display of the results. Data transfer between the units should proceed in a controlled manner and side effects should be avoided. Partitioning a program in such a way can reduce the programming effort considerably: Subroutines from mathematical libraries or other program collections may be employed. Certain types of subprograms can even be generated automatically whereby the speed of programming is increased and the number of errors reduced.

In this communication we describe the usage of the UNIX tools LEX-YACC to implement a command language and of the REDUCE tool GENTRAN to generate "C" subroutines.

2.0 THE LEX—YACC TOOL

The LEX-YACC programs [1] are well known as compiler generating tools in the UNIX world. One defines language rules as well as the associated code fragements and LEX-YACC produces "C"-code for the respective compiler. The LEX-YACC-program can equally well be employed to analyse input statements and to perfom the corresponding action. We used LEX-YACC to generate a command language parser for a general spectral analysis and display program named SPEC [2].

LEX-YACC [5,6] accepts as input a formal description of the command language for the SPEC program and generates the "C" code for the parser. The parser recognizes legal expressions and transforms them into actions. The syntax of the command language can be defined with BNF (Backus-Naur-form) production rules or with syntax graphs. The actual input for LEX-YACC must be described in a syntax similar to BNF. The parser produced is of the LR(1) type. Input is read from left to right and while interpreting the actual token the machine also knows the next one in line. Certain context dependent expressions can not be recognized. The LEX-YACC system also checks the syntax of the language thoroughly.

3.0 THE GENTRAN TOOL

The GENTRAN tool is part the computer algebra system REDUCE which allows to perform mathematical calculations or transformations in a symbolical manner. One can manipulate algebraic polynoms, differentiate or integrate functions or perform matrix computations. With the aid of the so called LET rules the algebraic properties of new operators may be defined. Additional rules for differentiation, integration, transformation and trigonometric simplification are thus created.

In the interactive mode only small problems can sensibly be solved. Larger problems are written as a REDUCE program.

Apart from its algebraic kernel, REDUCE offers modules such as an integration package for a special class of functions or the Gröbner package to handle systems of nonlinear equations, etc.. The GENTRAN module takes REDUCE code as input, performs symbolic calculations and translates the so modified REDUCE statments into a "C" or Fortran program.

4.0 APPLYING THE GENTRAN PROGRAM

4.1 Optimization And Curve Fitting.

One often encounters the task to fit experimental data to a theoretical model. Commonly the method of the least squares is applied to find optimal model parameters. If the model is nonlinear in the parameters, it is developed into a taylor series around an appropriate starting point. Then the model parameters are evaluated iteratively.

Algorithms to perform such calculations can be found in mathematical program libraries. We took the procedure "mrqmin" from the library "Numerical Recipes" [7] which is based on the algorithm developed by Levenberg and has later been improved by Marquardt. This subroutine requires as input, apart from the starting values for the parameters, a subroutine to compute the function value as well as all derivatives with respect to the parameters. Change of the model is simply achieved by a change of this subroutine. Therefore we implemented this subroutine in REDUCE code. The translation into "C" is performed in two steps: In the first, the function is differentiated analytically and, in the second, the actual code is produced. Introducing a new model means to insert a new function into the REDUCE version of the subroutine. By using an appropriate UNIX shell script or a "make-file" the whole process of model changing can be completely automated.

The basic functionality of the fitting program is shown in Fig. 4.1.

Input of the parameter
Graphical display of input data
Iterative fitting
Output of the results
Graphical display

Fig. 4.1: Basic tasks of the fitting program.

This fitting program was adapted to handle different problems, like analyzing DSC diagrams (differential scanning calorimetry) or resolving bandshapes of infrared and fluorescent spectra. As an example we describe how an experimental IR spectrum in the absorption mode can be resolved into a sum of standard spectral line shapes (Gauss, Lorentz, Maxwell etc.).

The REDUCE code for the model dependent subroutine is displayed in Fig. 4.2. To help understanding the code, we added more comment lines.

```
%
% program to produce the "C" subroutine "funk" for the fitting procedure
% mrqmin
%
    gentranlang!* :='c$          % translate into "C"
    clinelen!* := 72$
    on factor;                   % factorize expressions

    % gentranout "lorentz.c";    % name of "C" code output file

    operator a,yj,dyda;          % define symbolic functions
```

```
%-------------spectral line function------------------------

%-------------lorentzian: ---------------------------------

for all j,x let yj(j,x)=a(j)/(1+((x-a(j+1))/a(j+2))^2);
% the function is  defined by using the LET rule

%-------------gaussian-------------------------------------

% for all j,x, let yj(j,x)=a(j)*exp(-((x-a(j+1))/a(j+2))^2);

%----ma: number of coefficients a[i]-----------------------

ma:=3;           % 3 shapes are used
                 % in principle an arbitrary number is possible

gentran                         % calling gentran
procedure funk(x,a,y,dyda,ma);
begin
 declare
     <<
         x,y,sum,a(eval(ma)),dyda(eval(ma)) :float;
         ma,j : int;
         funk: void;              % declarations for the "C" code
     >>;

sum:=0.0;
for j:=1 step 3 until (ma-2) do
    sum:=: sum+yj(j,x);         % :=: means: evaluate first and then translate

!*y:=sum+a(ma-1)*x+a(ma);

for j:=1 step 3 until (ma-2) do

begin
    dyda(j)   :=: df(yj(j,x),a(j));
    dyda(j+1) :=: df(yj(j,x),a(j+1));
    dyda(j+2) :=: df(yj(j,x),a(j+2));
                                % all derivatives are calculated before the
                                % translation starts
end;

    dyda(ma-1):= x;             % derivatives for the baseline
    dyda(ma)  := 1.0;

end$
```

Fig. 4.2: REDUCE-GENTRAN program to produce a "C" subroutine.

In the file "lorentz.c" we find the "C" code generated (Fig. 4.3).

```
void funk(x,a,y,dyda,ma)
float x,*y,*a,*dyda;
int ma;
{
    float sum;
    int j;
    sum=0.0000000;
    for (j=1; j<=(ma-2); j+=3)
        sum=((power(a[j+1]-x,2)+power(a[j+2],2))*sum+power(a[j+2],2)*
        a[j])/(power(a[j+1]-x,2)+power(a[j+2],2));
    y=sum+a[ma-1]*x+a[ma];
    for (j=1;j<=(ma-2);j+=3)
        {
            dyda[j]=power(a[j+2],2)/(power(a[j+1]-x,2)+power(a[j+2],2));
            dyda[j+1]=-(2.0*(a[j+1]-x)*power(a[j+2],2)*a[j])/power(power
            (a[j+1]-x,2)+power(a[j+2],2),2);
            dyda[j+2]=2.0*power(a[j+1]-x,2)*a[j+2]*a[j]/power(power(a[j+
            1]-x,2)+power(a[j+2],2),2);
        }
    dyda[ma-1]=x;
    dyda[ma]=1.000;
}
```

Fig. 4.3: Generated "C" subroutine.

The spectrum may consist of an arbitrary number of lines. In the example only three were defined. A spectrum was simulated and gaussian noise added. Fig. 4.4 shows the result of the fitting procedure.

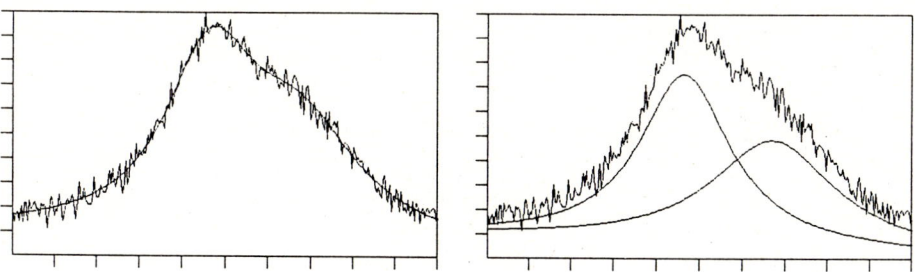

Fig. 4.4: Resolving a simulated bandshape into its parts.

4.2 Generation Of The Hamiltonian For A Spin 1/2 System

Analyzing complex line patterns is a typical problem encountered in NMR spectroscopy. To ease this task, large spin systems can be simplified by decoupling spins from each other. Decoupling is achieved by irradiating the sample with a monochromatic HF wave. Our goal is to simulate such an experiment.

The Hamiltonian for a n-spin-1/2 system is given by:

$$H = \sum_i \omega_i \cdot I_{zi} + \sum_{i<k} J_{ik} \cdot I_i \cdot I_k$$

To decouple spins of nuclei i from nuclei k we irradiate the sample with power P_i^* at frequency ω_i^*. This effect can be expressed by the operator:

$$H' = \sum_i (\omega_i - \omega_i^*) \cdot I_{zi} + \sum_{i<k} J_{ik} \cdot I_i \cdot I_k + \sum_i I_x P_i^*$$

For $\omega_i^* = 0$ und $P_i^* = 0$ the unperturbed hamiltonian results. To simplify the computation we neglect relaxation and assume that time dependent parts of H' do not influence I_{zi}. These terms are dropped. The spectrum itself is given as the Fourier transform (FT) of the trace of the time dependent density matrix multiplied by Ix:

$$I(\omega) = FT\ (\ \mathrm{Spur}\ (\ I_x \cdot e^{-iHt/h} \cdot I_y \cdot e^{iHt/h}\ )$$

With an appropriate basis, H' can be represented as a matrix. In order to express the exponential function of H' as a matrix as well we have to diagonalize H'. If we call

$$\sum_j \omega_j P_j$$

the spectral resolution of H', which means

$$\sum_j \omega_j \cdot P_j = U^{-1} \cdot H \cdot U\ ,$$

(ω_j: j^{th} eigenvalue, P_j: j^{th} projection operator)

we get:

$$I(\omega) \sim FT(\ \mathrm{Spur}(\ U^{-1} \cdot I_x \cdot U \cdot (\sum_j e^{-i\omega_j t} \cdot P_j) \cdot U^{-1} \cdot I_y \cdot U \cdot (\sum_k e^{i\omega_k} \cdot P_k)))$$

Computing the spectrum $I(\omega)$ can be reduced to the task of evaluating the eigenvalues and eigenvectors of a Hermitian matrix (in this case a real and symmetric one). For such calculations the necessary algorithms exist. We chose the Housholder method. The only problem left is to set up the Hamiltonian matrix. We use REDUCE for this purpose.

At first we calculate the spin matrices as direct products of the Pauli matrices. The Hamilton matrix "a" and the I_x matrix is set up. In the next step a REDUCE program including GENTRAN code is written to a file. This file is read in again and translated into "C" code. The whole process runs automatically.

```
%--------------Create Hamilton matrix---------------------------

        matrix ix,iy,iz,ee$
        operator v,c,freq,poww;
        array ax(5),ay(5),az(5);
        on comp;            % compiler on
        in dprod2;          % read in additional routines
        in spin;
        off comp;           % compiler off

        %n=number of spins
        n:=2;                              % 3,4 · · · · ·

        matrix xx;
        procedure ixx(n,k);
% ixx(n,k) generates a single spin matrix for spin k for n spins
% I_i(k) = 1 × 1 × 1 ···· × I_i × ···· × 1 × 1 × 1, there are n unit matrices 1 and the one in the
% k^th position is replaced by I_i
        begin
        scalar j;
        for j:=1:k-1 do
            xx:=ee#xx;       %ee is a 2 × 2 unit matrix, # indicates the direct product
        for j:=k+1:n do
            xx:=xx#ee;
        end;

        ee:=eins(2)$            % setting up the unit matrix
        ix:=matx(1/2)$          % creating the Pauli matrices
        iy:=maty(1/2)$
        iz:=matz(1/2)$
        for j:=1:n do
        <<
          xx:=ix;
          ixx(n,j);
          ax(j):=xx;
          xx:=iy;
          ixx(n,j);
          ay(j):=xx;
          xx:=iz;
          ixx(n,j);
          az(j):=xx;
        >>;
```

```
        matrix a(2**n,2**n);        % a is the Hamilton matrix
        for j:=1:n do
          <<
             xx:=az(j);
             a:=a+v(j)*xx;           % Zeeman term of the Hamiltonian
          >>;
        l:=1;
        matrix axj,axk,ayj,ayk,azj,azk;
        for j:=1:n do
          for k:=(j+1):n do
            <<
               axj:=ax(j); axk:=ax(k);
               ayj:=ay(j); ayk:=ay(k);
               azj:=az(j); azk:=az(k);
               a:=a+c(l)*(axj*axk+ayj*ayk+azj*azk);    %scalar coupling term
               l:=l+1;
            >>;
        clear axj,axk,ayj,ayk,azj,azk;
        matrix ix(2**n,2**n);
        for j:=1:n do
          <<
             xx:=ax(j);
             ix:=ix+200*xx;
             a:=a+poww(j)*xx;         %introducing the external irradiation
          >>;
        clear ax,ay,az,xx;
% -------------- we prepare a "GENTRAN" file -------------------------
        out ham3$                    % we write the Hamiltonian into a file
        write "gentranlang!*:='C$"$
        write "clinelen!*:=72$"$
        write "gentranout ham$"$
        write "gentran"$
        write "declare <<a(1,1),v(1),c(1),poww(1),freq(1):double$ ix(1,1) $"$
        write "gentran"$
        write "<<"$
        off nat$
        for j:=1:n do write "v(",j,"):=",v(j)-freq(j)$
        a:=a;
        ix:=ix;
        on nat$
        write ">>$"$
        write "end$"$
        write "end$"$
        shut ham3$
        load gentran$     %load the GENTRAN package
        in ham3$          %read in the Hamiltonian and produce the "C" code
        end;
```

Fig.4.5: REDUCE-GENTRAN program to set up a spin Hamiltonian to simulate spin decoupling.

One should mention that the spin matrices are stored as elements of a REDUCE array. This allows to access them through their index. The spin matrices in the product space of all spins is generated by the procedure ixx(n,k). The hamilton matrix "a" as well as the Ix matrix is stored in ham3 before being read in again to create the "C" code (Fig. 4.6).

```
DOUBLE A[2][2],V[2],C[2],POWW[2],FREQ[2];
INT IX[2][2];
{
    V[1]-=FREQ[1];
    V[2]-=FREQ[2];
    A[1][1]=(C[1]+2.0*V[2]+2.0*V[1])/4.0;
    A[1][2]=POWW[2]/2.0;
    A[1][3]=POWW[1]/2.0;
    A[1][4]=0.0;
    A[2][1]=POWW[2]/2.0;
    A[2][2]=-((C[1]+2.0*V[2]-(2.0*V[1]))/4.0);
    A[2][3]=C[1]/2.0;
    A[2][4]=POWW[1]/2.0;
    A[3][1]=POWW[1]/2.0;
    A[3][2]=C[1]/2.0;
    A[3][3]=-((C[1]-(2.0*V[2])+2.0*V[1])/4.0);
    A[3][4]=POWW[2]/2.0;
    A[4][1]=0.0;
    A[4][2]=POWW[1]/2.0;
    A[4][3]=POWW[2]/2.0;
    A[4][4]=(C[1]-(2.0*V[2])-(2.0*V[1]))/4.0;
    IX[1][1]=0.0;
    IX[1][2]=100.0;
    IX[1][3]=100.0;
    IX[1][4]=0.0;
    IX[2][1]=100.0;
    IX[2][2]=0.0;
    IX[2][3]=0.0;
    IX[2][4]=100.0;
    IX[3][1]=100.0;
    IX[3][2]=0.0;
    IX[3][3]=0.0;
    IX[3][4]=100.0;
    IX[4][1]=0.0;
    IX[4][2]=100.0;
    IX[4][3]=100.0;
    IX[4][4]=0.0;
}
```

Fig. 4.6 "C" array elements of the Hamilton matrix "a" and the Ix matrix.

In a slightly modified form was this code segment inserted into an existing NMR simulation program.

As an example we consider a 3-spin-system with chemical shifts at $\omega_1 = 20$ Hz, $\omega_2 = 50$ Hz and $\omega_3 = 80$ Hz and coupling constants at $J_{12} = 5$, $J_{13} = 12$ und $J_{23} = 15$ [all values in Hz].

The following spectra were obtained (Fig. 4.7 and Fig. 4.8).

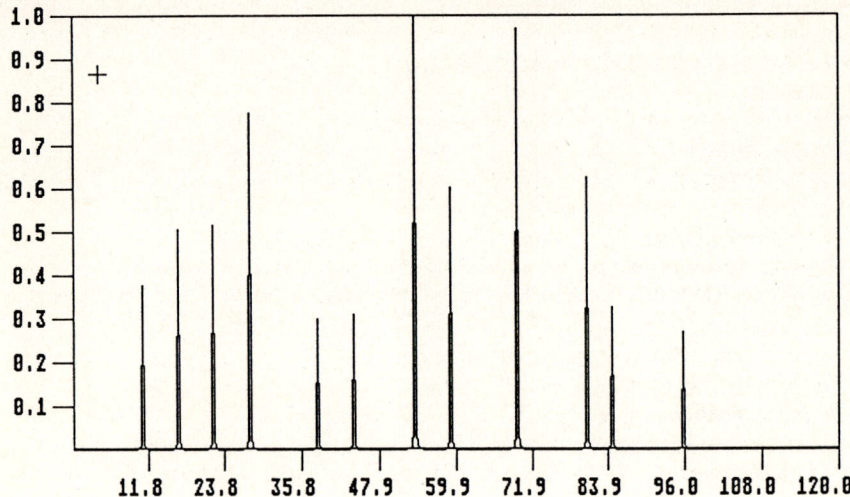

Fig. 4.7: Simulated 3-spin-1/2-system.

Decoupling at w2 = 50 Hz produces the spectrum of Fig. 4.8.

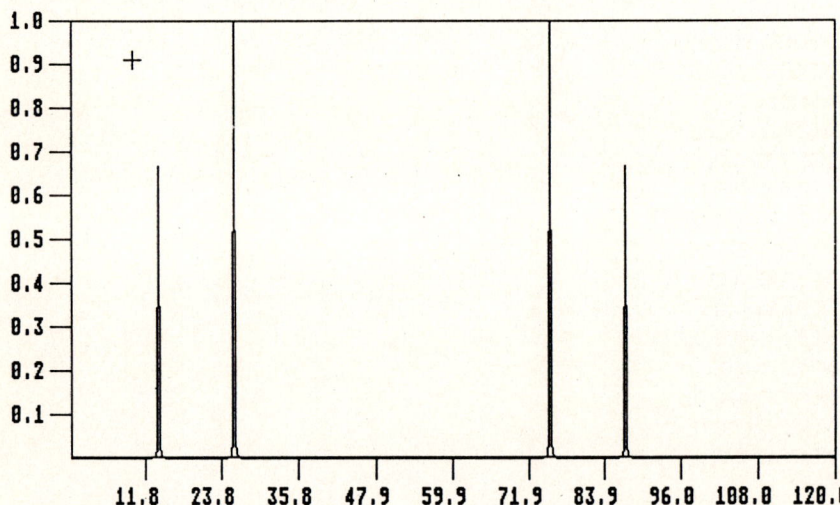

Fig. 4.8: Decoupling of one spin results in a pure 2-spin spectra.

5.0 REFERENCES

1   UNIX V.3 User Manual, Programmer's Guide, Part 1

2   Armitage H, Kolbe H und Ziessow D (1989) SPEC: Ein Programmsystem für die Spektroskopie, in: Gauglitz G (ed) Software-Entwicklung in der Chemie 3, Springer-Verlag Berlin Heidelberg New York

3   Armitage H (1989) Diplom-Arbeit, I-N-Stranski-Institut, TU-Berlin

4   Hearn A C (1987) REDUCE User's Manual, Version 3.3, RAND Publication CP78, Santa Monica
    Gates P L (1987) GENTRAN User's Manual, RAND Publication, Santa Monica
    Forster P, Khuen A, Ziessow D , Wasèn R (1989) Computer Algebra in Chemistry, in: Eckermannn R (ed) Computer Application in the Chemical Industry, Dechema Monograph 116: 613

5   Aho A V, Sethi R , Ullman J D (1988) Compilers: Principles, Techniques and Tools, Addison-Wesley, Reading, USA

6   Schreiner A T, Friedman G (1985), Compiler bauen mit UNIX, Hanser-Verlag, München Wien

7   Press W H, Flannery B P, Teukolsky S A and Vetterling W T (1986) Numerical Recipes, Cambridge University Press, Cambridge London

AUTOMATIC TRANSLATION
FROM FORTRAN TO "C"

Alfred Mechsner

2300 Kiel, Gärtnerstr. 4

Abstract: Programs written in FORTRAN can be translated to "C" with an automatic translator. It is shown, which techniques are used for this translation.

INTRODUCTION

During the last years the importance of the UNIX operating system as well as its programming language "C" has grown steadily. "C" is a sophisticated high level computer language which comprises also features near to the assembler level. Recently ANSI made "C" a standardized language. Nowadays "C" compilers are available for nearly every type of processor and operating system. New developements like "C++" add object orientated features, which will improve the capability of a programming language enormous.

"C" itself contains only a few basic language elements. For input/output no language constructs exists. For such and similar operations functions or macro definitions are used. A programmer can either employ quasi standardized functions from libraries or define functions according to his or her taste.

One is no longer forced to use implementation dependent I/O constructs like READ or WRITE statements of FORTRAN, but can define appropriate expressions without too much effort. This flexibility has made "C" a portable language. Programs written in "C" can be transferred fairly easy from one computer system to another, even if they rely on special hardware for graphics or data aquisition.

COMPARISION BETWEEN FORTRAN - "C".

In table 1 basic language elements which are present in one of the two languages are presented.

FORTRAN	"C"
Complex numbers	-
-	Structures
-	Pointer
-	Control structures like do $\cdot\cdot$ while, switch $\cdot\cdot$ case
-	prototypes of functions, which allows control of the parameters

–	good interface to the operating system
–	I/O can be programmed according to special needs

Table 1 : Comparison of FORTRAN and "C".

Even today FORTRAN is an important language and is widely used by the scientific community. The reason being that FORTRAN was the first mature computer language and many programs as well as subroutine libraries have been developed since. To be able to use a modern modular language like "C" without being forced to abondon the large treasure of existing FORTRAN programs, a FORTRAN to "C" converter (FORTOC) was developed.

WORKING OF THE TRANSLATOR

The translator is not much different from a common FORTRAN compiler. They differ only because instead of object code, "C" expressions are produced. The translation is done in two passes.

In the first stage the program is analyzed: all variables, functions, common blocks, labels and format statements are detected and a symboltable is set up containing this information. As a result of the first pass a cross reference listings and a "call tree" of the FORTRAN source can be obtained. Also function parameters can be checked, whether they will match the required type. This is of great value, when a FORTRAN program should be ported from one operating system to another or a FORTRAN program should be improved. In the second stage the actual translation is performed.

According to the following tree FORTRAN statements are transformed into the equivalent "C" expressions:

1. Prototyps of all functions (according to ANSI) are generated.
2. Common blocks are represented as global structures.
3. Arrays and matrices are initialized.
4. Translation of functionbodies (including the main program).

Point four is further subdivided:

4.1 Definition of the function, its parameters and variables.
4.2 Definition of local variables.
4.3 "define" preprocessor instructions are used for the format strings.
4.4 Translation of FORTRAN statements, including:

4.4.1 arithmetic expressions (real and complex),
4.4.2 string expressions,
4.4.3 comparison statements,
4.4.4 logical expressions.

4.5 Translation of control structures:

4.5.1 DO loops.
4.5.2 IF – THEN – ELSE – ENDIF expressions,
4.5.3 unconditional GOTO,

4.5.4 computed GOTO,
4.5.5 assigned GOTO,
4.5.6 arithmetic IF,
4.5.7 logical IF.

4.6 Translation of input/output expressions:

4.6.1 OPEN/CLOSE statement,
4.6.2 formatted READ/WRITE,
4.6.3 unformatted READ/WRITE,
4.6.4 READ/PRINT controlled by an iolist,
4.6.5 formatstrings.

GENERATING "C" - CODE

Translation Of Arrays And Matrices.

Translating arrays and matrices poses an interesting problem, since "C" allows several ways to create them. The following method turned out to be appropriate: Arrays and matrices are declared as pointer variables. Memory space is allocated by an usual array definition and not by the malloc function. In this way the task of freeing the memory, when leaving the function is avoided. Memory access is speeded up by using index arrays, so no index calculation for multidimensional arrays has to be done. An example will explain the method (figure 1).

FORTRAN array declarations like:
REAL X(10) , M(100,200) , Y(20:30)
are translated to:
float   __m[100][200];
float   *_m[100];
float   *m;

float   _x[10];
float   *x;

float   _y[11];
float   *y;

{ long d1,d2,d3,d4,d5,d6,d7;
  for (d1=0L; d1 < 100L; d1++)
    _m[d1]=   __m[d1]-1;
  x=_x-1;
  y= _y-20;
}
Fig. 1: Translation of arrays.

For the matrix m two new variables, _m and __m, are created. __m allocates space for all matrix elements and _m is neccessary to store the pointers to all rows of the matrix. m is now a pointer to the pointer array _m. To achieve in "C" the same pattern of access to array elements as in FORTRAN, the lower FORTRAN array boundary has to be subracted form the array adress, 1 in the case of array x and m and 20 in the case of y. Single array elements can now be accessed like

$$x(4) \longleftrightarrow x[4]$$
$$m(50,100) \longleftrightarrow m[50][100]$$

Using this method of array representation makes it possible to pass arrays with variable boundaries as parameters to subroutines. All declarations and definition as well as all initialization procedures, which are necessary to use arrays and matrices this way will be generated completely by the translator.

Passing Arguments To Functions.

In FORTRAN arguments are passed to functions by transfering the addresses of the variables, a way of communication which is called " call by reference".

"C" applies the principle "call by value": only the content of a variable is passed on. Changing the value of a parameter within a subroutine does not effect the calling program. In this way the so called "side effects" are avoided.

While a subroutine is translated, every variable of the argumentlist is checked, if its value is changed, if the variable is used as "lvalue". Such a usage of an argument is even recognized within deeply nested subroutines. All variables employed in this manner must pass their value back to the calling program. This is ensured by transfering the respective pointers to the subroutine.

An Example.

Figure 2 gives an example of a short FORTRAN program displaying how common blocks, functions, array declarations and arithmetic expressions are translated.

```
C       Example for usage of COMMON blocks,
C       function parameters and the initialisation of arrays
        common /test/ testa,testb,testarr(100,200)
        a=4.
        i=3
        c=i**a + a**i + i**i + a**a
        call funk1 (a,b,c)
        end

        subroutine funk1 (x,y,z)
        x=funk2 (y,z)
        return
        end

        function funk2 (e,f)
        common /test/ a,b,arr
        real arr (100,200)

        f=100.
        funk2= e*arr(50,50)
        return
        end
```
Fig. 2 : A short FORTRAN program.

The "C"-code produced by the translator is given in figure 3.

```c
#include <stdio.h>
#include <for.h>
/* Functionsprototypes */
void funk1 ( float*, float, float* );
float  funk2 ( float, float* );

/* Commonblocks as global structures */
struct CTEST
  {
float   testa;
float   testb;
float   __testarr[100][200];
float   *_testarr[100];
float   **testarr;
  } test;

#if 0
/* ##This block has to be included in main () */
{ long d1,d2,d3,d4,d5,d6,d7;
   for (d1=0;  d1<100;  d1++)
     test._testarr[d1]=test.__testarr[d1]-1;
     test.testarr=test._testarr-1;
}
/* ##End of initialisation of arrays of COMMON blocks  */
#endif

/*    Example for usage of COMMON blocks,
      function parameters and the initialisation of arrays
*/
 main ()
{
/* Local variables*/
float   a;
float   b;
float   c;
long    i;
/*------------------*/
       a=4.;
       i=3;
       c=ldpow(i,a)+dlpow(a,i)+llpow(i,i)+pow(a,a);
       funk1(&a,b,&c);
}
 void funk1(x,y,z)
/*################################*/
/* Parameter */
float   *x;
float   y;
float   *z;
{
/* Local variables */
/*------------------*/
```

```
        *x=funk2(y, z);
        return;
}
 float   funk2 (e,f)
/*################################*/
/* Parameter */
float   e;
float   *f;
{
/* Local variables*/
float   funk2;
/*------------------*/
        *f=100.;
        funk2=e*test.testarr[50L][50L];
        return ( funk2 );
}
```

Fig. 3: The "C"-code produced from the FORTRAN source code of fig. 2.

The REAL variable "a" of the main program receives a value in SUBROUTINE funk1 and to the variable "f" a value is assigned in FUNCTION funk2. Therefore both variables must be passed as addresses to the subroutines/functions. Their value is accessed by dereferencing them with the help of the operator *. Members of COMMON blocks are referenced only by their position within the COMMON block. So if in function func2 the variable name "arr" is used, the translation will insert the name "test.testarr", which is member of the structure "test".

The FORTRAN operator "**", which is used to compute powers, is translated into an appropriate function according to the type of the exponent. For expressions like "a ** b" the standard function pow (a,b) is used where as in the case of an integer exponent we use dlpow (a,i).

COMPLEX ARITHMETIC EXPRESSIONS

In "C" there is no variable type for complex numbers, while FORTRAN provides them. Complex numbers could be build by the help of structures, which contain the real and imaginary parts. Arithmetic expressions with complex numbers are translated to function calls. Both operants are passed to this function and the operation will be performed to real and imaginary part. The result is passed back as a complex structure. 'Real' complex arithmetic could by achieved by the use of a "C++" compiler, where operators like +-/* etc. can be overloaded by own functions.

CONCLUSION

With a tool like this the effort to convert FORTRAN source to "C" source is reduced to an absolute minimum, normally the generated source can be directly compiled by an "C" compiler.

References.
2 Press WH, Flannery BP et al. (1988) Numerical Recipes in C, Cambridge University Press
2 Kernighan BW, Ritchie DM (1978) The C-Programming Language, Prentice Hall
3 CDC Fortran Handbuch (1982) Rechenzentrum Niedersachsen, Hannover

LABORG

Laboratory optimisation by the elimination of weaknesses and the implementation of LIMS

Dr. Andreas Bielecki

KPMG Deutsche Treuhand-Unternehmensberatung GmbH
Hardefuststraße 1, D-5000 Köln 1

Summary

In the past few years the possibilities for quantitative improvements in laboratory services, as well as faster determination of results based on the introduction of information technology, have grown rapidly.

However, after the initial investemnt has been made, the expected increase in efficiency often does not occur to the extent planned. This is most often caused by the optimisation of isolated laboratory techniques. A comprehensive investigation of the company´s laboratories, and an associated system and development plan, create the efficient laboratory organisation which is a necessary condition for implementing of the Laboratory-Information-Management-System (LIMS).

Background

Through computerization, and the introduction of samples-changers and laboratory robots, investigative techniques can be highly optimized. However, the integration of these techniques into laboratory procediures is often based purely on task-oriented optimization.

The success of a laboratory-modernization is not only dependant on the automation of specific techniques. More important is the consideration of factors such as work-flows, dissemination of these ingredients must be combined in the concept of an integrated laboratory.

In many companies the necessary manpower and resources are not available for a laboratory study, nor to undertake the subsequent system development.

Weakness in processing laboratory contracts

Central to the modernization of a laboratory should be the samples and the results of the investigations, to which the necessary combinations of methods, laboratory organisation and the information and communication systems must be adepted.

The weakness of the existing laboratory administration are often not fully taken into account. Amongst the most important problems are:

- lengthy processing caused by:

    • sequential scheduling of techniques with several processing steps *(Fig. 1)*
    • lenghty waiting periods between tests carried out in separate laboratories *(Fig. 2 und 3)*

- incomplete description of the sample by the customer leading to *(Fig. 4)* :

    • labour-intensive supplementary testing
    • redundant and disorganized correlation of results

- difficulty in determing the status of the investigation:

    • information is only available at the end of the test sequence
    • multiple and redundant discussions and correlation of results, both within and outside the laboratory
    • testing is disrupted by frequent queries from customer

- missed deadlines in the laboratory processing *(Fig. 5)* , with the following results:

    • the results are no longer required - they came too late
    • the results are false - the sample has decomposed in the meanwhile

- lack of time is the main reason for inadequate methological development

Lengthy processing *(Fig. 3)* causes:

  . - false results caused by changes in the sample, or chemical or biological degradation
   - results which are not up-to-date, resulting in the following effects for the client:
    • "blind" production runs
    • disrupted production
    • unnecessary storage of products
    • customer complaints about quality
    • loss of image in the market

   - intensive laboratory coordination, which cannot be replicated.

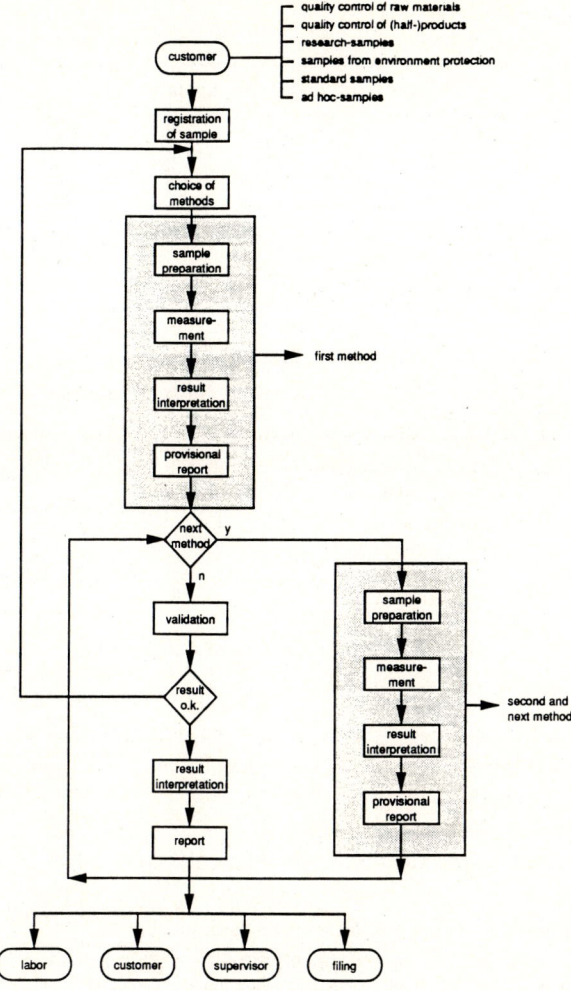

Fig. 1 Work-flow of typical, sequential laboratory processing

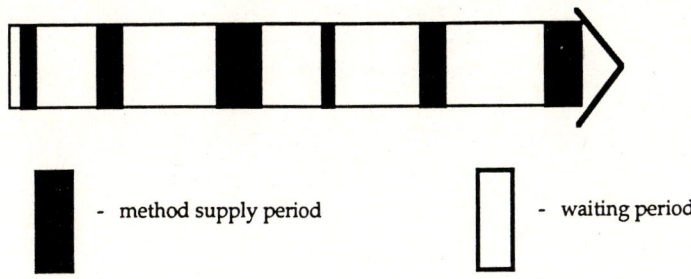

Fig. 2 Waiting periods iat the labor-processing

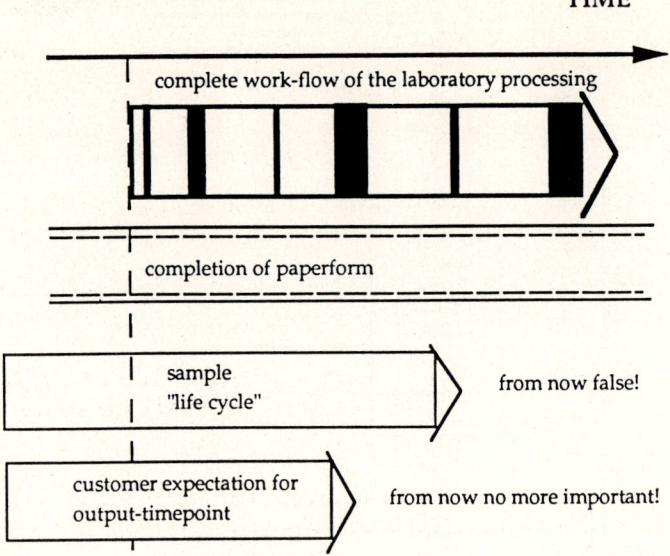

Fig. 3 Long run-time cycle causes, the labor-results will be uninteresting

Incomplete advance information about the customer´s samples often results in a range of weakness (*Fig. 4*), whose origin and consequences effect the customer rather than the laboratory.

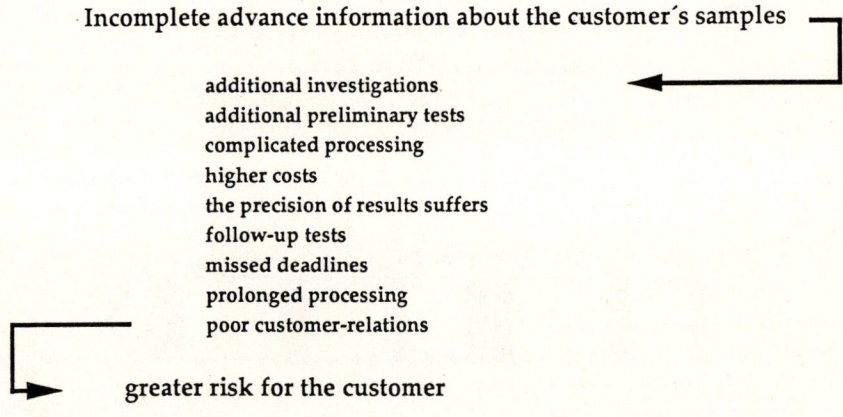

Fig. 4 A chain of weak passages of labor-process

Delays in delivering the test-results, and unnecessarily high laboratory costs can be viewed as the direct results of the weakness described above. The influence of these weakness on company-wide goals can only be assessed through an analysis of the connection between the interests of the client and the laboratory in question *(Fig. 5)*.

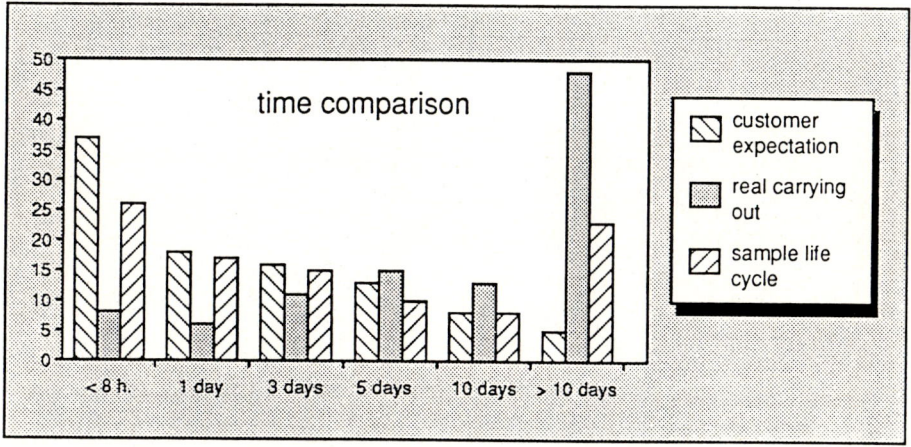

Fig. 5   Labor-run-time cycle in comparison with the sample-life cycle and the customer expectation

Some examples of these consequences are:

- delay in bringing products to the market
- incomplete quality control, based on faulty standards
- incomplete information available for inquiries, or in emergency situations

Efficiency in the laboratory with the help of LIMS-introduction

The well-planned and executed transmission from a sequential style laboratory to parallel procedures supported by a LIMS system adapted to the needs of the laboratory results in a considerable saving in laboratory processing times *(Fig. 6 and 7)*. The success of the introduction of the LIMS system depends on several factors, which can only be achieved by system planning and development independent of the manufacturer.

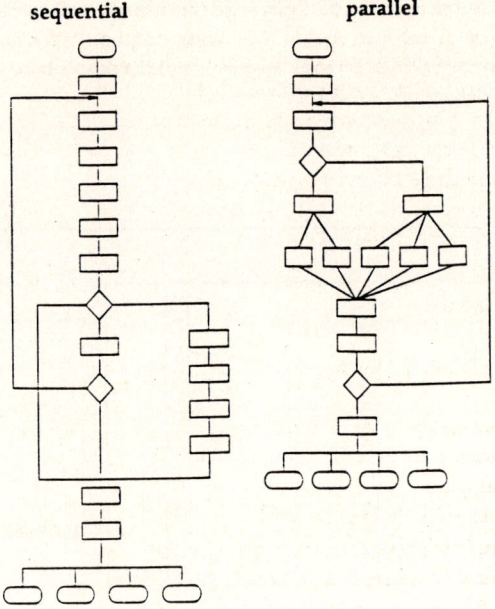

Fig. 6 Sequential and parallel labor-process

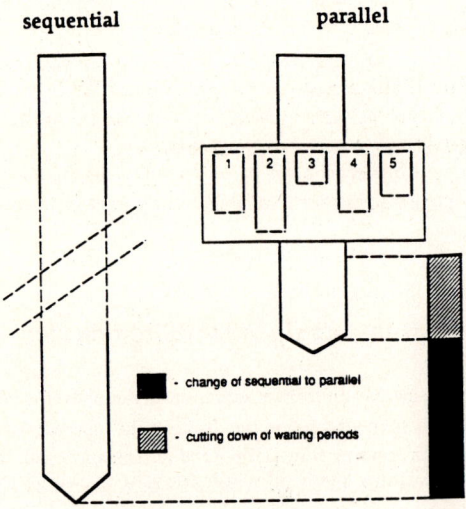

Fig. 7 Economy of run-time cycle

The well-planned and executed implementation of a LIMS system results in the following:

- control of information and communication
- improvement in information quality
- company-wide standards for laboratory methodology
- reproducable results
- uniform laboratory procedures
- improved efficiency in the laboratory and in the whole production area of the company

Scope of the LABORG study

Within the framework of the LABORG are investigated and optimized:

- laboratory organizational structure
- assignment of contracts and tasks
- sample and contract work-flows
- GLP-/GMP requirements
- study of elapsed-time for processing
- characteristics of the laboratory equipment
- integration of methodologies
- refereance-data and data-libraries
- data-modelling
- short, middle, and long-term reorganizational measures
- conceptual model for an integrated IV-Laboratory system
- handbook for the LIMS system
- support during system implementation

The manufacturer-independant implementation of the study and system development creates the best possible basis for an optimal improvement in the efficiency of the laboratories, matched to the specific conditions of the company.

C - LIT

Dipl.-Chem. Jürgen Kammerer

Albert-Ludwigs-Universität, Freiburg
Albertstraße 21, D-7800 Freiburg

Abstract: C-LIT is a literature data base program which helps chemists to organize their collection of literature. The features of this ShareWare program are described.

Every chemist is once faced with the problem of organizing his personal collection of literature. Conventional file card systems are inefficient, because they allow access to literature only by a limited number of keywords. An "electronic file card system", represented by a data base program, is more flexible. Each publication can be characterized with any number of keywords. These keywords can be used to extract a specific group of records based upon the user-defined criteria. Since professional data base programs require knowledge of the exact syntax, they are not easy to use for those chemists who are inexperienced in working with a computer. C-LIT was especially developed for this group; light bar menus and context specific help screens enable the user to work with C-LIT after a short time of practice. Commands are executed by highlighting the options using the arrows keys and then pressing <ENTER> or by typing the first letter of the option.

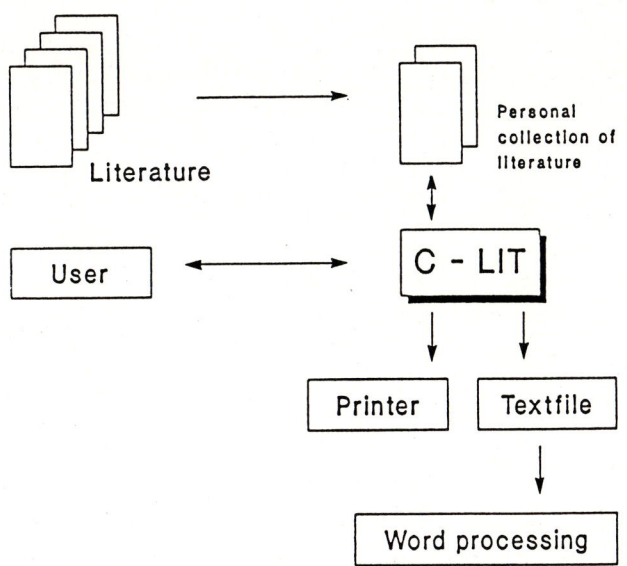

The structure of the database (Table 1) was adapted to the requirements of a chemist; a registration number (divided into 26 groups lettered from A to Z) will be attached to every publication in order to locate the reference within your collection of literature. Besides registration number and information about author and source, each record has 4 fields which contain keywords to describe the content of the publication.

Registration number	6 Characters
Author	100 Characters
Source	68 Characters
Volume	4 Characters
Publication year	4 Characters
Page	4 Characters
Title	136 Characters
Keywords	136 Characters
Structure1	68 Characters
Structure2	40 Characters
Reaction	68 Characters

Table 1: Data base structure of C-LIT

When adding a new record you press a letter key and the next free registration of this group number will be displayed (Figure 1). Strings (for example, titles of journals) can be connected with a function key. You only have to press a function key and the string appears on the screen. Each record can be extended or modified with the function "EDIT". It is possible to delete single records.

C-LIT has two possiblities to extract information from the database. The first one is to enter a registration number and the record appears on the screen at once. The second possibility is to select the data of a specific group of records based upon user-defined criteria. The user is asked to define a database field in which a certain string should occur. Two of these criteria can be joined. After selecting records the total number of records in the database and the number of selected records is displayed. You can display them on screen or printer. If the number of selected records is too large another selection is possible.

Another way to export data is to create a textfile with C-LIT. This textfile (ASCII or modified MS-Word format) contains the registration number and the correct citation of the publications. The two most common ways of citations are supported: (a) author, source, volume, page, year and (b) author, source, volume, year, page. This textfile can be used with any text processor. A makro for MS-Word 4.0 is available to print the volume in bold letters.

If more than one user works with C-LIT it is usefull to store each user's data base files in seperate subdirectories. C-LIT is able to create subdirectories and copy automatically an empty data base file in the new subdirectory. You can change the data base subdirectories without leaving the program. C-LIT database files are dBase III plus compatible, so that you can edit these files with any other program supporting this format. It is also very easy to make a back up with C-LIT; the program checks the destination disk and gives a warning if there isn't enough free disk space.

Hardware requirements: C-LIT runs on every 100 % compatible IBM PC/AT with more than 300 kB free RAM and MS-DOS higher 3.X. A hard disk is required. The program will be installed either in the monochrome or color version (for EGA or VGA) by an installation routine. All help screens and the documentation are in German language. The program is written in Pascal and compiled with Borlands Turbo Pascal 5.5 compiler.

C-LIT is distributed under the laws of ShareWare. The ShareWare concept permits you to try this program before you buy. You may freely copy and test C-LIT, but if you intend to continue the use of this program, then you need to register and obtain a license (license fee: 30.00 DM). For registration or copies of the program, please contact the author:

> Jürgen Kammerer
> In der Breite 2
> D-7801 Umkirch

```
┌─────────────────────────────────────────────────────────────┐
│ Reg. Nr.:  D 123                                            │
├─────────────────────────────────────────────────────────────┤
│ Autor(en):                                                  │
│                                                             │
│ Quelle:                                                     │
│ Band:              Jahr:              Seite:                │
│ Titel:                                                      │
├─────────────────────────────────────────────────────────────┤
│ Stichwort:                                                  │
│                                                             │
│ Struktur1:                                                  │
│                                                             │
│ Struktur2:                                                  │
│                                                             │
│ Reaktion:                                                   │
├─────────────────────────────────────────────────────────────┤
│   F1      F2     F3    F4     F5      F6     F7     F8     F9    F10 │
│  Angew   JACS   JOC   THL   ChComm  HelvCh ChemBe CanJC  ChRev  ChiuZ │
└─────────────────────────────────────────────────────────────┘
```

Figure 1: Screendump from C-LIT

References:

MS-Word 4.0 and MS-DOS are trademarks of Microsoft Cooperation. dBase III plus is trademark of Ashton – Tate.

POSSIBILITIES AND LIMITATIONS OF COMBINED DATABASE ENQUIRIES MODELLED ON "ENVIRONMENTAL SIGNIFICANCE OF ALUMINUM"

A. Müller and E. Striedl

SOFTEK Gesellschaft für Softwaretechnologie m.b.H.
Dept. *ChemInfo*
Landsbergerstraße 63a, D-8034 Germering/Mchn., FRG

Introduction

Environmental problems are discussed more and more, therefore it is obvious to use databases as comprehensive sources of mankind's knowledge in order to achieve unemotionally answers on holistic questions. The usefullness of databases in the area of research and development is well known, especially their importance as tools for rapid access to the demanded information. Nevertheless the possibility to act against an almost emotionally handled discussion concerning problems like "mankind and environment" remained to be seen until the present. A search for literary citations dealing with the "environmental significance of aluminum", a wide selection of primary references should be found. By starting an OnLine-search within the relevant databases, it could be expected to achieve a comprehensive point of view, because it is possible to avoid a biased representation of facts a priori.

Search terms and search strategy

For the purpose of including significant information as accurately and completely as possible, the search terms were arranged roughly by areas. The following number of hits was achieved by a search profile which was carried out within the databases below in order to evaluate the total costs:

HOST: STN-INTERNATIONAL
FILE CHEMICAL ABSTRACTS / 10.07.89

```
L1 =   627      (HAZARD? OR RISK? OR TOXIC? OR HEALTH OR BRAIN OR
HUMAN BODY OR HOMEOSTASIS OR MEDICIN? OR ALZHEIMER OR PHARMA?)/TI AND
(ALUMINIUM OR ALUMINUM OR AL)/TI

L2 =   723      (ENVIRONMENT OR SOIL OR WATER SUPPLY OR FOREST DECLINE
OR DIET OR FOOD OR ACID RAIN OR NATURAL RESOURCE? OR SAFE? OR
INDUSTRIAL HYGIENE)/TI AND (ALUMINIUM OR ALUMINUM OR AL)/TI

L3 = 1943       (PACKAGING OR TRAFFIC OR AUTOMOBIL? OR BUILD? OR
ARCHITECT? OR CONSTRUCT? OR ELECTRICAL ENGINEERING OR DOMESTIC
ARTICLE? OR RECYCLING OR ENERGY)/TI AND (ALUMINIUM OR ALUMINUM OR
AL)/TI

L4   3228  L1 OR L2 OR L3
```

FILE ULIDAT / 10.07.89
L5 =   39   (UMWELT? OR WALD? OR SAURER REGEN OR WASSER? OR
SICHER? OR NAHRUNG? OR ERNAEHR? OR RUECK? OR WIEDER? OR ENERGIE? OR
BODEN? OR GEFAHR? OR GEFAEHR? OR GESUND? OR HAUSHALT?)/TI AND
(ALUMINIUM OR AL)/TI

FILE UFORDAT / 10.07.89
L6 =   29   (UMWELT? OR WALD? OR SAURER REGEN OR WASSER? OR
SICHER? OR NAHRUNG? OR ERNAEHR? OR RUECK? OR WIEDER? OR ENERGIE? OR
BODEN? OR GEFAHR? OR GEFAEHR? OR GESUND? OR HAUSHALT?)/TI   AND
(ALUMINIUM OR AL)/TI

HOST: DIALOG

File   5:**BIOSIS PREVIEWS** 69-89/JUL BA8803;RRM3703

File 240:**PAPERCHEM** - 67-89/JUN

File 203:**AGRIS INTERNATIONAL** 74-89/APR

File 50:**CAB ABSTRACTS** - 1984-89/JUN

File 53:**CAB ABSTRACTS** 1972-1983

File 51:**FSTA** - 69-89/JUL

File 155:**MEDLINE** 66-89/AUG (890802)

File 252:**PACKAGING SCIENCE & TECHNOLOGY ABSTRACTS**/1989 MAR

File 151:**Health Planning and Administration** - 1975-89/Aug

File 33:**WORLD ALUMINUM ABSTRACTS** - 68-89/JULY

File 161:**Occupational Safety & Health (NIOSH)** -73-88/Dec

TOTAL: FILES 5,240,203 ... 161

```
S1    506357   (HAZARD? OR RISK? OR TOXIC? OR HEALTH OR BRAIN)/TI
S2    173394   (HUMAN(W)BODY OR HOMEOSTASIS OR MEDICIN? OR PHARMA?)/TI
S3    372429   (ALZHEIMER? OR ENVIRONMENT OR SOIL OR DIET OR FOOD)/TI
S4      4501   (WATER(W)SUPPLY OR FOREST(W)DECLINE OR ACID(W)RAIN)/TI
S5     63711   (NATURAL(W)RESOURCE? OR SAFE? OR PACKAGING OR BUILD?)/TI
S6      8077   (INDUSTRIAL(W)HYGIENE OR TRAFFIC OR AUTOMOBIL?)/TI
S7    107650   (ARCHITECT? OR CONSTRUCT? OR ENERGY OR RECYCLING)/TI
S8        59   (ELECTRICAL(W)ENGINEERING OR DOMESTIC(W)ARTICLE?)/TI
S9    112532   (ALUMINIUM OR ALUMINUM OR AL)/TI

S10     7488   (S1 OR S2 OR S3 OR S4 OR S5 OR S6 OR S7 OR S8) AND S9
```

Within the **Phytomed** database/<u>**DIMDI**</u> and the **ACID RAIN** database/<u>**ESA-IRS**</u> the same search strategy done on **12.07.89** resulted in **159** hits respectively **26 hits** on **11.07.89**.

Because of the enormous number of literary citations, which was not manageable within the project's scope, the limited number of hits from a few specialized databases was downloaded in order to plan further conceivable steps. Among the displayed hits - all of them verifying the relevance of the "emotive word" aluminum - one leaps to the eye because it repeats the projects objective in a really provocative manner:

```
L6     ANSWER 1 OF 29
AN     25121    UFORDAT       DN   UFOKAT CH*T16
TI     Ist Aluminium ein Umweltgift? (Arbeitstitel).
TT     Is aluminum an environmental poison? (working title)
SF     Projektleiter: Lietha, R., Dr.
CSP    Institut fuer angewandte Biologie
       Zuercherstr. 146
       CH-8640 Rapperswil
       Schweiz
DB     01 Jan 1983
DE     31 Dez 1985
AB     Aluminium-Belastung des Organismus durch Gebrauchsmittel des
       Alltags (Haushaltgeraete, Kosmetika, Medikamente,
       Lebensmittelzusaetze) insb. auch Aluminium-Gehalt von
       Getraenken.
CC     *CH21 Umweltchemikalien/Schadstoffe: Wirkung auf Menschen
       (incl. menschbezogene Tierversuche); CH10
       Umweltchemikalien/Schadstoffe: Herkunft, Verhalten, Ausbreitung
       und Verbleib in der Umwelt; LF20 Wirkungen und Rueckwirkungen
       von Belastungen auf die Land- und Forstwirtschaft, Fischerei,
       Nahrungsmittel; LF55 Umweltaspekte der Land- und
       Forstwirtschaft, Fischerei, Nahrungsmittel:
       Nahrungsmitteltechnologie; CH20 Umweltchemikalien/Schadstoffe:
       Wirkung von Belastungen
CT     Aluminium; Arzneimittel; Kosmetika; Getraenk;
       Lebensmittelzusatz; Gesundheitsgefaehrdung; Schadstoffwirkung;
       Haushaltsgeraet; Umweltchemikalien; Schadstoffbelastung;
       Lebensmittel
```

An attempt to obtain the original article concerning the citation mentioned above failed, because according to phone messages of the project leader, this project was never ended (or even started at all?).

Just from this incident, questions arise about the reliability of database contents and of existing or perhaps yet to be developed producer-independent control mechanisms.

Summary

"The holistic approach" is a slogan used more and more, especially in discussions on environmental problems. Until the present, that kind of thinking solved problems on a basic level only, because it is enormous time consuming and cost expensive to solve such problems in a really holistic way. Therefore integral online searches will remain infeasible. The vast costs of truly comprehensive enquiries are evidence of that fact. A desirable scientific and neutral assessment of advantageous and of adverse impacts of individual substances on our environment seems not to be achievable for the lay public in this way.

G. Gauglitz, Universität Tübingen (Hrsg.)

Software-Entwicklung in der Chemie 3

Proceedings des 3. Workshops „Computer in der Chemie" Tübingen, 16.–18. November 1988

1989. XII, 472 S. 188 Abb. 36 Tab. Brosch. DM 88,- ISBN 3-540-50673-X

Dieser Band enthält die Beiträge des 3. Workshops „Computer in der Chemie" (16.–18. November 1988 in Tübingen). Das Meeting wurde von der Fachgruppe Chemie-Information der GDCh veranstaltet und enthält Beiträge für folgende Gebiete:
- Chemometrie
- Molekülmodellierung
- Syntheseplanung
- Prozeßsteuerung
- Design und Aufbau von Datenbanken, z. B. in der Toxikologie
- Spektrenbibliotheken
- Datenerfassung in der Analytik
- Röntgenstrukturanalyse
- Simulation elektrochemischer Prozesse

Hierzu lieferbar:
J. Gasteiger (Hrsg.), **Software-Entwicklung in der Chemie 1.** 1987. Brosch. DM 54,- ISBN 3-540-18465-1

J. Gasteiger (Hrsg.), **Software-Entwicklung in der Chemie 2.** 1988. Brosch. DM 79,- ISBN 3-540-18696-4

Springer-Verlag Berlin Heidelberg New York London Paris Tokyo Hong Kong

W. A. Warr, ICI Pharmaceuticals Division, Macclesfield, Cheshire (Ed.)

Chemical Structures

The International Language of Chemistry

1988. XII, 472 pp. 213 figs. 18 tabs. Hardcover DM 168,- ISBN 3-540-50143-6

From the contents: Progress in chemical information science and future trends – In-house chemical structure databases and related property data – MACCS – Substructure searching methodology – Generic structure search – Online databases – Spectral databases – Computer-aided library search systems – Expert systems for structure analysis – Hardware and software developments – Managing personal databases – Publishing on CD-ROM – Molecular model building – Parallel processing techniques – Chemical reaction retrieval and synthesis planning – Chemical nomenclature and grammar in chemical indexing languages.

Springer-Verlag Berlin Heidelberg New York London Paris Tokyo Hong Kong

NOV 1 3 1990